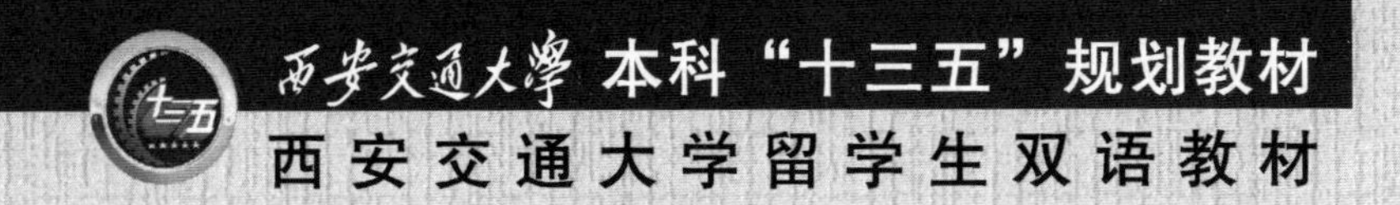

西安交通大学留学生双语教材

高等数学
(汉英对照)
Advanced Mathematics

主编 吴慧卓 副主编 温广瑞 赵 斌

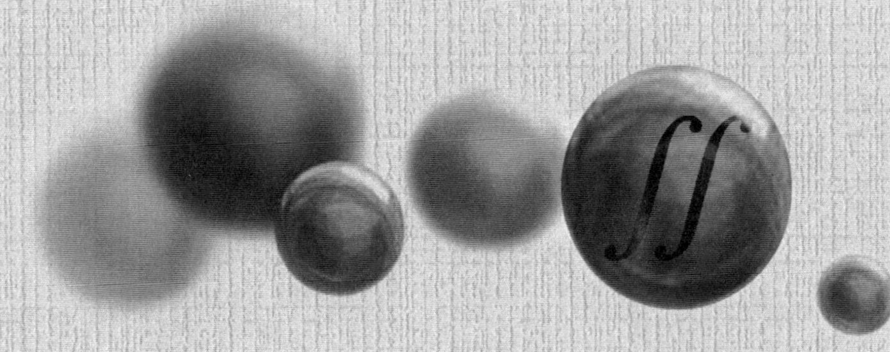

西安交通大学出版社
XI'AN JIAOTONG UNIVERSITY PRESS
国家一级出版社
全国百佳图书出版单位

图书在版编目(CIP)数据

高等数学 :汉英对照 / 吴慧卓主编. — 西安 :西安交通大学出版社,2021.9
ISBN 978-7-5693-1819-7

Ⅰ. ①高… Ⅱ. ①吴… Ⅲ. ①高等数学-高等学校-教材-汉、英 Ⅳ. ①O13

中国版本图书馆 CIP 数据核字(2020)第 189296 号

书 名 高等数学(汉英对照)
GAODENG SHUXUE(HAN YING DUIZHAO)
主 编 吴慧卓
副 主 编 温广瑞 赵 斌
责任编辑 王 欣
责任校对 陈 昕

出版发行 西安交通大学出版社
(西安市兴庆南路 1 号 邮政编码 710048)
网 址 http://www.xjtupress.com
电 话 (029)82668357 82667874(市场营销中心)
(029)82668315(总编办)
传 真 (029)82668280
印 刷 西安日报社印务中心

开 本 787 mm×1092 mm 1/16 印张 18.625 字数 532 千字
版次印次 2021 年 9 月第 1 版 2021 年 9 月第 1 次印刷
书 号 ISBN 978-7-5693-1819-7
定 价 56.00 元

如发现印装质量问题,请与本社市场营销中心联系。
订购热线:(029)82665248 (029)82665249
投稿热线:(029)82664954 QQ:1410465857
读者信箱:1410465857@qq.com

前　言

随着经济全球化的发展和“一带一路”倡议的提出，我国高等教育的国际化程度越来越高，来华留学生的规模不断扩大。但是留学生教育背景的差异给教学带来了很大的困难。作者长期从事留学生的工科数学教学工作，积累了一定的教学经验，在调研的基础上根据留学生的实际知识结构、学习习惯、数学基础水平、学习现状及汉语水平，结合国内的教学大纲，参考国内外高等院校高等数学教材编写了留学生用双语教材《高等数学》。本教材适合来自官方语言为英语国家的理工科及经济学、管理学等专业背景的留学生使用，同时也为任课教师的教学带来便利。

本教材共分九章。第 1 章至第 5 章属于一元函数微积分学及其应用，包含极限与连续、导数、导数的应用、不定积分、定积分及其应用。第 6 章至第 8 章属于多元函数微积分学及其应用，包含多元函数微分学及其应用、重积分及其及应用、曲线积分与曲面积分。第 9 章是常微分方程。本教材编写时加入了大量的例题和习题，借助几何直观帮助学生理解抽象概念，弱化了抽象理论推导，将重点放在计算和应用上，有利于提高学生的学习兴趣，巩固学生的基础知识。例如，在一元函数微积分学部分，删去中值定理，借助几何图形给出了单调性和极值的判别方法。在多元函数微分部分，不加证明地直接给出极值的判定方法。在多元函数积分学部分，从物质块质量的角度出发，给出三重积分三种不同的计算方法，即“先单后重”法、“先重后单”法和楔形法，使得内容更简洁，学生更容易掌握。

本教材是西安交通大学本科“十三五”规划教材，由西安交通大学吴慧卓（主编）、温广瑞（副主编）、湖北工业大学赵斌（副主编）编写。感谢西安交通大学教务处给予的资助和大力支持，感谢西安交通大学出版社编辑的辛勤工作。由于时间仓促，书中难免出现不妥之处，请广大读者批评指正。

作　者

2020 年 9 月 29 日于西安交通大学

Preface

With the development of economic globalization and the proposing of the Belt and Road Initiative(BRI), higher education in China is becoming more and more internationalized, and the enrollment of international students is constantly increasing. But the difference of educational background of the students has brought great difficulty to teaching. Owing to the long-term teaching experince in engineering mathematics, the authours compiles this textbook, *Advanced Mathematics*, which is based on the actual knowledge structure, learning babits, mathematical knowledge level and the difference of Chinese proficiency of the overseas students. This textbook is suitable for the students with science, engineering, economics, management, and other professional background.

This textbook contains nine chapters. The first chapter to the fifth chapter are about the calculus with one variable and its application, including limit and continuity, derivative, the application of derivatives, indefinite integral, definite integral and its application. The sixth chapter to the eighth chapter are about the calculus with several variables and its application, including the differentiation of functions, multivariable integrals and their applications, curve integral and surface integral. The last chapter is the ordinary differential equation. In order to strengthen the basic knowledge of the students, a large number of examples and exercises are compiled in this textbook. Many figures are used to help students to understand abstract concepts. Some contents concerning abstract theoretical derivation have been weakened while we focus on calculation and application. For example, the mean value theorems are deleted, monotonicity and extreme value are introduced by means of some geometric figures in differentiation of one variable. And the method of extrem value in differentiation of several variables is given directly without proof. From the point of view of mass of material block, three different calculation methods of triple integral are given, that is, strip cutting method, slice method and wedge-shape method, which makes the content more concise and easier for students to learn.

According to the investigation on the learning situation of the overseas students in China, this Chinese-English version textbook is compiled in combination with the calculus textbook used in the colleges and universities of America and other countries. Therefore, it is not only suitable for international students from English-speaking conutries, but also for those with adequate ability in Chinese communication. It also brings great convenience to teachers.

This textbook is for the 13th Five-Year Plan of Xi'an Jiaotong University. The textbook is written by Wu Huizhuo (chief editor), Wen Guangrui and Zhao Bin (deputy chief editors). The authors express their thanks to the academic affairs office of Xi'an Jiaotong University for the finacial support, and editors of Xi'an Jiaotong University Press for their hard work.

Authors

June 2020 at Xi'an Jiaotong University

目　录

中文部分

English Section

中文部分

第 1 章　极限与连续

1.1　函数

极限是研究函数的工具，利用极限可以研究函数的连续性、可导性、可微性与可积性. 在给出极限的定义前，有必要回顾函数的概念.

1.1.1　函数的定义

设 A 和 B 是两个非空实数集合，则对于 A 中的每个元素，按照对应法则 f，都有 B 中的唯一元素与之相对应，即对于任意的 $x\in A$，都有唯一的 $f(x)\in B$，可以写成 $y=f(x)$，记号 $f: A\to B$ 表示 A 到 B 上的函数. 元素 $f(x)$称为 x 在函数 f 下的值.

1.1.2　函数值

设 f 是集合 A 到 B 上的函数，$x=a$ 是 f 的定义域中的元素，则与 $x=a$ 相对应的像 $f(a)$称为函数在 $x=a$ 处的函数值. 如果函数 $f(x)$在 $x=a$ 处的函数值 $f(a)$是实数，则称 $f(x)$在 $x=a$ 处有定义，否则称为没有定义.

例如：

(1) $y=f(x)=3x+5$ 在 $x=2$ 处有定义，$f(2)=3\times 2+5=11$，$f(2)\in \mathbf{R}$ 是一个有限数.

(2) 因为 $f(1)=\frac{1}{0}\to\infty$，$f(1)\notin\mathbf{R}$，所以 $y=f(x)=\frac{1}{x-1}$ 在 $x=1$ 处没有定义.

(3) 考虑函数 $y=f(x)=\frac{x^2-1}{x-1}$和 $x=1$. 当 $x=1$ 时，$y=f(1)=\frac{0}{0}$，这是一个不确定的数，它仅仅意味着分子、分母均为 0，我们把形如$\frac{0}{0}$的表达式称为**未定式**. 还有一些其他形式的未定式，如$\frac{\infty}{\infty}$，$\infty-\infty$，1^∞ 和 0^∞.

1.2　极限

1.2.1　变量趋于有限值 $x\to x_0$

我们先考察 $x\to 2$ 的含义. 令变量 x 取值 1.9，1.99，1.999，1.9999，…，随着 9 的个数增加，x 越来越接近 2 但永远不等于 2. 这种情况下 x 与 2 的差充分小. 因此，当 x 比 2 大或比 2 小，但它们之间的差充分小时，称 x 趋近于 2，写作 $x\to 2$.

一般地，设 x 是一个变量，x_0 是一个常数，$x\to x_0$ 表示 x 与 x_0 的差充分小.

1.2.2 极限的直观认识

在这一节中，我们通过一些例子来理解极限．考虑一个圆内接正多边形，如图 1－1 所示．保持圆是固定的，当多边形边数增加时，其面积(或周长)也会增加．但是，无论边数有多少，多边形的面积(或周长)都不能大于圆的面积(或周长)．从而可通过取足够大的边的数量，使多边形的面积(或周长)与圆的面积(或周长)的差足够小．这种情况下，我们说圆的面积(或周长)是多边形的面积(或周长)的极限．

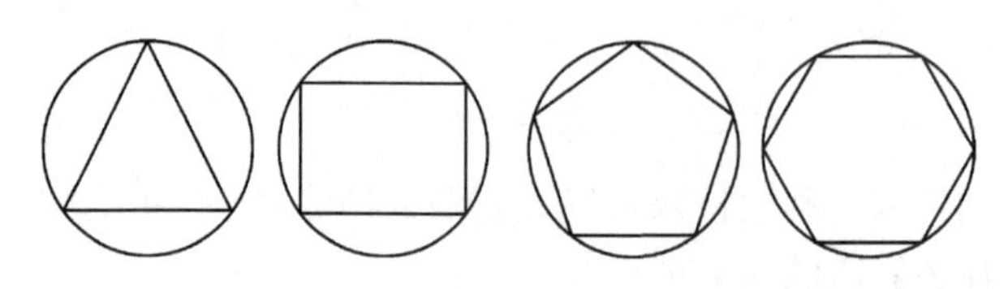

图 1－1

若将上面的例子用函数表示，可将多边形的面积 A_n 表示为边数 n 的函数 $f(n)=A_n$，$A_n(n>2)$是一个递增数列，当 n 趋于无穷时，$f(n)$或 A_n 的极限就是圆的面积，记作 $\lim\limits_{n\to\infty}f(n)$ 或$\lim\limits_{n\to\infty}A_n$．

再考虑一个数列：0.9，0.99，0.999，0.9999，…，在这个数列中，这些数逐渐增加，但始终小于 1．随着项数的逐渐增加，数列中的数充分接近 1，或者使 1 和这个数之间的差足够小(或者任意小)．这种情况下，我们说数列趋向于极限值 1．

上面的数列可以通过函数来表示．

$$
\begin{aligned}
&f(1)=0.9 && \text{第 1 项}\\
&f(2)=0.99 && \text{第 2 项}\\
&f(3)=0.999 && \text{第 3 项}\\
&\quad\vdots && \quad\vdots\\
&f(n)=0.99\cdots9 && \text{第 } n \text{ 项}
\end{aligned}
$$

当 n 趋于无穷时，$f(n)$与 1 几乎相等．所以 $f(n)$的极限是 1，并记作

$$\lim_{n\to\infty}f(n)=1$$

再看一个数列的例子，这个数列记作 a_n，有

$$1,\ 1-\frac{1}{2},\ 1-\frac{1}{2^2},\ \cdots,\ 1-\frac{1}{2^{n-1}},\ \cdots$$

显然，当 n 充分大时，$\frac{1}{2^{n-1}}$充分小．所以当 n 趋于无穷时，数列的极限为 1，记作

$$\lim_{n\to\infty}a_n=1$$

1.2.3 函数的极限

我们通过数列极限的概念来理解函数极限的概念．

首先，考虑函数 $y=f(x)=2x+3$．

令 x 的取值为 0.5，0.75，0.9，0.99，0.999，0.9999，…，即 x 从 1 的左侧越来越靠近

1，我们将发现 $f(x)$的值分别为 4，4.5，4.8，4.98，4.998，4.9998，…，越来越接近 5.

如果让 x 的取值为 2，1.5，1.25，1.1，1.01，1.001，1.0001，…，即 x 从 1 的右侧越来越接近 1，我们将发现 $f(x)$的值分别为 7，6，5.5，5.2，5.02，5.002，5.0002，…，越来越接近 5，也就是 $x\to1$，$f(x)\to5$，记作

$$\lim_{x\to1}f(x)=\lim_{x\to1}(2x+3)=5$$

下面我们给出函数极限的定义.

定义 1.1　如果当 x 充分靠近 x_0 时，$f(x)$与常数 a 的差充分小，即当 $x\to x_0$ 时 $f(x)\to a$，则称常数 a 是函数 $f(x)$当 $x\to x_0$ 时的极限，记作

$$\lim_{x\to x_0}f(x)=a$$

按照此定义，不难得到$\lim\limits_{x\to1}(x+1)=2$，$\lim\limits_{x\to0}\sin x=0$，$\lim\limits_{x\to0}\cos x=1$.

另外，根据极限的定义，可以得到如下定理.

定理 1.1

$$\lim_{x\to x_0}f(x)=a\Leftrightarrow\lim_{x\to x_0^-}f(x)=\lim_{x\to x_0^+}f(x)=a$$

例如，考察下列函数的极限

$$u(t)=\begin{cases}1, & t\geqslant0\\0, & t<0\end{cases}$$

我们可以先画出函数的图像，如图 1－2 所示.

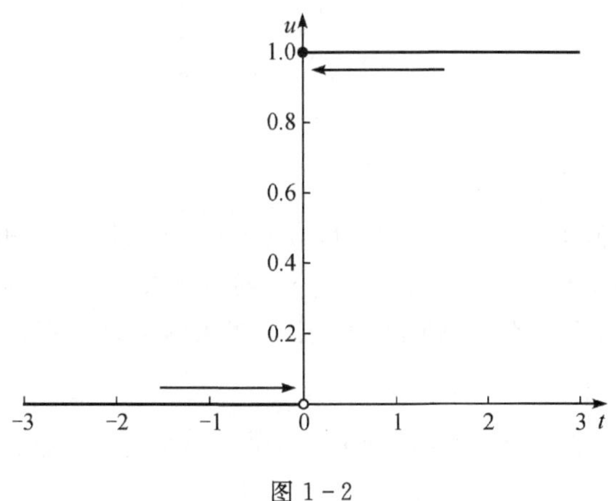

图 1－2

观察图像可以看出，当 t 从 0 的右侧趋近 0 时，所对应的函数值无限趋近于 1. 当 t 从 0 的左侧趋近于 0 时，所对应的函数值无限趋近于 0. 从而有

$$\lim_{t\to0^+}u(t)=1,\quad\lim_{t\to0^-}u(t)=0$$

所以$\lim\limits_{t\to0}u(t)$不存在.

再如，考察函数$\lim\limits_{x\to0}\sin\dfrac{\pi}{x}$. 利用作图软件画出函数的图像，如图 1－3 所示. 但从图像中很难看出函数的极限是否存在.

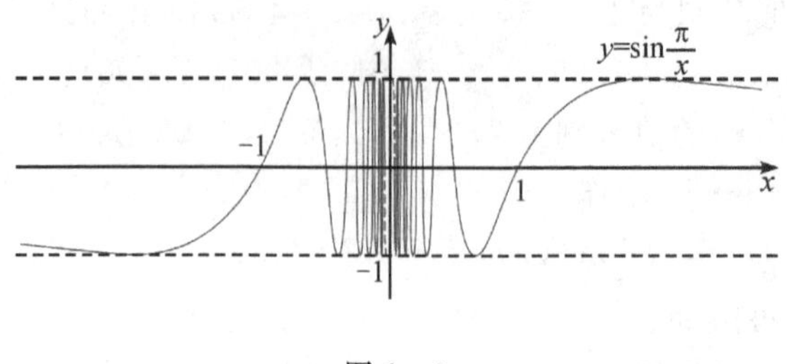

图 1-3

我们可以计算函数在 $x=0$ 附近的函数值来观察函数的变化趋势.

$$f(1)=\sin\pi=0,\ f\left(\frac{1}{2}\right)=\sin 2\pi=0,\ f\left(\frac{1}{3}\right)=\sin 3\pi=0,\ f\left(\frac{1}{4}\right)=\sin 4\pi=0$$

$$f(0.1)=\sin 10\pi=0,\ f(0.01)=\sin 100\pi=0,\ \cdots,\ f(0.001)=f(0.0001)=0$$

但如果就此猜想 $\lim\limits_{x\to 0}\sin\dfrac{\pi}{x}=0$，则这个猜想结果是错的.

因为虽然在 x 靠近 0 处取 $\dfrac{1}{n}$ 时(n 为任意的正整数)，$f\left(\dfrac{1}{n}\right)=\sin n\pi=0$，但是在 x 靠近 0 处取 $\dfrac{1}{n+\frac{1}{2}}$ 时，$f\left(\dfrac{1}{n+\frac{1}{2}}\right)=\sin(n+\frac{1}{2})\pi=\pm 1$.

图 1-3 中虚线表示当 $x\to 0$ 时，函数 $\sin\dfrac{\pi}{x}$ 在 -1 与 1 之间无限次振荡，所以该极限不存在.

1.2.4　无穷大

考虑函数 $y=f(x)=\dfrac{1}{x}$.

如果令 x 取一系列值 1，0.5，0.1，0.01，0.001，0.0001，…，即越来越接近 0，相应的函数值 $f(x)$ 为 1，2，10，100，1000，10000，…，逐渐增大. 如果 x 取得足够小，则 $f(x)$ 足够大. 令 x 从 0 的右侧充分靠近 0，则 $f(x)$ 大于任意大的正数. 这种情况下我们说，x 趋于 0 时 $f(x)$ 趋于无穷大，记作

$$f(x)\to\infty,\ x\to 0^{+}\quad 或\quad \lim_{x\to 0^{+}}\frac{1}{x}=\infty$$

这里的 ∞ 实际上是 $+\infty$.

定义 1.2　如果当 x 充分靠近 x_0 时，所对应的函数值大于任意一个事先给定的正数，则称当 x 趋近 x_0 时 $f(x)$ 的极限是正无穷大，记作

$$\lim_{x\to x_0}f(x)=+\infty$$

类似地，可以定义负无穷.

1.2.5　$x\to\infty$

考虑函数 $f(x)=\dfrac{1}{x^2}$，观察 x 取一系列值时相应的函数值的变化情况.

x	1	10	100	1000	…
$f(x)$	1	0.01	0.0001	0.000001	…

从上面的表格中我们看到，当 x 增加时，相应的函数值 $f(x)$减少. 当 x 充分大时，相应的函数值 $f(x)$足够小. 也就是说，当 x 大到可以大于一个任意大的正数，相应的函数值 $f(x)$可以充分靠近 0. 这种情况下我们说，当 x 趋于正无穷时，$f(x)$趋于 0，记作 $f(x)\to 0$，$x\to+\infty$，或

$$\lim_{x\to+\infty} f(x) = \lim_{x\to+\infty}\frac{1}{x^2} = 0$$

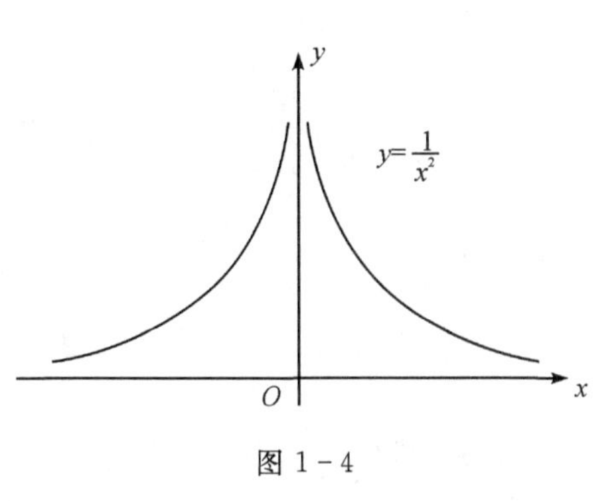

图 1－4

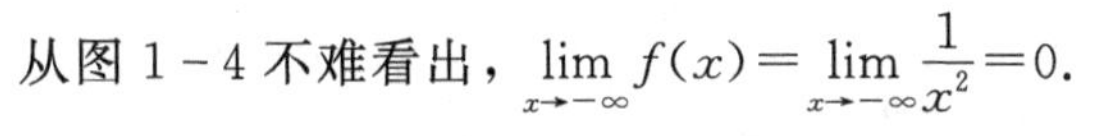

从图 1－4 不难看出，$\lim\limits_{x\to-\infty} f(x)=\lim\limits_{x\to-\infty}\frac{1}{x^2}=0$.

定义 1.3 如果当 x 趋于$+\infty$时，所对应的函数值 $f(x)$充分靠近某个常数 a，则称 a 是当 x 趋于$+\infty$时 $f(x)$的极限，记作

$$\lim_{x\to+\infty} f(x) = a$$

类似地，可以定义

$$\lim_{x\to-\infty} f(x) = a$$

显然，

$$\lim_{x\to-\infty} f(x) = \lim_{x\to+\infty} f(x) = a \Leftrightarrow \lim_{x\to\infty} f(x) = a$$

极限运算法则

设 $\lim\limits_{x\to x_0} f(x)=a$，$\lim\limits_{x\to x_0} g(x)=b$，则

(1) $\lim\limits_{x\to x_0}[f(x)\pm g(x)]=\lim\limits_{x\to x_0} f(x)\pm\lim\limits_{x\to x_0} g(x)=a\pm b$.

(2) $\lim\limits_{x\to x_0}[f(x)\cdot g(x)]=\lim\limits_{x\to x_0} f(x)\cdot\lim\limits_{x\to x_0} g(x)=a\cdot b$.

(3) $\lim\limits_{x\to x_0}\frac{f(x)}{g(x)}=\frac{\lim\limits_{x\to x_0} f(x)}{\lim\limits_{x\to x_0} g(x)}=\frac{a}{b}$，其中 $b\neq 0$.

(4) $\lim\limits_{x\to x_0}[kf(x)]=k\lim\limits_{x\to x_0} f(x)=ka$.

(5) $\lim\limits_{x\to x_0}[f(x)]^n=[\lim\limits_{x\to x_0} f(x)]^n=a^n$，$n\in\mathbf{Z}^+$.

(6) $\lim\limits_{x\to x_0}\sqrt[n]{f(x)}=\sqrt[n]{\lim\limits_{x\to x_0} f(x)}=\sqrt[n]{a}\ (a>0)$，$n\in\mathbf{Z}^+$.

例 1.1 求$\lim\limits_{x\to 3}(3x-2)$.

解 当 x 趋近于 3 时，$3x$ 趋近于 $3\times 3=9$. 于是 $3x-2$ 趋近于 $9-2=7$. 因此，

$$\lim_{x\to 3} f(x) = \lim_{x\to 3}(3x-2) = 7$$

例 1.2 计算$\lim\limits_{x\to 2}(3x^2-5x+6)$.

解 $\lim\limits_{x\to 2}(3x^2-5x+6)=\lim\limits_{x\to 2}3x^2-\lim\limits_{x\to 2}5x+\lim\limits_{x\to 2}6=3\lim\limits_{x\to 2}x^2-5\lim\limits_{x\to 2}x+6$

$$=3\times 4-5\times 2+6=8$$

例 1.3 计算$\lim\limits_{x\to 3}\dfrac{4x-5}{2x+3}$.

解 $\lim\limits_{x\to 3}\dfrac{4x-5}{2x+3}=\dfrac{\lim\limits_{x\to 3}(4x-5)}{\lim\limits_{x\to 3}(2x+3)}=\dfrac{12-5}{6+3}=\dfrac{7}{9}$.

例 1.4 计算$\lim\limits_{x\to 0}\dfrac{5x^2+3x}{x}$.

解 $\lim\limits_{x\to 0}\dfrac{5x^2+3x}{x}=\lim\limits_{x\to 0}\dfrac{x(5x+3)}{x}=\lim\limits_{x\to 0}(5x+3)=3$.

例 1.5 求$\lim\limits_{x\to a}\dfrac{x^5-a^5}{x^4-a^4}$.

解 $\lim\limits_{x\to a}\dfrac{x^5-a^5}{x^4-a^4}=\lim\limits_{x\to a}\dfrac{(x-a)(x^4+x^3a+x^2a^2+xa^3+a^4)}{(x-a)(x^3+x^2a+xa^2+a^3)}$

$$=\lim\limits_{x\to a}\frac{x^4+x^3a+x^2a^2+xa^3+a^4}{x^3+x^2a+xa^2+a^3}$$

$$=\frac{5a^4}{4a^3}=\frac{5}{4}a.$$

例 1.6 计算 $\lim\limits_{x\to a}\dfrac{x^{\frac{1}{3}}-a^{\frac{1}{3}}}{x^{\frac{1}{2}}-a^{\frac{1}{2}}}$.

解 $\lim\limits_{x\to a}\dfrac{x^{\frac{1}{3}}-a^{\frac{1}{3}}}{x^{\frac{1}{2}}-a^{\frac{1}{2}}}=\lim\limits_{x\to a}\dfrac{(x^{1/6})^2-(a^{1/6})^2}{(x^{1/6})^3-(a^{1/6})^3}=\lim\limits_{x\to a}\dfrac{(x^{1/6}-a^{1/6})(x^{1/6}+a^{1/6})}{(x^{1/6}-a^{1/6})(x^{2/6}+x^{1/6}a^{1/6}+a^{2/6})}$

$$=\lim\limits_{x\to a}\frac{x^{1/6}+a^{1/6}}{x^{2/6}+x^{1/6}a^{1/6}+a^{2/6}}=\frac{2a^{1/6}}{3a^{1/3}}=\frac{2}{3a^{1/6}}.$$

例 1.7 计算 $\lim\limits_{x\to a}\dfrac{\sqrt{x+a}-\sqrt{3x-a}}{x-a}$.

解 $\lim\limits_{x\to a}\dfrac{\sqrt{x+a}-\sqrt{3x-a}}{x-a}=\lim\limits_{x\to a}\dfrac{(\sqrt{x+a}-\sqrt{3x-a})(\sqrt{x+a}+\sqrt{3x-a})}{(x-a)(\sqrt{x+a}+\sqrt{3x-a})}$

$$=\lim\limits_{x\to a}\frac{x+a-3x+a}{(x-a)(\sqrt{x+a}+\sqrt{3x-a})}$$

$$=\lim\limits_{x\to a}\frac{-2(x-a)}{(x-a)(\sqrt{x+a}+\sqrt{3x-a})}$$

$$=\lim\limits_{x\to a}\frac{-2}{\sqrt{x+a}+\sqrt{3x-a}}=\frac{-2}{\sqrt{2a}+\sqrt{2a}}=\frac{-1}{\sqrt{2a}}$$

例 1.8 求 $\lim\limits_{x\to +\infty}(\sqrt{x+a}-\sqrt{x})$.

解 $\lim\limits_{x\to +\infty}(\sqrt{x+a}-\sqrt{x})=\lim\limits_{x\to +\infty}\dfrac{(\sqrt{x+a}-\sqrt{x})(\sqrt{x+a}+\sqrt{x})}{\sqrt{x+a}+\sqrt{x}}$

$$=\lim\limits_{x\to +\infty}\frac{x+a-x}{\sqrt{x+a}+\sqrt{x}}=\lim\limits_{x\to +\infty}\frac{a}{\sqrt{x+a}+\sqrt{x}}=0$$

1.2.6　函数极限的两个重要结论及两个重要极限

下面不加证明地给出关于函数极限的两个重要结论.

定理 1.2　如果在点 x_0 附近(x_0 除外)，$f(x)\leqslant g(x)$，且 $\lim\limits_{x\to x_0}f(x)$ 与 $\lim\limits_{x\to x_0}g(x)$ 存在，则

$$\lim_{x\to x_0}f(x)\leqslant\lim_{x\to x_0}g(x)$$

定理 1.3(夹逼定理)　如果在点 x_0 附近(x_0 除外)，$f(x)\leqslant g(x)\leqslant h(x)$，且 $\lim\limits_{x\to x_0}f(x)=\lim\limits_{x\to x_0}h(x)=L$，则

$$\lim_{x\to x_0}g(x)=L$$

重要极限 1　$\lim\limits_{x\to 0}\dfrac{\sin x}{x}=1$.

证明　在如图 1-5 所示的单位圆中，因为 $\triangle OAB$ 的面积$\leqslant$扇形 OAB 的面积$\leqslant\triangle OPA$，故有

$$\frac{1}{2}OA\cdot BC\leqslant\frac{1}{2}OA^2\cdot\theta\leqslant\frac{1}{2}OA\cdot AP$$

又因为 $OA=OB=1$，$BC=\sin\theta$，$AP=\tan\theta$，从而有

$$\sin\theta\leqslant\theta\leqslant\tan\theta$$

进一步有

$$1\leqslant\frac{\theta}{\sin\theta}\leqslant\frac{1}{\cos\theta}$$

即

$$\cos\theta\leqslant\frac{\sin\theta}{\theta}\leqslant 1$$

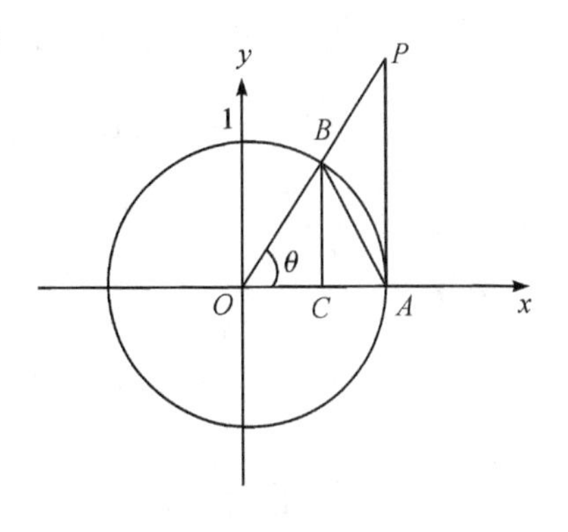

图 1-5

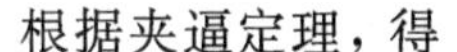

根据夹逼定理，得

$$\lim_{x\to 0^+}\frac{\sin x}{x}=1$$

由于 $\dfrac{\sin x}{x}$ 是偶函数，故有

$$\lim_{x\to 0^-}\frac{\sin x}{x}=1$$

综上，有

$$\lim_{x\to 0}\frac{\sin x}{x}=1$$

重要极限 2　$\lim\limits_{x\to\infty}\left(1+\dfrac{1}{x}\right)^x=\mathrm{e}$.

由于该结论的证明超出了本教材的教学要求，因此我们通过一个近似的方法去验证上面的结论，见表 1-1.

表 1-1

x	1	10	100	1000	10000	100000	1000000
$\left(1+\frac{1}{x}\right)^x$	≈2.000000	≈2.563742	≈2.704814	≈2.716924	≈2.718146	≈2.718268	≈2.718280

通过变量替换，不难得到

$$\lim_{x\to 0}(1+x)^{\frac{1}{x}}=\mathrm{e}$$

例 1.9 求下列函数的极限.

(1) $\lim\limits_{x\to 0}\frac{\ln(1+x)}{x}$；　　(2) $\lim\limits_{x\to 0}\frac{\mathrm{e}^x-1}{x}$；

(3) $\lim\limits_{x\to 0}\frac{a^x-1}{x}$；　　(4) $\lim\limits_{x\to 0}\frac{\mathrm{e}^{3x}-1}{x\cdot 5^x}$；

(5) $\lim\limits_{x\to 0}\frac{a^x-b^x}{x}$；　　(6) $\lim\limits_{x\to 0}\frac{\tan x}{x}$；

(7) $\lim\limits_{x\to 0}\frac{1-\cos 3x}{3x^2}$；　　(8) $\lim\limits_{x\to a}\frac{\sqrt{x}-\sqrt{a}}{\tan(x-a)}$.

解 (1) $\lim\limits_{x\to 0}\frac{\ln(1+x)}{x}=\lim\limits_{x\to 0}\frac{1}{x}\ln(1+x)=\lim\limits_{x\to 0}\ln(1+x)^{\frac{1}{x}}=\ln\lim\limits_{x\to 0}(1+x)^{\frac{1}{x}}=\ln\mathrm{e}=1$.

(2) 令 $\mathrm{e}^x-1=t\Rightarrow x=\ln(1+t)$，而且当 $x\to 0$ 时，$t\to 0$. 所以，$\lim\limits_{x\to 0}\frac{\mathrm{e}^x-1}{x}=\lim\limits_{t\to 0}\frac{t}{\ln(1+t)}=\lim\limits_{t\to 0}\frac{1}{\frac{\ln(1+t)}{t}}=\frac{1}{1}=1$.

(3) 令 $a^x-1=t\Rightarrow x\ln a=\ln(1+t)\Rightarrow x=\frac{\ln(1+t)}{\ln a}$，而且当 $x\to 0$ 时，$t\to 0$. 所以 $\lim\limits_{x\to 0}\frac{a^x-1}{x}=\lim\limits_{t\to 0}\frac{t}{\frac{\ln(1+t)}{\ln a}}=\ln a$.

(4) $\lim\limits_{x\to 0}\frac{\mathrm{e}^{3x}-1}{x\cdot 5^x}=\lim\limits_{x\to 0}\frac{\mathrm{e}^{3x}-1}{3x}\times 3\times\frac{1}{5^x}=3\lim\limits_{x\to 0}\frac{\mathrm{e}^{3x}-1}{3x}\cdot\lim\limits_{x\to 0}\frac{1}{5^x}=3$.

(5) $\lim\limits_{x\to 0}\frac{a^x-b^x}{x}=\lim\limits_{x\to 0}\frac{(a^x-1)-(b^x-1)}{x}=\lim\limits_{x\to 0}\frac{a^x-1}{x}-\lim\limits_{x\to 0}\frac{b^x-1}{x}=\ln\frac{a}{b}$.

(6) $\lim\limits_{x\to 0}\frac{\tan x}{x}=\lim\limits_{x\to 0}\frac{\sin x}{\cos x}\frac{1}{x}=\lim\limits_{x\to 0}\frac{\sin x}{x}\cdot\lim\limits_{x\to 0}\frac{1}{\cos x}=1$.

(7) $\lim\limits_{x\to 0}\frac{1-\cos 3x}{3x^2}=\lim\limits_{x\to 0}\frac{2\left(\sin\frac{3x}{2}\right)^2}{3\times\frac{4}{9}\left(\frac{3x}{2}\right)^2}=\frac{3}{2}\lim\limits_{x\to 0}\left(\frac{\sin\frac{3x}{2}}{\frac{3x}{2}}\right)^2=\frac{3}{2}$.

(8) $\lim\limits_{x\to a}\frac{\sqrt{x}-\sqrt{a}}{\tan(x-a)}=\lim\limits_{x\to a}\left[\frac{\sqrt{x}-\sqrt{a}}{\tan(x-a)}\cdot\frac{\sqrt{x}+\sqrt{a}}{\sqrt{x}+\sqrt{a}}\right]$，利用题(6)的结论，$\lim\limits_{x\to 0}\frac{\tan x}{x}=1\Rightarrow\lim\limits_{x\to 0}\frac{x}{\tan x}=1\Rightarrow\lim\limits_{x\to a}\frac{x-a}{\tan(x-a)}=1$，所以 $\lim\limits_{x\to a}\frac{\sqrt{x}-\sqrt{a}}{\tan(x-a)}=\lim\limits_{x\to a}\frac{x-a}{\tan(x-a)}\cdot\lim\limits_{x\to a}\frac{1}{\sqrt{x}+\sqrt{a}}=\frac{1}{2\sqrt{a}}$.

1.3　连续性

1.3.1　在一点的连续性

在$[a, b]$上连续函数的几何直观是，在这个区间上函数 f 的图像是一条光滑没有断点的曲线，这条曲线可以通过连续地移动铅笔，且不离开纸面画出来. 类似地，不连续函数的图像是一条不连续的曲线. 看图 1－6 并讨论它们的本质.

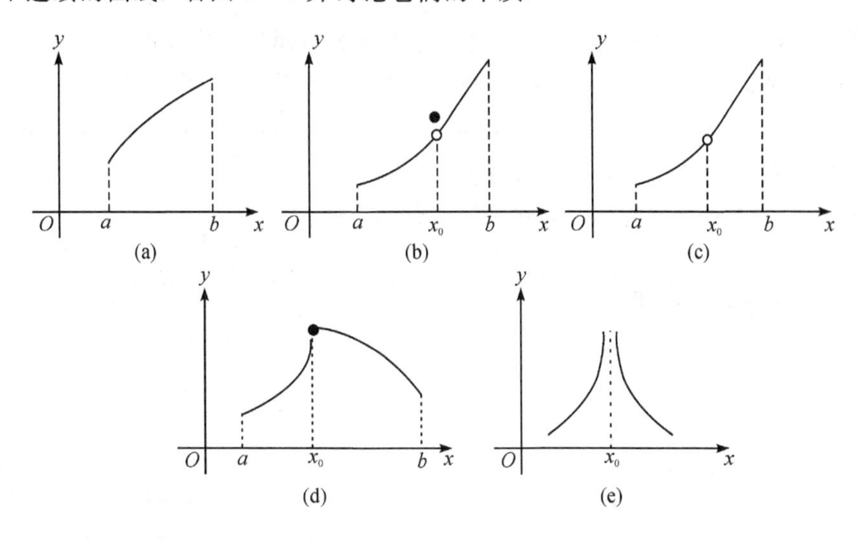

图 1－6

显然，除了图 1－6(a)中的函数以外，其他几个图中的函数都是不连续的. 在图 1－6(b)中，$\lim\limits_{x\to x_0}f(x)$存在且 $f(x_0)$有定义，但 $\lim\limits_{x\to x_0}f(x)\neq f(x_0)$. 在图 1－6(c)中，$\lim\limits_{x\to x_0}f(x)$存在但 $f(x)$在 $x=x_0$ 无定义. 在图 1－6(d)中，$f(x)$在 $x=x_0$ 有定义，但 $\lim\limits_{x\to x_0}f(x)$不存在. 在图 1－6(e)中，$\lim\limits_{x\to x_0}f(x)$不存在而且 $f(x)$在 $x=x_0$ 无定义. 从而$\lim\limits_{x\to x_0}f(x)=f(x_0)$可以作为函数在点 $x=x_0$ 处连续的充分必要条件.

定义 1.4　如果$\lim\limits_{x\to x_0}f(x)=f(x_0)$，则称函数 $f(x)$在 $x=x_0$ 处连续.

函数 $f(x)$在 $x=x_0$处连续需要同时满足三个条件：

(1) $\lim\limits_{x\to x_0}f(x)$存在，即 $\lim\limits_{x\to x_0^-}f(x)$和 $\lim\limits_{x\to x_0^+}f(x)$ 存在且相等；

(2) $f(x)$在 $x=x_0$ 处有定义；

(3) $\lim\limits_{x\to x_0}f(x)=f(x_0)$.

因此，如果 $\lim\limits_{x\to x_0^-}f(x)=\lim\limits_{x\to x_0^+}f(x)=f(x_0)$，则 $f(x)$在 $x=x_0$处连续. 如果上述条件中任何一个不满足，则称函数在那个点是间断的.

1.3.2　间断点的类型

间断点包含下列几种类型.

(1) 如果$\lim\limits_{x\to x_0}f(x)$不存在，且$\lim\limits_{x\to x_0^-}f(x)\neq\lim\limits_{x\to x_0^+}f(x)$，则称$x=x_0$是函数$f(x)$的跳跃间断点.

(2) 如果$\lim\limits_{x\to x_0}f(x)\neq f(x_0)$或函数在$x=x_0$无定义，则称$x=x_0$是函数的可去间断点. 这种类型的间断点可以通过重新定义函数值而变为连续点.

(3) 如果$\lim\limits_{x\to x_0}f(x)=\infty$，则称$x=x_0$是函数$f(x)$的无穷间断点.

下面的例子有助于我们理解连续和间断.

(1) 图 1 - 7 中的函数是连续的，它在每一点都连续.

(2) 设$f(x)=\begin{cases}1, & x>0\\ 2, & x<0\end{cases}$，图像如图 1 - 8 所示. 函数在$x=0$处间断，$x=0$是跳跃间断点.

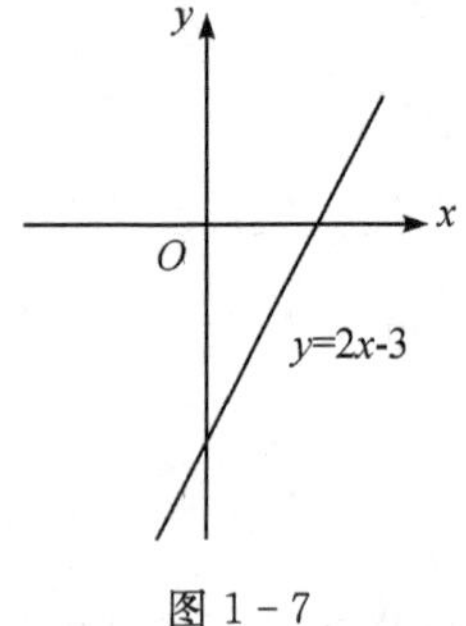

图 1 - 7

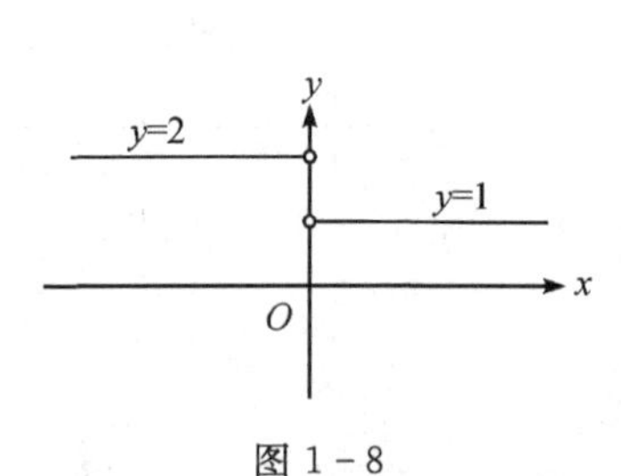

图 1 - 8

(3) 函数$f(x)=\dfrac{x^2-4}{x-2}$的图像如图 1 - 9 所示. 但$f(2)=\dfrac{0}{0}$为未定式，所以$f(2)$没有定义，$x=2$是函数的可去间断点.

(4) 函数$f(x)$的图像如图 1 - 10 所示. 由于$\lim\limits_{x\to 2^-}f(x)=-\infty$，$\lim\limits_{x\to 2^+}f(x)=+\infty$，所以$x=2$是函数的无穷间断点.

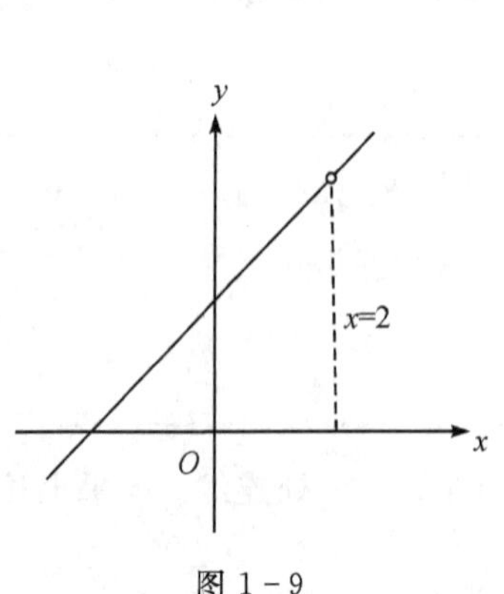

图 1 - 9

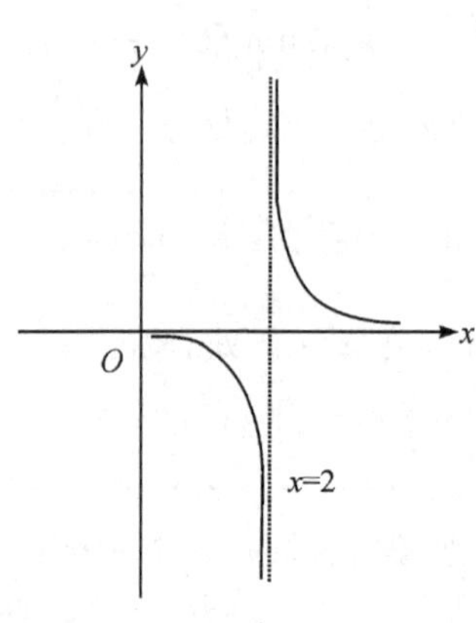

图 1 - 10

例 1.10　通过计算左右极限及函数值，判定下列函数在给定点处是否连续.

(1) $f(x)=2x^2-3x+10$，$x=1$；

(2) $f(x)=\dfrac{1}{x-2}$，$x=2$.

解　(1) $\lim\limits_{x\to1^-}f(x)=\lim\limits_{x\to1^-}(2x^2-3x+10)=9$，$\lim\limits_{x\to1^+}f(x)=\lim\limits_{x\to1^+}(2x^2-3x+10)=9$，又因为 $f(1)=9$，$\lim\limits_{x\to1}f(x)=f(1)$，所以 $f(x)$在 $x=1$ 处是连续的.

(2) 因为 $\lim\limits_{x\to2^-}f(x)=\lim\limits_{x\to2^-}\dfrac{1}{x-2}=-\infty$，说明函数在 $x=2$ 处极限不存在. 故 $f(x)$在 $x=2$ 处不连续.

例 1.11　设 $f(x)$定义如下

$$f(x)=\begin{cases}2x+3, & x<1\\4, & x=1\\6x-1, & x>1\end{cases}$$

问函数在 $x=1$ 处是否连续？如果不连续，怎样可以使函数在该点连续？

解　因为 $\lim\limits_{x\to1^-}f(x)=\lim\limits_{x\to1^-}(2x+3)=5$，$\lim\limits_{x\to1^+}f(x)=\lim\limits_{x\to1^+}(6x-1)=5$，所以 $\lim\limits_{x\to1}f(x)=5$. 但 $f(1)=4$，所以 $\lim\limits_{x\to1}f(x)\neq f(1)$. 因此 $f(x)$在 $x=1$ 是间断的，$x=1$ 是函数的可去间断点.

如果把所给函数在 $x=1$ 处的函数值改为 5

$$f(x)=\begin{cases}2x+3, & x<1\\5, & x=1\\6x-1, & x>1\end{cases}$$

则所给函数在重新定义后在 $x=1$ 处连续.

例 1.12　设函数 $f(x)$定义如下

$$f(x)=\begin{cases}x^2-1, & x<3\\2kx, & x\geqslant3\end{cases}$$

求使得函数在 $x=3$ 处连续的 k 值.

解　由于 $\lim\limits_{x\to3^-}f(x)=\lim\limits_{x\to3^-}(x^2-1)=8$，$\lim\limits_{x\to3^+}f(x)=\lim\limits_{x\to3^+}2kx=6k$，$f(3)=6k$，要使 $f(x)$在 $x=3$ 处连续，必有 $6k=8\Rightarrow k=\dfrac{4}{3}$.

1.3.3　区间上连续

定义 1.5　如果函数 $f(x)$在(a,b)内的每一点都连续，则称函数在开区间(a,b)内连续. 如果函数 $f(x)$在开区间(a,b)内连续，且在 $x=a$ 处右连续，在 $x=b$ 处左连续，即 $\lim\limits_{x\to a^+}f(x)=f(a)$且 $\lim\limits_{x\to b^-}f(x)=f(b)$，则称函数 $f(x)$在闭区间$[a,b]$上连续.

习题 1

1. 求下列极限.

(1) $\lim\limits_{x\to2}(2x^2+3x-14)$；　　(2) $\lim\limits_{x\to1}\dfrac{3x^2+2x-4}{x^2+5x-4}$；

(3) $\lim\limits_{x\to 0}\dfrac{4x^3-x^2+2x}{3x^2+4x}$;

(4) $\lim\limits_{x\to a}\dfrac{x^{\frac{2}{3}}-a^{\frac{2}{3}}}{x-a}$;

(5) $\lim\limits_{x\to 2}\dfrac{x^2-5x+6}{x^2-x-2}$;

(6) $\lim\limits_{x\to a}\dfrac{\sqrt{3x}-\sqrt{2x+a}}{2(x-a)}$;

(7) $\lim\limits_{x\to 1}\dfrac{\sqrt{2x}-\sqrt{3-x^2}}{x-1}$;

(8) $\lim\limits_{x\to 0}\dfrac{\sqrt[6]{x}-2}{\sqrt[3]{x}-4}$;

(9) $\lim\limits_{x\to 1}\dfrac{x-\sqrt{2-x^2}}{2x-\sqrt{2+2x^2}}$.

2. 计算下列极限.

(1) $\lim\limits_{x\to\infty}\dfrac{3x^2-4}{4x^2}$;

(2) $\lim\limits_{x\to\infty}\dfrac{5x^2+2x-7}{3x^2+5x+2}$;

(3) $\lim\limits_{x\to+\infty}(\sqrt{x-a}-\sqrt{x-b})$;

(4) $\lim\limits_{x\to+\infty}\sqrt{x}(\sqrt{x}-\sqrt{x-a})$.

3. 计算下列极限.

(1) $\lim\limits_{x\to 0}\dfrac{\sin mx}{\sin nx}$;

(2) $\lim\limits_{x\to 0}\dfrac{\tan ax}{\tan bx}$;

(3) $\lim\limits_{x\to 0}\dfrac{\sin ax\cos bx}{\sin cx}$;

(4) $\lim\limits_{x\to 0}\dfrac{1-\cos 2x}{x^2}$;

(5) $\lim\limits_{x\to 0}\dfrac{\sin ax-\sin bx}{x}$;

(6) $\lim\limits_{x\to 0}\dfrac{\tan x-\sin x}{x^3}$;

(7) $\lim\limits_{x\to\frac{\pi}{4}}\dfrac{\sec^2 x-2}{\tan x-1}$;

(8) $\lim\limits_{x\to y}\dfrac{\sin x-\sin y}{x-y}$;

(9) $\lim\limits_{\theta\to\frac{\pi}{4}}\dfrac{\cos\theta-\sin\theta}{\theta-\dfrac{\pi}{4}}$;

(10) $\lim\limits_{x\to c}\dfrac{\sqrt{x}-\sqrt{c}}{\sin x-\sin c}$.

4. 求下列极限.

(1) $\lim\limits_{x\to 0}\dfrac{e^{6x}-1}{x}$;

(2) $\lim\limits_{x\to 0}\dfrac{e^{2x}-1}{x\cdot 2^{x+1}}$;

(3) $\lim\limits_{x\to 0}\dfrac{e^{ax}-e^{bx}}{x}$;

(4) $\lim\limits_{x\to 0}\dfrac{a^x+b^x-2}{x}$;

(5) $\lim\limits_{x\to 2}\dfrac{x-2}{\ln(x-1)}$;

(6) $\lim\limits_{x\to\frac{\pi}{2}}\dfrac{\cos x}{\ln\left(x-\dfrac{\pi}{2}+1\right)}$.

5. 通过计算左右极限及函数值,判定下列函数在给定点处是否连续.

(1) $f(x)=2-3x^2$, $x=1$;

(2) $f(x)=\dfrac{1}{2x}$, $x=0$;

(3) $f(x)=\dfrac{x^2-9}{x-3}$, $x=3$;

(4) $f(x)=\dfrac{|x-2|}{x-2}$, $x=2$.

6. 讨论函数在给定点处的连续性.

(1) $f(x)=\begin{cases}2x^2+1, & x\leqslant 2\\ 4x+1, & x>2\end{cases}$, $x=2$;

(2) $f(x)=\begin{cases}2x+1, & x<1\\ 2, & x=1\\ 3x, & x>1\end{cases}$, $x=1$

7. (1)设函数 $f(x)$定义如下

$$f(x)=\begin{cases}x^2+2, & x<5\\ 20, & x=5\\ 3x+12, & x>5\end{cases}$$

证明 $x=5$ 是函数 $f(x)$的可去间断点.

(2) 设函数 $f(x)$定义如下

$$f(x)=\begin{cases}2x-3, & x<2\\ 2, & x=2\\ 3x-5, & x>2\end{cases}$$

该函数 $f(x)$在 $x=2$ 连续吗？如果不连续，怎样使得函数在 $x=2$ 连续？

8. (1)设函数 $f(x)$定义如下

$$f(x)=\begin{cases}kx+3, & x\geqslant 2\\ 3x-1, & x<2\end{cases}$$

求 k,使得函数 $f(x)$在 $x=2$ 连续.

(2) 设函数 $f(x)$定义如下

$$f(x)=\begin{cases}\dfrac{2x^2-18}{x-3}, & x\neq 3\\ k, & x=3\end{cases}$$

求 k,使得函数 $f(x)$在 $x=3$ 连续.

第2章　导数

微分学的产生来源于对两个问题的求解，一个问题是求曲线的切线，一个问题是计算变速直线运动的瞬时速度．在这两个问题中，所涉及的曲线为连续曲线，采用的是极限的思想．因此，微分学的研究对象是连续函数，这两个问题是由牛顿（英国，1642—1727）和莱布尼茨（德国，1646—1716）在同一时期从不同的角度解决的，在此过程中微分学创立了．

2.1　曲线的切线

设连续曲线 AB 的方程为 $y=f(x)$，P，Q 是曲线上的任意两点．点 P 和 Q 的坐标分别为 (x_0, y_0) 和 (x, y)．当点 P 沿曲线移动到点 Q 时，产生的水平位移为 PR，产生的垂直位移为 RQ，如图 2-1 所示．

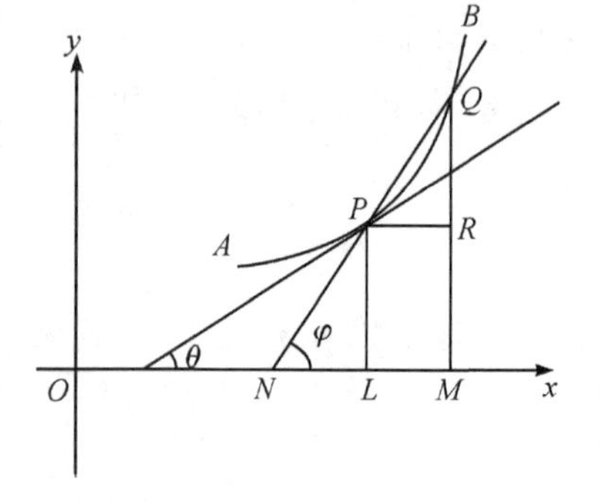

图 2-1

$$PR=LM=OM-OL=x-x_0$$

$$RQ=QM-RM=y-PL=y-y_0$$

$x-x_0$ 和 $y-y_0$ 分别为 x 的改变量和 y 的改变量，记作 Δx 和 Δy，即

$$\Delta x = x - x_0,\ \Delta y = y - y_0$$

$$\Delta y = y - y_0 = f(x) - f(x_0) = f(x_0 + \Delta x) - f(x_0)$$

连接点 P 和点 Q，得到割线 PQ 与 x 轴的夹角为 φ，即 $\angle QPR=\angle QNM=\varphi$，且割线 PQ 的斜率为

$$\tan\varphi = \frac{QR}{PR} = \frac{\Delta y}{\Delta x}$$

当 Q 沿着曲线移动趋近于 P 时，割线绕着 P 转动．当 Q 最终与 P 重合时，割线转到了极限位置，即转到了点 P 的切线处，此时切线与 x 轴的夹角为 θ，所以

$$\tan\theta = \lim_{\Delta x\to 0}\tan\varphi = \lim_{\Delta x\to 0}\frac{\Delta y}{\Delta x} \text{ 或 } \tan\theta = \lim_{\Delta x\to 0}\frac{\Delta y}{\Delta x} = \lim_{\Delta x\to 0}\frac{f(x_0+\Delta x)-f(x_0)}{\Delta x}$$

因此 $\lim\limits_{\Delta x\to 0}\dfrac{\Delta y}{\Delta x}$ 或 $\lim\limits_{\Delta x\to 0}\dfrac{f(x_0+\Delta x)-f(x_0)}{\Delta x}$ 就是曲线 $y=f(x)$ 在点 P 处的切线的斜率．

2.2　瞬时速度

设质点沿直线 AB 运动，则距离随着时间增加．因此距离 s 可以看成时间 t 的函数 f，设为 $s=f(t)$．

A　　P　　Q　　B

在时刻 t 质点位于点 P；在时刻 $t+\Delta t$ 质点位于点 Q，则 $AP=s$，$AQ=s+\Delta s$．于是，

$$PQ = AQ - AP = s + \Delta s - s = \Delta s$$

且有，

$$\Delta s = s + \Delta s - s = f(t + \Delta t) - f(t)$$

从而，质点在时间区间$[t, t+\Delta t]$的平均速度$\bar{v}$为

$$\bar{v} = \frac{\Delta s}{\Delta t} = \frac{f(t + \Delta t) - f(t)}{\Delta t}$$

当$\Delta t \to 0$时，Q趋近P. 此时质点在时刻t的瞬时速度v为当$\Delta t \to 0$时平均速度$\bar{v}$的极限.

$$v = \lim_{\Delta t \to 0} \frac{\Delta s}{\Delta t} = \lim_{\Delta t \to 0} \frac{f(t + \Delta t) - f(t)}{\Delta t}$$

2.3　导数

上述两个问题需要我们计算出当自变量趋于零时函数的改变量之比的极限. 这种求极限的方法叫作函数**微分法**. 记作$\frac{\mathrm{d}}{\mathrm{d}x}$，其结果称为**导数**.

定义 2.1　设函数f定义在区间(a, b)内，$x_0 \in (a, b)$. 若$\lim\limits_{\Delta x \to 0} \frac{f(x_0 + \Delta x) - f(x_0)}{\Delta x}$存在，则称函数$f(x)$在$x = x_0$处可导，并把该极限值称为函数$f(x)$在$x = x_0$的导数，记作$f'(x_0)$，$\frac{\mathrm{d}y}{\mathrm{d}x}\Big|_{x=x_0}$，$y'\big|_{x=x_0}$，即

$$f'(x_0) = \lim_{\Delta x \to 0} \frac{f(x_0 + \Delta x) - f(x_0)}{\Delta x}$$

如果记$x = x_0 + \Delta x$，则$\Delta x = x - x_0$，于是上述表达式也可写成

$$f'(x_0) = \lim_{x \to x_0} \frac{f(x) - f(x_0)}{x - x_0}$$

如果函数在区间(a, b)的每一点都可导，则称$f(x)$为区间(a, b)上的可导函数，从而有

$$f'(x) = \lim_{\Delta x \to 0} \frac{f(x + \Delta x) - f(x)}{\Delta x}$$

f的导函数记作

$$f'(x),\ \frac{\mathrm{d}f(x)}{\mathrm{d}x},\ y',\ \frac{\mathrm{d}y}{\mathrm{d}x}$$

如果$f(x)$在区间上可导，则$f'(x_0) = f'(x)\big|_{x=x_0}$.

利用记号，可以把上述两个问题表述为：

(1) 函数$y = f(x)$在点(x, y)处切线的斜率等于函数$y = f(x)$对x的导数，即$f'(x)$或$\frac{\mathrm{d}f(x)}{\mathrm{d}x}$.

(2) 质点在时刻t的瞬时速度等于位移函数$s = f(t)$对t的导数，即s'或$\frac{\mathrm{d}f(t)}{\mathrm{d}t}$.

定义 2.2　如果函数$y = f(x)$可导，则称$f'(x)\mathrm{d}x$为该函数的微分，记作$\mathrm{d}y = f'(x)\mathrm{d}x$.

根据微分的定义，$\mathrm{d}x = \Delta x$. 图 2-2 显示了微分的几何意义，即曲线在x_0处的切线的纵坐标的改变量，即SR.

$\triangle PRS$ 称为弧微分三角形，它表明在 P 点附近，可以用切线的长度 PS 近似代替曲线的长度$\overset{\frown}{PQ}$，即

$$\overset{\frown}{PQ} \approx \mathrm{d}s = \sqrt{(\mathrm{d}x)^2 + (\mathrm{d}y)^2}$$

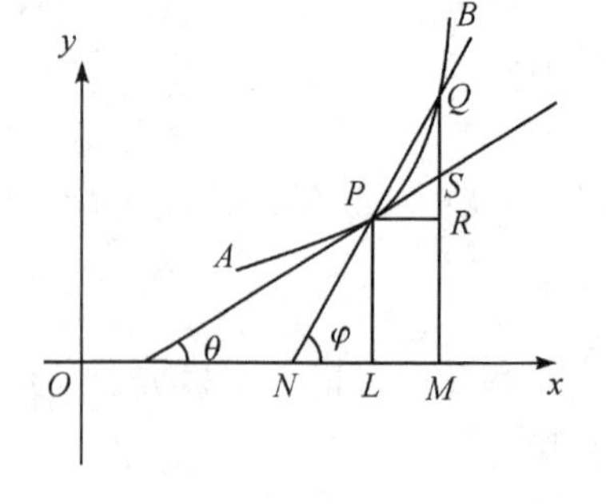

图 2-2

现在，我们计算常数函数 C 和 x^n ($n \in \mathbf{Z}^+$)的导数.

(1) 设 $y=f(x)=C$，C 为常数.

于是 $y+\Delta y=C$，从而 $\Delta y=C-y=C-C=0 \Rightarrow \dfrac{\Delta y}{\Delta x}=0$，所以

$$\frac{\mathrm{d}y}{\mathrm{d}x} = \lim_{\Delta x \to 0} \frac{\Delta y}{\Delta x} = 0$$

(2) 设 $y=f(x)=x^n$ ($n \in \mathbf{Z}^+$)，则

$$\Delta y = (x + \Delta x)^n - x^n$$

$$\frac{\mathrm{d}y}{\mathrm{d}x} = \lim_{\Delta x \to 0} \frac{\Delta y}{\Delta x} = \lim_{\Delta x \to 0} \frac{(x + \Delta x)^n - x^n}{(x + \Delta x) - x} = nx^{n-1}$$

注意: 后面将会证明这个公式对 n 为任何实数都成立.

定理 2.1 可导与连续的关系：

如果 $f(x)$在 $x=x_0$ 可导，则 $f(x)$在 $x=x_0$ 处连续.

证明 因为 $f(x)$在 $x=x_0$ 可导，所以 $f'(x_0)=\lim\limits_{\Delta x \to 0}\dfrac{\Delta y}{\Delta x}$存在，从而

$$\lim_{\Delta x \to 0} \Delta y = \lim_{\Delta x \to 0} \frac{\Delta y}{\Delta x} \Delta x = \lim_{\Delta x \to 0} \frac{\Delta y}{\Delta x} \lim_{\Delta x \to 0} \Delta x = 0$$

根据连续的定义知，函数 $f(x)$在 $x=x_0$ 处连续.

2.4 微分法

下面推导求导公式.

2.4.1 和的导数

如果 $f(x)$和 $g(x)$是任意两个可导函数，则

$$\frac{\mathrm{d}}{\mathrm{d}x}[f(x) \pm g(x)] = \frac{\mathrm{d}}{\mathrm{d}x} f(x) \pm \frac{\mathrm{d}}{\mathrm{d}x} g(x)$$

或

$$[f(x) \pm g(x)]' = f'(x) \pm g'(x)$$

证明 令 $h(x)=f(x)+g(x)$，则

$$\begin{aligned} h'(x) &= \lim_{\Delta x \to 0} \frac{h(x + \Delta x) - h(x)}{\Delta x} \\ &= \lim_{\Delta x \to 0} \frac{[f(x + \Delta x) \pm g(x + \Delta x)] - [f(x) \pm g(x)]}{\Delta x} \\ &= \lim_{\Delta x \to 0} \left[\frac{f(x + \Delta x) - f(x)}{\Delta x} \pm \frac{g(x + \Delta x) - g(x)}{\Delta x} \right] \end{aligned}$$

$$= \lim_{\Delta x \to 0}\left[\frac{f(x+\Delta x)-f(x)}{\Delta x}\right] \pm \lim_{\Delta x \to 0}\left[\frac{g(x+\Delta x)-g(x)}{\Delta x}\right]$$
$$= f'(x) \pm g'(x)$$

和的导数公式可以推广到多个函数的和，例如

$$[f(x)+g(x)+p(x)]' = f'(x)+g'(x)+p'(x)$$

例 2.1　求函数 $5x^3+4x^2-2x+7$ 的导数.

解　令 $y=5x^3+4x^2-2x+7$，有

$$y' = (5x^3+4x^2-2x+7)' = (5x^3)'+(4x^2)'+(-2x)'+(7)'$$
$$= 5(x^3)'+4(x^2)'-2(x)' = 15x^2+8x-2$$

2.4.2　乘积的求导法则

设 $f(x)$和 $g(x)$均可导，则

$$\frac{\mathrm{d}}{\mathrm{d}x}[f(x)g(x)] = f(x)\frac{\mathrm{d}}{\mathrm{d}x}[g(x)] + g(x)\frac{\mathrm{d}}{\mathrm{d}x}[f(x)]$$

或

$$[f(x)g(x)]' = f'(x)g(x)+f(x)g'(x)$$

证明　令 $h(x)=f(x)g(x)$，则

$$h'(x) = \lim_{\Delta x \to 0}\frac{h(x+\Delta x)-h(x)}{\Delta x}$$
$$= \lim_{\Delta x \to 0}\frac{f(x+\Delta x)g(x+\Delta x)-f(x)g(x)}{\Delta x}$$

为了计算这个极限，我们像证明和的导数公式一样，通过在分子中加减 $f(x+\Delta x)g(x)$把函数拆开.

$$h'(x) = \lim_{\Delta x \to 0}\frac{f(x+\Delta x)g(x+\Delta x)-f(x+\Delta x)g(x)+f(x+\Delta x)g(x)-f(x)g(x)}{\Delta x}$$
$$= \lim_{\Delta x \to 0}\left[f(x+\Delta x)\frac{g(x+\Delta x)-g(x)}{\Delta x}+g(x)\frac{f(x+\Delta x)-f(x)}{\Delta x}\right]$$
$$= \lim_{\Delta x \to 0}f(x+\Delta x)\cdot\lim_{\Delta x \to 0}\frac{g(x+\Delta x)-g(x)}{\Delta x}+\lim_{\Delta x \to 0}g(x)\cdot\lim_{\Delta x \to 0}\frac{f(x+\Delta x)-f(x)}{\Delta x}$$
$$= f(x)g'(x)+g(x)f'(x)$$

注意：极限$\lim\limits_{\Delta x \to 0}g(x)$的变量是 Δx，所以 $g(x)$可看作常数，故$\lim\limits_{\Delta x \to 0}g(x)=g(x)$. 又因为 f 在 x 是可导的，所以也是连续的，因此

$$\lim_{\Delta x \to 0}f(x+\Delta x) = f(x)$$

根据乘积的求导法则，不难得到，当 k 为任意实数时，有

$$[kf(x)]' = kf'(x)$$

例 2.2　设 $f(x)=(3x^2-5x)(2x+3)$，求 $f'(x)$.

解

$$f'(x) = (3x^2-5x)'(2x+3)+(3x^2-5x)(2x+3)'$$
$$= (6x-5)(2x+3)+2(3x^2-5x) = 18x^2-2x-15$$

2.4.3　商的求导法则

设 $u(x)$和 $v(x)$是任意两个可导函数，令 $h(x)=\dfrac{u(x)}{v(x)}$，则

$$h'(x)=\left[\frac{u(x)}{v(x)}\right]'=\frac{u'(x)v(x)-u(x)v'(x)}{v^2(x)}$$

证明 根据定义，有

$$\begin{aligned}
h'(a)&=\lim_{x\to a}\frac{h(x)-h(a)}{x-a}=\lim_{x\to a}\frac{\dfrac{u(x)}{v(x)}-\dfrac{u(a)}{v(a)}}{x-a}\\
&=\lim_{x\to a}\frac{v(a)u(x)-u(a)v(x)}{(x-a)v(x)v(a)}\\
&=\lim_{x\to a}\frac{v(a)u(x)-v(a)u(a)+v(a)u(a)-u(a)v(x)}{(x-a)v(x)v(a)}\\
&=\lim_{x\to a}\frac{1}{v(x)v(a)}\left[v(a)\frac{u(x)-u(a)}{(x-a)}-u(a)\frac{v(x)-v(a)}{(x-a)}\right]\\
&=\frac{1}{v(a)v(a)}\left[v(a)u'(a)-u(a)v'(a)\right]\\
&=\frac{v(a)u'(a)-u(a)v'(a)}{v^2(a)}
\end{aligned}$$

一般地

$$h'(x)=\left[\frac{u(x)}{v(x)}\right]'=\frac{u'(x)v(x)-u(x)v'(x)}{v^2(x)}$$

或

$$\frac{\mathrm{d}}{\mathrm{d}x}\left[\frac{u(x)}{v(x)}\right]=\frac{v(x)\dfrac{\mathrm{d}}{\mathrm{d}x}u(x)-u(x)\dfrac{\mathrm{d}}{\mathrm{d}x}v(x)}{v^2(x)}$$

例 2.3 求$\dfrac{4x^2+3}{3x^2-2}$的导数.

解 令 $u(x)=4x^2+3$，$v(x)=3x^2-2$，则

$$\begin{aligned}
\frac{\mathrm{d}}{\mathrm{d}x}\left[\frac{u(x)}{v(x)}\right]&=\frac{v(x)\dfrac{\mathrm{d}}{\mathrm{d}x}u(x)-u(x)\dfrac{\mathrm{d}}{\mathrm{d}x}v(x)}{v^2(x)}\\
&=\frac{(3x^2-2)\times 8x-(4x^2+3)\times 6x}{(3x^2-2)^2}=\frac{-34x}{(3x^2-2)^2}
\end{aligned}$$

例 2.4 求$\dfrac{\sqrt{x}+a}{\sqrt{x}+b}$的导数.

解 $\left(\dfrac{\sqrt{x}+a}{\sqrt{x}+b}\right)'=\dfrac{(\sqrt{x}+a)'(\sqrt{x}+b)-(\sqrt{x}+a)(\sqrt{x}+b)'}{(\sqrt{x}+b)^2}=\dfrac{b-a}{2\sqrt{x}(\sqrt{x}+b)^2}$

2.4.4 链式法则

定理 2.2 如果 $y=f(u)$，$u=g(x)$，且 f 和 g 是可导的，则 $y=f[g(x)]$是可导的，且

$$\frac{\mathrm{d}y}{\mathrm{d}x}=\frac{\mathrm{d}y}{\mathrm{d}u}\cdot\frac{\mathrm{d}u}{\mathrm{d}x}=f'(u)u'(x)$$

证明　设 Δx 是 x 的改变量，相应的 u 的改变量为 Δu，则

$$\Delta u = g(x+\Delta x) - g(x)$$

由于 $g(x)$是可导的，所以 $g(x)$连续，因此

$$\lim_{\Delta x \to 0} \Delta u = \lim_{\Delta x \to 0} [g(x+\Delta x) - g(x)] = g(x) - g(x) = 0$$

故

$$\frac{dy}{dx} = \lim_{\Delta x \to 0} \frac{\Delta y}{\Delta x} = \lim_{\Delta x \to 0} \left[\frac{\Delta y}{\Delta u} \frac{\Delta u}{\Delta x}\right] = \lim_{\Delta u \to 0} \frac{\Delta y}{\Delta u} \lim_{\Delta x \to 0} \frac{\Delta u}{\Delta x} = \frac{dy}{du} \cdot \frac{du}{dx}$$

例 2.5　设 $y=4u^2-3u+5$，$u=2x^2-3$，求$\dfrac{dy}{dx}$.

解　因为$\dfrac{dy}{du}=8u-3$，$\dfrac{du}{dx}=4x$，所以

$$\frac{dy}{dx} = \frac{dy}{du} \cdot \frac{du}{dx} = (8u-3)4x = [8(2x^2-3)-3]4x = 64x^3 - 108x.$$

在熟悉链式法则后，可以跳过对中间变量的求导，即$\dfrac{dy}{du}$，从复合函数的外层向内层直接求导.

例 2.6　求函数$(2-3x)^{1/2}$的导数.

解　$[(2-3x)^{1/2}]'=\dfrac{1}{2}(2-3x)^{\frac{1}{2}-1}(2-3x)'=\dfrac{-3}{2\sqrt{2-3x}}$.

例 2.7　求$\sqrt{\dfrac{1-x}{1+x}}$的导数.

解

$$\left(\sqrt{\frac{1-x}{1+x}}\right)' = \frac{1}{2\sqrt{\frac{1-x}{1+x}}} \cdot \left(\frac{1-x}{1+x}\right)'$$

$$= \frac{1}{2\sqrt{\frac{1-x}{1+x}}} \cdot \frac{(1-x)'(1+x)-(1-x)(1+x)'}{(1+x)^2}$$

$$= -\frac{1}{(1+x)\sqrt{1-x^2}}.$$

例 2.8　求函数$\dfrac{1}{x+\sqrt{x^2-a^2}}$的导数.

解

$$\left(\frac{1}{x+\sqrt{x^2-a^2}}\right)' = -\frac{1}{(x+\sqrt{x^2-a^2})^2}(x+\sqrt{x^2-a^2})'$$

$$= -\frac{1}{(x+\sqrt{x^2-a^2})^2}\left[1+\frac{1}{2\sqrt{x^2-a^2}}(x^2-a^2)'\right]$$

$$= -\frac{1}{(x+\sqrt{x^2-a^2})\sqrt{x^2-a^2}}$$

求导法则

1. 设 $f(x)$ 和 $g(x)$ 均可导，则

(1) $[f(x)\pm g(x)]'=f'(x)\pm g'(x)$；

或 $\mathrm{d}[f(x)\pm g(x)]=f'(x)\mathrm{d}x\pm g'(x)\mathrm{d}x$；

(2) $[f(x)g(x)]'=f'(x)g(x)+f(x)g'(x)$；

或 $\mathrm{d}[f(x)g(x)]=g(x)\mathrm{d}f(x)+f(x)\mathrm{d}g(x)$；

(3) $\left[\dfrac{u(x)}{v(x)}\right]'=\dfrac{u'(x)v(x)-u(x)v'(x)}{v^2(x)}$，$v(x)\neq 0$；

或 $\mathrm{d}\left[\dfrac{u(x)}{v(x)}\right]=\dfrac{v(x)\mathrm{d}u(x)-u(x)\mathrm{d}v(x)}{v^2(x)}$.

2. 设 $y=f(u)$，$u=g(x)$，其中 f 和 g 均可导，则 $y=f[g(x)]$ 可导，且

$$\frac{\mathrm{d}y}{\mathrm{d}x}=\frac{\mathrm{d}y}{\mathrm{d}u}\cdot\frac{\mathrm{d}u}{\mathrm{d}x}=f'(u)u'(x)$$

2.5 高阶导数

定义 2.3 设函数 $y=f(x)$ 可导，则称 $\dfrac{\mathrm{d}y}{\mathrm{d}x}=f'(x)$ 为 $f(x)$ 关于 x 的一阶导数，如果再对 x 求导，$\dfrac{\mathrm{d}}{\mathrm{d}x}\left(\dfrac{\mathrm{d}y}{\mathrm{d}x}\right)$ 可以写成 $\dfrac{\mathrm{d}^2y}{\mathrm{d}x^2}$ 或 $f''(x)$，称为关于 x 的**二阶导数**. 按照此方法，可以定义更高阶的导数.

例 2.9 求函数 $y=5x^4-3x^2+11$ 的二阶和高阶导数.

解 $\dfrac{\mathrm{d}y}{\mathrm{d}x}=y'=20x^3-6x$，$\dfrac{\mathrm{d}^2y}{\mathrm{d}x^2}=y''=60x^2-6$，$\dfrac{\mathrm{d}^3y}{\mathrm{d}x^3}=y'''=120x$，$\dfrac{\mathrm{d}^4y}{\mathrm{d}x^4}=y^{(4)}=120$，

$\dfrac{\mathrm{d}^5y}{\mathrm{d}x^5}=y^{(5)}=0$，其他阶导数为 0.

如果 $y=f(x)=a_nx^n+a_{n-1}x^{n-1}+\cdots+a_1x+a_0$，则

$$\frac{\mathrm{d}^ny}{\mathrm{d}x^n}=f^{(n)}(x)=n!\cdot a_n$$

$$\frac{\mathrm{d}^{n+1}y}{\mathrm{d}x^{n+1}}=f^{(n+1)}(x)=0$$

2.6 隐函数及隐函数微分法

定义 2.4 设 $F(x, y)$ 为任意的含有两个变量 x 和 y 的函数，如果 $y=f(x)$ 是由方程 $F(x, y)=0$ 所确定的，则称 $F(x, y)=0$ 为一个**隐函数**.

注意：从方程 $F(x, y)=0$ 可以解 y，也可以不解 y 直接逐项对 x 求导，解出 $\dfrac{\mathrm{d}y}{\mathrm{d}x}$. 不解 y 求 $\dfrac{\mathrm{d}y}{\mathrm{d}x}$ 的过程称为**隐函数微分法**.

例 2.10　设 $2x^2-3y^2=16$，利用隐函数微分法求$\frac{dy}{dx}$.

解　方程两端直接对 x 求导，得到

$$4x-6y\cdot\frac{dy}{dx}=0$$

$$\frac{dy}{dx}=\frac{2x}{3y}.$$

例 2.11　设 $x^3+y^3-3axy=0$，求$\frac{dy}{dx}$.

解　方程两端直接对 x 求导，得到

$$3x^2+3y^2\frac{dy}{dx}-3a\frac{d}{dx}(xy)=0$$

$$3x^2+3y^2\frac{dy}{dx}-3a\left(y+x\frac{dy}{dx}\right)=0$$

$$\frac{dy}{dx}=\frac{ay-x^2}{y^2-ax}$$

例 2.12　设 $xy^2=(x+2y)^3$，求$\frac{dy}{dx}$.

解　方程两端直接对 x 求导，得到

$$y^2+x\cdot 2y\frac{dy}{dx}=3(x+2y)^2\frac{d}{dx}(x+2y)$$

$$y^2+x\cdot 2y\frac{dy}{dx}=3(x+2y)^2\left(1+2\frac{dy}{dx}\right)$$

$$\frac{dy}{dx}=\frac{y}{x}\quad[\text{因为 } xy^2=(x+2y)^3]$$

2.7　参数方程和其所确定的函数的微分法

定义 2.5　设 $x=x(t)$，$y=y(t)$均可导，且$\frac{dx}{dt}\neq 0$，则称$\begin{cases}x=x(t)\\y=y(t)\end{cases}$为**参数方程**.

注意：可以消去 t 也可以不消去 t，通过把 t 看成中间变量然后利用链式法则求$\frac{dy}{dx}$.

例 2.13　如果 $x=t+\frac{1}{t}$，$y=t-\frac{1}{t}$，求$\frac{dy}{dx}$.

解　$\frac{dy}{dx}=\frac{dy}{dt}\cdot\frac{dt}{dx}=\frac{y'(t)}{x'(t)}=\frac{1+\frac{1}{t^2}}{1-\frac{1}{t^2}}=\frac{t^2+1}{t^2-1}.$

例 2.14　求函数$(3x-1)^2$ 对 $2x+1$ 的导数.

解　令 $y=(3x-1)^2$，$u=2x-1$，则

$$\frac{dy}{du}=\frac{dy/dx}{du/dx}=\frac{6(3x-1)}{2}=3(3x-1)$$

2.8 三角函数的导数

2.8.1 sin$x$ 和 cos$x$ 的导数

设
$$y=\sin x \tag{2.1}$$
令 Δx 为 x 的改变量，Δy 为相应的 y 的改变量，则
$$y+\Delta y=\sin(x+\Delta x) \tag{2.2}$$
式(2.2)－式(2.1)得
$$\Delta y=\sin(x+\Delta x)-\sin x=2\sin\frac{\Delta x}{2}\cos\frac{2x+\Delta x}{2}$$
所以
$$\frac{\mathrm{d}y}{\mathrm{d}x}=\lim_{\Delta x\to 0}\frac{\Delta y}{\Delta x}=\lim_{\Delta x\to 0}\left(\frac{\sin\frac{\Delta x}{2}}{\frac{\Delta x}{2}}\cos\frac{2x+\Delta x}{2}\right)=\cos x$$
用类似的方法可以得到
$$\frac{\mathrm{d}\cos x}{\mathrm{d}x}=-\sin x \text{ 或 } (\cos x)'=-\sin x$$

2.8.2 tan$x$ 和 cot$x$ 的导数

利用商的求导法则，我们有
$$\frac{\mathrm{d}\tan x}{\mathrm{d}x}=\left(\frac{\sin x}{\cos x}\right)'=\frac{(\sin x)'\cos x-\sin x\,(\cos x)'}{\cos^2 x}$$
$$=\frac{\cos^2 x+\sin^2 x}{\cos^2 x}=\sec^2 x$$
用类似的方法可以得到
$$\frac{\mathrm{d}\cot x}{\mathrm{d}x}=-\csc^2 x$$

2.8.3 sec$x$ 和 csc$x$ 的导数

因为
$$\sec x=\frac{1}{\cos x}$$
利用商的求导法则和链式求导法则，我们有
$$(\sec x)'=\left(\frac{1}{\cos x}\right)'=-\frac{1}{\cos^2 x}(\cos x)'=\frac{\sin x}{\cos x\cdot\cos x}=\tan x\sec x$$
所以
$$(\sec x)'=\tan x\sec x$$
用类似的方法可以得到
$$(\csc x)'=-\cot x\csc x$$

三角函数的导数公式

$(\sin x)'=\cos x$　　$(\cos x)'=-\sin x$

$(\tan x)'=\sec^2 x$　　$(\cot x)'=-\csc^2 x$

$(\sec x)'=\tan x\sec x$　　$(\csc x)'=-\cot x\csc x$

例 2.14　求下列函数的导数.

(1) $\sin(ax+b)$；　　(2) $\sin(ax^2-b)$；

(3) $\sqrt{\tan 2x}$；　　(4) $x^2\sec(ax-b)$；

(5) $\dfrac{1+\sin x}{1-\sin x}$；　　(6) $\dfrac{1-\tan x}{\sec x}$.

解　(1) $[\sin(ax+b)]'=\cos(ax+b)\cdot(ax+b)'=a\cos(ax+b)$.

(2) $[\sin(ax^2-b)]'=\cos(ax^2-b)\cdot(ax^2-b)'=2ax\cos(ax^2-b)$.

(3) $(\sqrt{\tan 2x})'=\dfrac{1}{2\sqrt{\tan 2x}}(\tan 2x)'=\dfrac{\sec^2 2x}{\sqrt{\tan 2x}}$.

(4) $[x^2\sec(ax-b)]'=2x\sec(ax-b)+ax^2\sec(ax-b)\tan(ax-b)$.

(5) $\left(\dfrac{1+\sin x}{1-\sin x}\right)'=\dfrac{\cos x(1-\sin x)-(1+\sin x)(-\cos x)}{(1-\sin x)^2}=\dfrac{2\cos x}{(1-\sin x)^2}$.

(6) $\left(\dfrac{1-\tan x}{\sec x}\right)'=[\cos x(1-\tan x)]'=-\sin x(1-\tan x)+\cos x(-\sec^2 x)$

$=\sin x\tan x-\sin x-\cos x\sec^2 x$.

例 2.15　设 $x-y=\sin xy$，求$\dfrac{\mathrm{d}y}{\mathrm{d}x}$.

解　等式两端对 x 求导

$$1-y'=\cos xy\ (xy)'=\cos xy\cdot(y+xy')$$

$$y'=\frac{1-y\cos xy}{1+x\cos xy}$$

例 2.16　设 $y=2\theta-\tan\theta$，$x=\tan\theta$，求$\dfrac{\mathrm{d}y}{\mathrm{d}x}$.

解　$\dfrac{\mathrm{d}y}{\mathrm{d}x}=\dfrac{y'(\theta)}{x'(\theta)}=\dfrac{2-\sec^2\theta}{\sec^2\theta}=2\cos^2\theta-1=\cos 2\theta$.

2.9　反三角函数的导数

2.9.1　arcsin*x* 和 arccos*x* 的导数

令 $y=\arcsin x$，则 $x=\sin y$，等式两端对 y 求导，得

$$\frac{\mathrm{d}x}{\mathrm{d}y}=\cos y=\sqrt{1-\sin^2 y}=\sqrt{1-x^2}$$

$$\frac{\mathrm{d}y}{\mathrm{d}x}=\frac{1}{\sqrt{1-x^2}}$$

类似地，可以得到

$$\frac{\mathrm{d}}{\mathrm{d}x}(\arccos x)=-\frac{1}{\sqrt{1-x^2}}$$

2.9.2 arctan$x$ 和 arccot$x$ 的导数

令 $y=\arctan x$，则 $x=\tan y$，等式两端对 y 求导，得

$$\frac{\mathrm{d}x}{\mathrm{d}y}=\sec^2 y=1+\tan^2 y=1+x^2$$

$$\frac{\mathrm{d}y}{\mathrm{d}x}=\frac{1}{1+x^2}$$

类似地，可得

$$\frac{\mathrm{d}}{\mathrm{d}x}(\operatorname{arccot} x)=-\frac{1}{1+x^2}$$

反三角函数的导数

$$(\arcsin x)'=\frac{1}{\sqrt{1-x^2}} \qquad (\arccos x)'=-\frac{1}{\sqrt{1-x^2}}$$

$$(\arctan x)'=\frac{1}{1+x^2} \qquad (\operatorname{arccot} x)'=-\frac{1}{1+x^2}$$

例 2.17 设 $y=x^3\operatorname{arccot} x$，求 $\frac{\mathrm{d}y}{\mathrm{d}x}$.

解 $\frac{\mathrm{d}y}{\mathrm{d}x}=3x^2\operatorname{arccot} x+x^3\left(-\frac{1}{1+x^2}\right)=3x^2\operatorname{arccot} x-\frac{x^3}{1+x^2}$

例 2.18 设 $y=\arctan\frac{2x}{1-x^2}$，求 $\frac{\mathrm{d}y}{\mathrm{d}x}$.

解
$$\begin{aligned}y'=\left(\arctan\frac{2x}{1-x^2}\right)'&=\frac{1}{1+\left(\frac{2x}{1-x^2}\right)^2}\left(\frac{2x}{1-x^2}\right)'\\&=\frac{(1-x^2)^2}{(1+x^2)^2}\frac{[2(1-x^2)-2x\ (1-x^2)']}{(1-x^2)^2}\\&=\frac{2}{1+x^2}\end{aligned}$$

2.10 对数函数与指数函数的导数

我们知道，指数函数与对数函数互为反函数，故如果设 $y=f(x)=a^x$，则有 $\log_a y=x$.

有一个特殊的指数函数 e^x，其中 $\mathrm{e}=\lim\limits_{n\to\infty}\left(1+\frac{1}{n}\right)^n$. e 的值介于 2 和 3 之间，约等于 2.718. 以 e 为底数的对数函数称为自然对数函数，记作 $\ln x$.

进一步，由

$$\mathrm{e}=\lim_{x\to\infty}\left(1+\frac{1}{x}\right)^x$$

且当 $x\to\infty$ 时 $t=\frac{1}{x}\to 0$，所以

$$\mathrm{e}=\lim_{x\to\infty}\left(1+\frac{1}{x}\right)^{x}=\lim_{t\to 0}(1+t)^{\frac{1}{t}}$$

我们应该熟悉对数运算的一些性质.

(1) $\log_a xy=\log_a x+\log_a y$；　(2) $\log_a \dfrac{x}{y}=\log_a x-\log_a y$；

(3) $\log_a x^n=n\log_a x$；　(4) $\log_a a=1$；

(5) $\log_a b=\dfrac{\log_m b}{\log_m a}=\log_a m\cdot\log_m b$；　(6) $\log_a 1=0$.

现在，我们通过定义利用下列极限结果来求指数函数和对数函数的导数.

(1) $\lim\limits_{x\to 0}\dfrac{\ln(1+x)}{x}=1$；　(2) $\lim\limits_{x\to 0}\dfrac{\mathrm{e}^x-1}{x}=1$；　(3) $\lim\limits_{x\to 0}\dfrac{a^x-1}{x}=\ln a$.

2.10.1　自然对数函数的导数

设 $y=\ln x$，则

$$(\ln x)'=\lim_{\Delta x\to 0}\frac{\Delta y}{\Delta x}=\lim_{\Delta x\to 0}\frac{\ln(x+\Delta x)-\ln x}{\Delta x}$$

$$=\lim_{\Delta x\to 0}\frac{\ln\left(1+\dfrac{\Delta x}{x}\right)}{x\dfrac{\Delta x}{x}}=\frac{1}{x}$$

例 2.19　求函数 $y=\ln|x|$ 的导数.

解　当 $x>0$ 时，$y=\ln x\Rightarrow y'=\dfrac{1}{x}$；

当 $x<0$ 时，$y=\ln(-x)\Rightarrow y'=\dfrac{1}{-x}(-x)'=\dfrac{1}{x}$；

所以，$(\ln|x|)'=\dfrac{1}{x}$.

2.10.2　e^x 的导数

令 $y=\mathrm{e}^x$，则

$$(\mathrm{e}^x)'=\lim_{\Delta x\to 0}\frac{\Delta y}{\Delta x}=\lim_{\Delta x\to 0}\frac{\mathrm{e}^{x+\Delta x}-\mathrm{e}^x}{\Delta x}$$

$$=\lim_{\Delta x\to 0}\mathrm{e}^x\frac{(\mathrm{e}^{\Delta x}-1)}{\Delta x}=\mathrm{e}^x$$

2.10.3　对数函数的导数

令 $y=\log_a x$，则

$$y=\log_a x=\frac{\ln x}{\ln a}\text{（把底换成 e）}$$

$$\frac{\mathrm{d}y}{\mathrm{d}x}=\frac{1}{\ln a}\frac{\mathrm{d}(\ln x)}{\mathrm{d}x}=\frac{1}{x\ln a}$$

2.10.4 a^x 的导数

令 $y=a^x$，则

$$(a^x)'=\lim_{\Delta x\to 0}\frac{\Delta y}{\Delta x}=\lim_{\Delta x\to 0}\frac{a^{x+\Delta x}-a^x}{\Delta x}$$

$$=\lim_{\Delta x\to 0}a^x\frac{(a^{\Delta x}-1)}{\Delta x}=a^x\ln a$$

基本初等函数的导数

1. $(x^n)'=nx^{n-1}(n\in\mathbf{Z}^+)$，实际上这个公式对任意的实数均成立.
2. $(a^x)'=a^x\ln a$，特殊地，$(e^x)'=e^x$.
3. $(\log_a x)'=\dfrac{1}{x\ln a}$，特殊地，$(\ln x)'=\dfrac{1}{x}$.
4. $(\sin x)'=\cos x$，$(\cos x)'=-\sin x$；
   $(\tan x)'=\sec^2 x$，$(\cot x)'=-\csc^2 x$；
   $(\sec x)'=\sec x\tan x$ $(\csc x)'=-\csc x\cot x$.
5. $(\arcsin x)'=\dfrac{1}{\sqrt{1-x^2}}$，$(\arccos x)'=-\dfrac{1}{\sqrt{1-x^2}}$；
   $(\arctan x)'=\dfrac{1}{1+x^2}$，$(\text{arccot}\, x)'=-\dfrac{1}{1+x^2}$.

2.10.5 $f(x)^{g(x)}(f(x)>0)$的导数

定义 2.5 $f(x)^{g(x)}(f(x)>0)$称为**幂指函数**.

方法 1 令 $y=f(x)^{g(x)}$，则 $\ln y=g(x)\ln f(x)$. 等式两端对 x 求导，得

$$\frac{1}{y}\frac{dy}{dx}=g'(x)\ln f(x)+\frac{g(x)}{f(x)}f'(x)$$

所以 $\dfrac{dy}{dx}=y\left[g'(x)\ln f(x)+\dfrac{g(x)}{f(x)}f'(x)\right]=f(x)^{g(x)}\left[g'(x)\ln f(x)+\dfrac{g(x)}{f(x)}f'(x)\right]$. 这个方法称为**对数求导法**.

方法 2 利用指数函数的性质和链式法则，我们有

$$y=f(x)^{g(x)}=e^{\ln f(x)^{g(x)}}=e^{g(x)\ln f(x)}$$

所以

$$\frac{dy}{dx}=f(x)^{g(x)}\left[g'(x)\ln f(x)+\frac{g(x)}{f(x)}f'(x)\right]$$

例 2.20 求 $y=\ln(5x^2+6)$的导数.

解 $\dfrac{dy}{dx}=[\ln(5x^2+6)]'=\dfrac{10x}{5x^2+6}$.

例 2.21 证明$(x^\alpha)'=\alpha x^{\alpha-1}(x>0)$，其中 α 为任意实数.

证明 因为 $x^\alpha=e^{\alpha\ln x}$,所以

$$(x^\alpha)'=(e^{\alpha\ln x})'=e^{\alpha\ln x}\alpha\frac{1}{x}=\alpha x^{\alpha-1}$$

例 2.22 求 $y=\sin 3x\ln(ax+b)$ 的导数.

解 $\frac{dy}{dx}=[\sin 3x\ln(ax+b)]'=3\cos 3x\ln(ax+b)+\frac{a\sin 3x}{ax+b}$.

例 2.23 求 $y=\frac{\sin x}{\ln(3x+4)}$ 的导数.

解 $\frac{dy}{dx}=\left[\frac{\sin x}{\ln(3x+4)}\right]'=\frac{\cos x\ln(3x+4)-\frac{3\sin x}{3x+4}}{[\ln(3x+4)]^2}=\frac{(3x+4)\cos x\ln(3x+4)-3\sin x}{(3x+4)[\ln(3x+4)]^2}$.

例 2.24 求 $y=e^{5x+4}\sin 6x$ 的导数.

解 $\frac{dy}{dx}=(e^{5x+4}\sin 6x)'=5e^{5x+4}\sin 6x+6e^{5x+4}\cos 6x=e^{5x+4}(5\sin 6x+6\cos 6x)$.

例 2.25 设 $x^y=y^x(x>0,\ y>0)$，求$\frac{dy}{dx}$.

解 利用对数函数的性质，有

$$y\ln x=x\ln y$$

则

$$\frac{dy}{dx}\ln x+\frac{y}{x}=\ln y+\frac{x}{y}\frac{dy}{dx}$$

因此

$$\frac{dy}{dx}=\frac{\ln y-\frac{y}{x}}{\ln x-\frac{x}{y}}=\frac{xy\ln y-y^2}{xy\ln x-x^2}$$

例 2.26 设 $y=\sqrt{\frac{a-x}{a+x}}$，求$\frac{dy}{dx}$.

解 利用对数函数的性质，有

$$\ln y=\frac{1}{2}\left(\ln|a-x|-\ln|a+x|\right)$$

等式两端对 x 求导，有

$$\frac{1}{y}y'=\frac{1}{2}\left(\frac{-1}{a-x}-\frac{1}{a+x}\right)=\frac{a}{x^2-a^2}$$

所以

$$y'=\frac{a}{x^2-a^2}\sqrt{\frac{a-x}{a+x}}$$

习题 2

1. 利用导数定义，求下列函数的导数.

(1) x^2-2； (2) $\frac{1}{x}$；

(3) $x^{\frac{1}{2}}$； (4) $\frac{2x+3x^{\frac{3}{4}}+x^{\frac{1}{2}}+1}{x^{\frac{1}{4}}}$；

(5) $(3x^4+5)(4x^5-3)$； (6) $\frac{x^2-2x}{x+1}$.

2. 利用链式法则求导.

(1) $y=2u^2-3u+1$ 和 $u=2x^2$; (2) $y=(2u^2+3)^{\frac{1}{3}}$ 和 $u=\sqrt{2x+1}$;

(3) $(3x^2+2x-1)^4$; (4) $\sqrt{8-5x}$;

(5) $\dfrac{1}{\sqrt{ax^2+bx+c}}$; (6) $\sqrt{\dfrac{x^2+a^2}{x^2-a^2}}$.

3. 求下列函数的二阶导数.

(1) $y=7x^2+6x-5$; (2) $y=3x^4-x^2+1$;

(3) $y=\dfrac{3}{x^2}$; (4) $\dfrac{1}{2x+1}$.

4. 利用隐函数求导法求下列函数的导数.

(1) $x^2+y^2=16$; (2) $\dfrac{x^2}{a^2}-\dfrac{y^2}{b^2}=1$;

(3) $x^2+2x^2y=y^3$; (4) $x^3y^6=(x+y)^9$.

5. 求下列函数的导数.

(1) $\cos^2 2x$; (2) $\sqrt{\sin 2x}$;

(3) $\tan(5x^2+6)$; (4) $\sec^2\dfrac{1}{x}$;

(5) $\tan(\cos 5x)$; (6) $\csc^3(\cot 4x)$;

(7) $\sec^2\left(\tan\sqrt{x}\right)$; (8) $\sin^2(\cos 6x)$.

6. 求下列函数的导数.

(1) $(x^2+3x)\sin 5x$; (2) $(x+\sin 2x)\sec 3x^2$;

(3) $\dfrac{1}{\sqrt{x}}\sin\sqrt{x}$; (4) $\dfrac{\sec nx}{ax-b}$;

(5) $\sin 2mx\sin 2nx$; (6) $\sin 3x\cos 5x$.

7. 求下列函数的导数.

(1) $\dfrac{1-2\sin^2\dfrac{x}{2}}{\cos^2 x}$; (2) $\dfrac{\sin 2nx}{\cos 2nx}$;

(3) $\dfrac{1-\cos x}{1+\cos x}$; (4) $\sqrt{\dfrac{1-\sin x}{1+\sin x}}$;

(5) $\dfrac{\cos 2x}{1-\sin 2x}$; (6) $\dfrac{1+\cos x}{1-\cos x}$;

(7) $\dfrac{\sec x+\tan x}{\sec x-\tan x}$; (8) $\dfrac{1+\tan x}{1-\tan x}$.

8. 求下列反三角函数的导数.

(1) $\arcsin(3x-4)$; (2) $\arccos\dfrac{3x^2-2}{2}$;

(3) $\arccos\dfrac{1-x^2}{1+x^2}$; (4) $\arctan\dfrac{1}{1-x^2}$;

(5) $\arctan\dfrac{\sin 2x}{1+\cos 2x}$; (6) $\operatorname{arcsec}\dfrac{1}{\sqrt{1-x^2}}$.

9. 求下列隐函数的导数.

(1) $x+y=\cos(x-y)$；　　(2) $x^2+y^2=\sin xy$；

(3) $x^2y^2=\tan(ax+by)$；　　(4) $xy=\tan(x^2+y^2)$.

10. 求由参数方程所确定的函数的导数.

(1) $x=a\cos^2\theta$，$y=b\sin^2\theta$；　　(2) $x=2a\sin t\cos t$，$y=b\cos 2t$；

(3) $x=2a\tan\theta$，$y=a\sec^2\theta$；　　(4) $x=a(\cos t+t\sin t)$，$y=a(\sin t-t\cos t)$；

(5) $x=a(\tan t-t\sec^2 t)$，$y=a\sec^2 t$.

11. 求下列函数的导数.

(1) $y=\ln\sin x$；　　(2) $y=\ln(x+\tan x)$；

(3) $y=\ln(1+e^{5x})$；　　(4) $y=\ln\ln x$；

(5) $y=\ln\sec x$；　　(6) $y=\ln(1+\sin^2 x)$；

(7) $y=\ln(e^{ax}+e^{-ax})$；　　(8) $y=\ln(\sqrt{a^2+x^2}+b)$.

12. 求下列函数的导数.

(1) $y=e^{\sqrt{\cos x}}$；　　(2) $y=e^{(1+\ln x)}$；

(3) $y=e^{\sin(\ln x)}$；　　(4) $y=\tan(\ln x)$；

(5) $y=\cos(\ln\sec x)$；　　(6) $y=\sec(\ln\tan x)$.

13. 求下列函数的导数.

(1) $y=x^y$；　　(2) $e^{\sin x}+e^{\sin y}=1$；

(3) $x^y\cdot y^x=1$；　　(4) $x^{\sin y}=y^{\sin x}$.

第 3 章　导数的应用

导数在科学、工程、经济、商业等领域发挥着十分重要的作用。例如，利用导数可以研究函数的性态，解决瞬时速度问题、收益最大成本最小问题等.

3.1　函数的单调性

函数有多种表现形式，其中最重要的一种是函数的图像. 首先从函数的增减性来了解函数的性态；其次，利用增减性求最大和最小值. 在这一节，我们应用导数来判定函数在区间上的增减性.

图 3-1 是函数 $y=f(x)$的图像，设 x_1，x_2，x_3 和 x_4 分别为点 A，B，C 和 D 的横坐标，相应地，$PA=f(x_1)$，$QB=f(x_2)$，$RC=f(x_3)$，$SD=f(x_4)$.

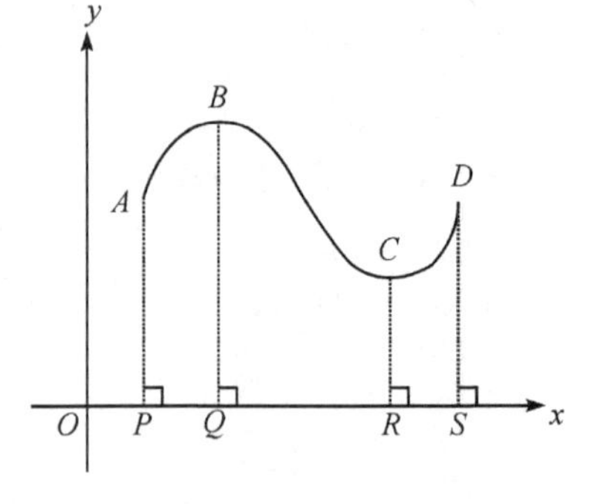

图 3-1

从图中可以看出，曲线从 A 到 B 是上升的，从 B 到 C 是下降的，从 C 到 D 又是上升的. 也就是说，

在 AB 部分，$x_2>x_1 \Rightarrow f(x_2)>f(x_1)$

在 BC 部分，$x_3>x_2 \Rightarrow f(x_3)<f(x_2)$

在 CD 部分，$x_4>x_3 \Rightarrow f(x_4)>f(x_3)$

下面我们给出函数单调增加和单调减少的定义.

3.1.1　单调增加

定义 3.1　对任意的 x_1，$x_2\in(a,b)$，如果当 $x_1<x_2$ 时，则有 $f(x_1)<f(x_2)$，则称函数 $y=f(x)$ 在区间(a,b)上是单调增加的.

这意味着随着 x 的增加，y(即 $f(x)$)也增加. 所以，曲线在任意一点处切线的斜率是正的，即如果$\dfrac{\mathrm{d}y}{\mathrm{d}x}=f'(x)>0$，则 $y=f(x)$是单调增的，如图 3-2 所示.

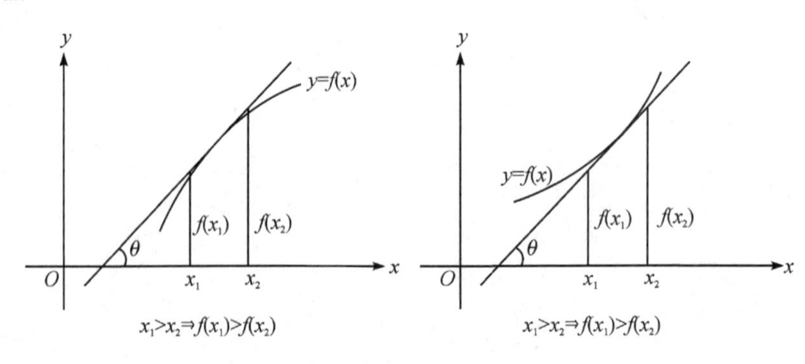

图 3-2

3.1.2　单调减少

定义 3.2　对任意的 x_1，$x_2 \in (a, b)$，如果当 $x_1 < x_2$ 时，有 $f(x_1) > f(x_2)$，则称函数 $y=f(x)$ 在区间 (a, b) 上是单调减少的.

这意味着随着 x 的增加，y（即 $f(x)$）减少，所以，曲线在任意一点处的斜率是负的，即如果 $\frac{\mathrm{d}y}{\mathrm{d}x}=f'(x)<0$，则 $y=f(x)$ 是单调减少的，如图 3－3 所示.

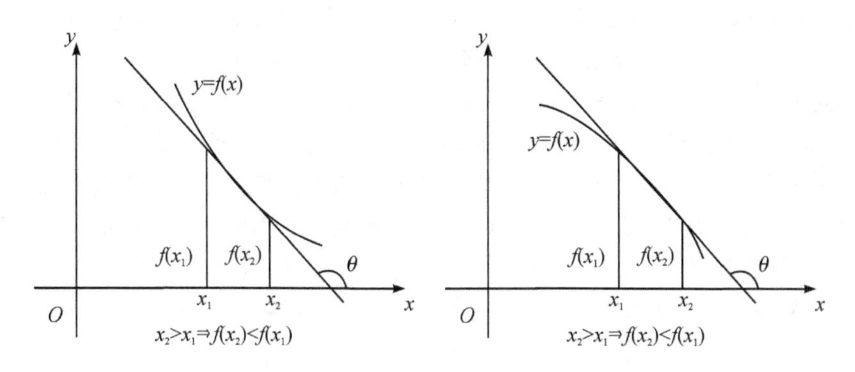

图 3－3

下面给出了如何利用导数的符号来判定函数的单调性.

利用导数符号判定函数的单调性

$f'(x)>0 \Rightarrow$ 函数单调增加

$f'(x)<0 \Rightarrow$ 函数单调减少

例 3.1　证明函数 $f(x)=\frac{1}{2}x^2-3x$ 在区间 $(3, +\infty)$ 上单调增加，在区间 $(-\infty, 3)$ 上单调减少.

证明　由 $f(x)=\frac{1}{2}x^2-3x \Rightarrow f'(x)=x-3$. 因为 $x>3$ 时 $f'(x)>0$，$x<3$ 时 $f'(x)<0$，所以 $f(x)$ 在区间 $(3, +\infty)$ 上单调增加，在区间 $(-\infty, 3)$ 上单调减少.

例 3.2　设 $f(x)=2x^3-15x^2+36x+1$，求函数的增减区间.

解　$f'(x)=6x^2-30x+36=6(x^2-5x+6)=6(x-2)(x-3)$.

令 $f'(x)=0 \Rightarrow x=2$ 和 $x=3$.

$x>3$ 时，$f'(x)>0$，$f(x)$ 在区间 $(3, +\infty)$ 上单调增加；$x<2$ 时，$f'(x)>0$，$f(x)$ 在区间 $(-\infty, 2)$ 上单调增加；$2<x<3$ 时，$f'(x)<0$，$f(x)$ 在区间 $2<x<3$ 上单调减少.

于是，$f(x)$ 在区间 $(-\infty, 2)\cup(3, \infty)$ 上单调增加，在 $x\in(2, 3)$ 上单调减少.

例 3.3　证明函数 $f(x)=x-\frac{1}{x}$ 在其定义域上单调增加.

证明　$f(x)=x-\frac{1}{x}$ 的定义域为 $(-\infty, 0)\cup(0, +\infty)$，则 $f'(x)=1+\frac{1}{x^2}>0\ (x\neq 0)$. 所以 $f(x)$ 在 $(-\infty, 0)\cup(0, +\infty)$ 上单调增加.

3.2　极值与最值

3.2.1　极值

定义 3.3　如果 $f(x_0 \pm \Delta x) < f(x_0)$，则称 $f(x)$ 在 $x = x_0$ 处取得极大值；如果 $f(x_0 \pm \Delta x) > f(x_0)$，则称 $f(x)$ 在 $x = x_0$ 处取得极小值.

为了正确地理解这个问题，有必要知道函数的局部最大值或局部最小值点附近的曲线的性态. 曲线在某一点上的性态是通过研究其邻近点切线的斜率得到的.

函数曲线上一点切线的斜率可以是正的、负的或零. 图 3-4 中，在区间 (a, c) 内斜率为正，即 $f'(x) > 0$，则切线的倾角为锐角，且切线从左下向右上倾斜(或上山)，函数是递增的. 在区间 (c, b) 内斜率为负，即 $f'(x) < 0$，则切线的倾角为钝角，且切线从左上向右下倾斜(或下山)，函数是递减的. 在点 c 处，$f'(x) = 0$，该点处的切线既不向上倾斜也不向下倾斜，而是平行于 x 轴. 我们把满足 $f'(x) = 0$ 的点称为函数的**驻点**. 曲线在驻点处有水平切线. 图 3-4 中，在驻点的左边邻域，函数单调增；在驻点的右边邻域，函数单调减，所以函数在这点取得极大值.

另一种情况如图 3-5 所示，函数在 c' 处取得极小值.

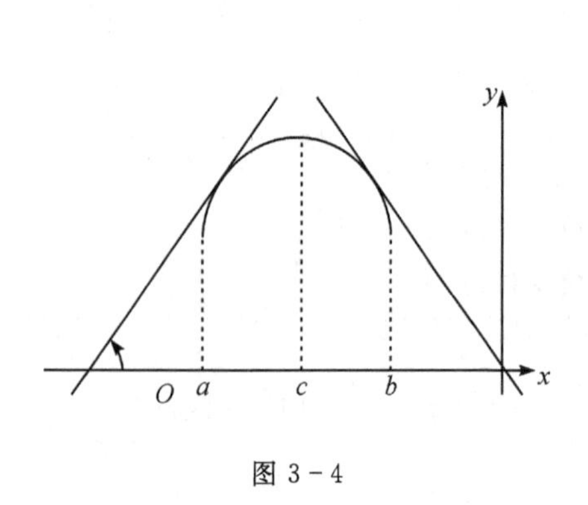

图 3-4

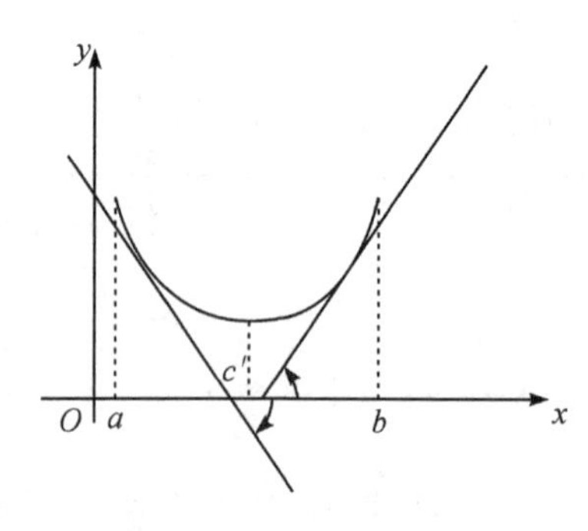

图 3-5

$f''(x)$ 是 $f'(x)$ 的导数，所以，如果 $f''(x) > 0$，则 $f'(x)$ 单调增加. 对于图 3-6(a)，斜率为增大的负值，对于图 3-6(b)，斜率为增大的正值. 图 3-6(a) 和图 3-6(b) 结合在一起构成图 3-6(c)，该曲线是凹的. 所以，曲线为凹的条件是函数的二阶导数 $f''(x)$ 大于 0.

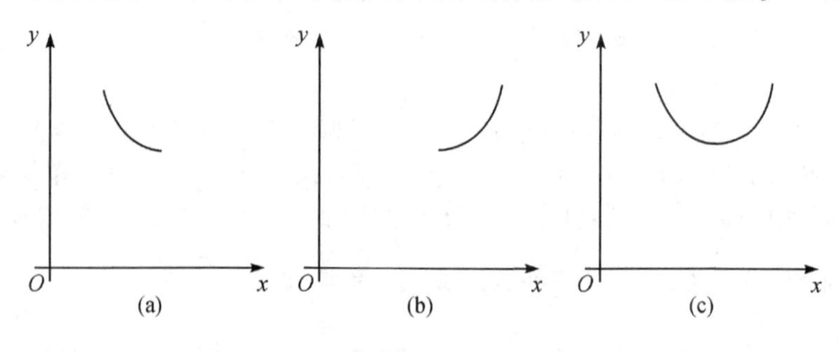

图 3-6

类似地，所以如果 $f''(x)<0$，则 $f'(x)$单调减. 对于图 3－7(a)，斜率为减小的正值，对于图 3－7(b)，斜率为减小的负值. 图 3－7(a)和图 3－7(b)结合在一起构成图 3－7(c)，该曲线是凸的. 所以，曲线为凸的条件是函数的二阶导数 $f''(x)$小于 0.

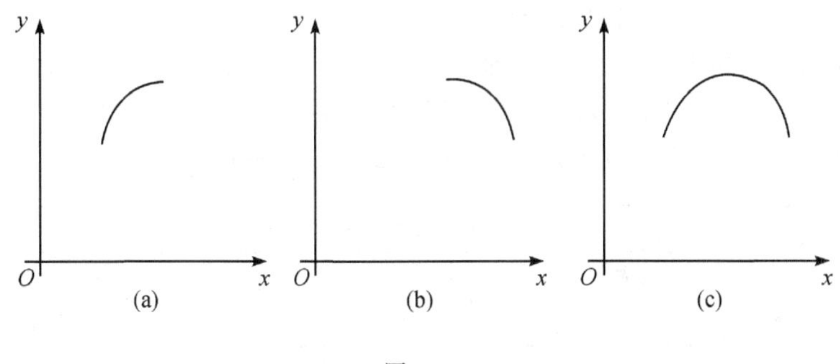

图 3－7

考虑图 3－8 中的函数图. 在区间(a, c)内，曲线是下凹的，因此 $f''(x)>0$，而在区间(c, b)内，曲线是上凸的，因此 $f''(x)<0$. 所以很自然地，在这个区间内一定存在一个点，在该点处 $f''(x)=0$，这个点很重要，它把下凹的部分和上凸的部分分开，这样的点称为**拐点**.

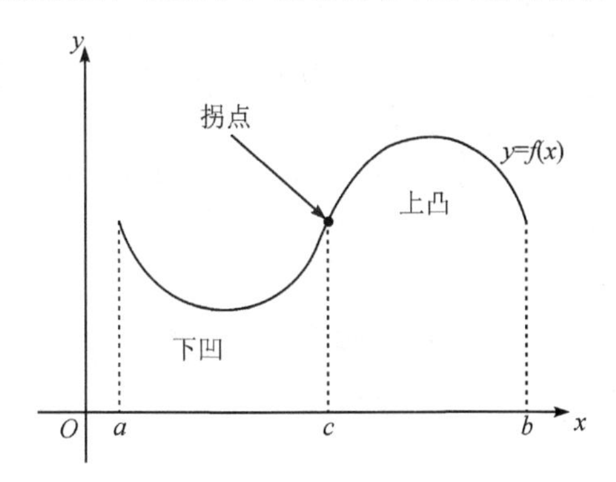

图 3－8

3.2.2　求极大值和极小值的步骤

设函数 $y=f(x)$，求极大值和极小值的步骤如下：

(1) 求 $f'(x)$ 或$\frac{\mathrm{d}y}{\mathrm{d}x}$；

(2) 令 $f'(x)=0$，解方程得到所有驻点 $x=x_i(i=1, 2, 3, \cdots, k)$；

(3) 确定当 x 从 $x_i-\Delta x$ 变到 $x_i+\Delta x$ 时$\frac{\mathrm{d}y}{\mathrm{d}x}$的符号；

(4) 如果$\frac{\mathrm{d}y}{\mathrm{d}x}$由正变负，则 $f(x)$在 $x=x_i$ 处取得极大值；如果$\frac{\mathrm{d}y}{\mathrm{d}x}$由负变正，则 $f(x)$ 在 $x=x_i$处取得极小值；如果$\frac{\mathrm{d}y}{\mathrm{d}x}$不变号，则 $f(x)$在 $x=x_i$ 处不取得极值.

3.2.3 求极大值和极小值的另一种方法

(1) 求 $f'(x)$，$f''(x)$；

(2) 令 $f'(x)=0$，求出所有驻点 $x=x_i(i=1, 2, \cdots, k)$；

(3) 计算 $f''(x_i)$. 如果 $f''(x_i)<0$，则 $f(x)$在 $x=x_i$ 处取得极大值 $f(x_i)$；

如果 $f''(x_i)>0$，则 $f(x)$在 $x=x_i$ 处取得极小值 $f(x_i)$；

如果 $f''(x_i)=0$，$f'''(x_i)\neq 0$，则 $f(x)$在 $x=x_i$ 处不取得极值.

项目	$y=f(x)$是否能取得极值的条件			
	极大值	极小值	非极值	
一阶导数	$f'(x)=0$	$f'(x)=0$	$f'(x)=0$	$f'(x)\neq 0$
二阶导数	$f''(x)<0$	$f''(x)>0$	$f''(x)=0$	
三阶导数			$f'''(x)\neq 0$	

例 3.4 求函数 $f(x)=2x^3-3x^2-36x$ 的极大值和极小值.

解 $f'(x)=6x^2-6x-36$，$f''(x)=12x-6$，令 $f'(x)=0 \Rightarrow 6x^2-6x-36=0 \Rightarrow x=3$ 或 $x=-2$.

(1) 当 $x=3$ 时，$f''(x)=36-6=30>0$，所以 $f(x)$在 $x=3$ 处取得极小值. 极小值为

$$f(3)=54-27-108=-81$$

(2) 当 $x=-2$ 时，$f''(x)=-24-6=-30<0$，所以 $f(x)$在 $x=-2$ 处取得极大值. 极大值为

$$f(-2)=-16-12+72=44$$

例 3.5 证明 $f(x)=x^3-3x^2+6x+4$ 既没有极大值也没有极小值.

证明 $f'(x)=3x^2-6x+6=3(x^2-2x+2)=3[(x-1)^2+1]>0$. 所以 $f(x)$既没有极大值也没有极小值.

3.2.4 最大值和最小值

函数 $y=f(x)$在 $x=x_0$ 处取得的最大值是函数 $y=f(x)$在定义域内所有函数值中最大的一个，即对于任意的 $x\in D(f)$，$f(x)\geqslant f(x_0)$.

如图 3-9 所示. 函数的极大值和极小值未必是函数的最大值和最小值，而只需是点附近的最大值和最小值. 一个函数可能有多个极大值和极小值，极大值未必大于极小值. 最大值在极大值点或端点处取得，最小值在极小值点或端点处取得.

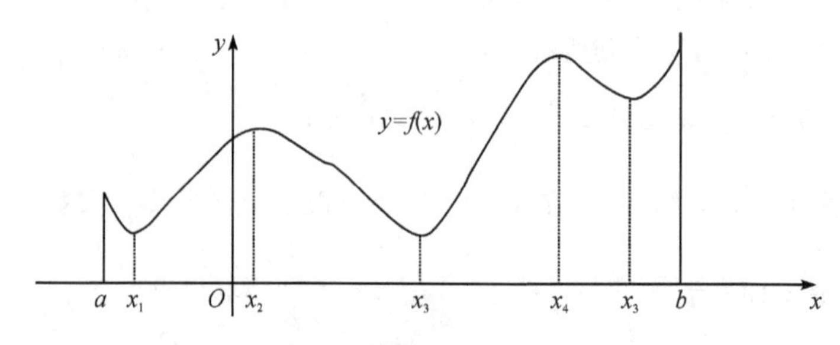

图 3-9

有时函数的最大值和最小值在区间的端点处取得，如图 3 - 10 所示.

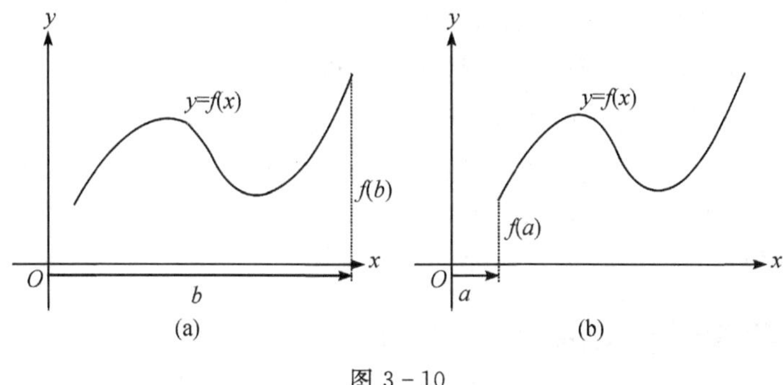

图 3 - 10

图 3 - 10(a)中，函数在右端点 $x=b$ 处取得最大值 $f(b)$. 图 3 - 10(b)中，函数在左端点 $x=a$ 处取得最小值 $f(a)$.

3.2.5　求最大值和最小值的步骤

设 $y=f(x)$定义在区间$[a, b]$上.

(1) 求 $f'(x)$.

(2) 令 $f'(x)=0$，求出所有驻点 $x=x_i(i=1, 2, \cdots, k)$.

(3) 计算 $f(x_i)$，$f(a)$，$f(b)$，从而

$$\max_{a\leqslant x\leqslant b} f(x)=\max\{f(x_i), f(a), f(b)\}, \quad \min_{a\leqslant x\leqslant b} f(x)=\min\{f(x_i), f(a), f(b)\}$$

例 3.6　求 $f(x)=2x^3-9x^2+12x+20$ 在$[-1, 5]$的最大值和最小值.

解　$f'(x)=6x^2-18x+12$，令 $f'(x)=0 \Rightarrow (x-1)(x-2)=0$，得驻点为 $x=1, 2$. 又因为

$$f(-1)=2(-1)^3-9(-1)^2+12\times(-1)+20=-3$$
$$f(5)=2\times5^3-9\times5^2+12\times5+20=105$$
$$f(1)=2\times1^3-9\times1^2+12\times1+20=25$$
$$f(2)=2\times2^3-9\times2^2+12\times2+20=24$$

所以最大值为 105，最小值为-3.

例 3.7　求长度为 60 m 的绳子所围成的矩形面积的最大值.

解　设矩形的长和宽分别为 x 和 y. 由题意知

$$2x+2y=60 \Rightarrow y=30-x$$

矩形面积为 $A=xy=x(30-x)=30x-x^2$，有

$$\frac{\mathrm{d}A}{\mathrm{d}x}=30-2x \Rightarrow \text{驻点为 } x=15$$

又因为$\frac{\mathrm{d}^2A}{\mathrm{d}x^2}=-2<0$，所以当 $x=15$ 时，面积取得最大值，最大值为 $A=15\text{ m}\times15\text{ m}=225\text{ m}^2$.

3.2.6　求函数凹凸区间的步骤

设 $y=f(x)$，求函数凹凸区间的步骤如下：

(1) 求 $f''(x)$；

(2) 求 $f''(x)>0$ 的区间，可得函数凹的区间；

(3) 求 $f''(x)<0$ 的区间，可得函数凸的区间.

例 3.8 设 $f(x)=2x^3-6x^2+5$，求函数的凹凸区间和拐点.

解 $f'(x)=6x^2-12x$，$f''(x)=12x-12=12(x-1)$，$f''(x)=0 \Rightarrow x=1$.

$x>1$ 时，$f''(x)>0$，函数在区间上是凹的.

$x<1$ 时，$f''(x)<0$，函数在区间上是凸的.

所以 (1，1)是曲线的拐点.

3.3 应用导数度量变化率

设 $y=f(x)$是连续函数，当 x 变化时，y 也随之变化. 当 x 和 y 产生微小改变 Δx 和 Δy 时，则

$$\frac{\Delta y}{\Delta x}=\frac{f(x+\Delta x)-f(x)}{\Delta x}$$

表示 x 每变化一个单位，y 的变化，因此，它刻画了在区间$[x, x+\Delta x]$上 y 关于 x 的平均变化率. 当 $\Delta x \to 0$ 时，平均变化率的极限就是瞬时变化率. 因此，

$$\lim_{\Delta x \to 0}\frac{\Delta y}{\Delta x}\bigg|_{x=x_0}=\lim_{\Delta x \to 0}\frac{f(x_0+\Delta x)-f(x_0)}{\Delta x}$$

即

$$\frac{\mathrm{d}y}{\mathrm{d}x}\bigg|_{x=x_0}=f'(x_0)=\lim_{\Delta x \to 0}\frac{f(x_0+\Delta x)-f(x_0)}{\Delta x}$$

是在 $x=x_0$ 处，y 关于 x 的瞬时变化率.

质点的平均速度和瞬时速度是平均变化率和瞬时变化率最常见的例子，如果时间的改变量为 Δt，相应的位移的改变量为 Δs，则$\frac{\Delta s}{\Delta t}$表示位移的平均变化率，$\lim\limits_{\Delta t \to 0}\frac{\Delta s}{\Delta t}=\frac{\mathrm{d}s}{\mathrm{d}t}$表示质点在时刻 t 的瞬时速度. 类似地，$\frac{\mathrm{d}v}{\mathrm{d}t}=\frac{\mathrm{d}^2 s}{\mathrm{d}t^2}$表示质点在时刻 t 的加速度.

例 3.9 设一个质点从给定点处开始沿直线运动，t 时刻后的位移为 $s=3+5t+t^3$. 求

(1) 在 $2\frac{1}{4}$ s 末的速度；(2) 在 $3\frac{2}{3}$ s 末的加速度；(3) 在第 5 s 内的平均速度.

解 $\frac{\mathrm{d}s}{\mathrm{d}t}=5+3t^2$，$\frac{\mathrm{d}^2 s}{\mathrm{d}t^2}=6t$.

(1) 当 $t=2\frac{1}{4}$ s 时，$\frac{\mathrm{d}s}{\mathrm{d}t}=5+3\times\left(\frac{9}{4}\right)^2=20\frac{3}{16}$ m/s.

(2) 当 $t=3\frac{2}{3}$ s 时，$\frac{\mathrm{d}^2 s}{\mathrm{d}t^2}=6\times\frac{11}{3}=22$ m/s^2.

(3) 平均速度$=\frac{\Delta s}{\Delta t}=\frac{s(5)-s(4)}{5-4}=\frac{153-87}{1}=66$ m/s.

在商业和经济领域，某些量的变化率常常为了解各种经济系统提供了可靠的数据. 例如，制造商不仅仅对某一生产水平的总成本感兴趣，也对不同生产水平的成本变化率感兴趣.

在经济学中，“边际”这个词指的是变化的程度，也就是导数. 因此，如果

$$C(x)=\text{产品数量为 } x \text{ 时的总成本}$$

则

$$C'(x) = \text{边际成本}$$
$$= \text{总成本 } C(x) \text{ 随产品数量 } x \text{ 变化的瞬时变化率}$$

例 3.10(边际成本)　假设每年制造 x 艘帆船的总成本函数为

$$C(x)=575+25x-0.25x^2 \text{(千元)}$$

(1) 求生产水平为每年 x 艘船时的边际成本.

(2) 求生产水平为每年 40 艘船时的边际成本，并对结果进行解释.

解　(1) $C'(x)=25-0.5x$.

(2) $C'(40)=25-0.5\times40=5$，即每艘船 5000 元.

当生产水平为每年 40 艘帆船时，每多生产一艘帆船，总成本将增加 5000 元.

例 3.11(销售分析)　一个家庭电子游戏推出 t 个月后的总销量 S (以千件计)为

$$S(t)=\frac{125t^2}{t^2+100}$$

(1) 求 $S'(t)$.

(2) 求 $S(10)$和 $S'(10)$，并对结果进行解释.

(3) 用(2)中的结果估计 11 个月后的总销量.

解　(1) $S'(t)=\dfrac{(t^2+100)(125t^2)'-125t^2\ (t^2+100)'}{(t^2+100)^2}=\dfrac{25000t}{(t^2+100)^2}$.

(2) $S(10)=62.5$，$S'(10)=6.25$.

10 个月后，生产游戏量为 62500 件. 在此基础上，每多生产一个月，游戏销量增加 6250 件.

(3) 11 月游戏销量将增加 6250 件，即 11 个月后预计总销量为

$$62500+6250=68750$$

例 3.12　一个金属圆盘受热膨胀，半径的膨胀率为 0.25 cm/s，求当半径为 7 cm 时，面积的膨胀率.

解　设 r 和 s 分别表示金属圆盘的半径和面积，则

$$s=\pi r^2$$

$$\frac{\mathrm{d}s}{\mathrm{d}t}=\frac{\mathrm{d}}{\mathrm{d}t}(\pi r^2)=2\pi r\frac{\mathrm{d}r}{\mathrm{d}t}$$

当 $r=7$ cm，

$$\frac{\mathrm{d}s}{\mathrm{d}t}=2\times\pi\times7\times0.25=3.5\pi\ \mathrm{cm^2/s}$$

例 3.13　某日下午两点有两架飞机在城市上空飞行，一架飞机以 300 km/h 的速度向东飞行，另一架以 400 km/h 的速度向北飞行。两架飞机之间的距离在下午四点时以什么速度变化？

解　如图 3-11 所示，设起飞后 t 时刻两架飞机的位置为 A 和 B. 此时两飞机间的距离为 $AB=x(t)$. 从而有

$$OA=300t,\ OB=400t$$

因为，$AB^2=OA^2+OB^2$，于是

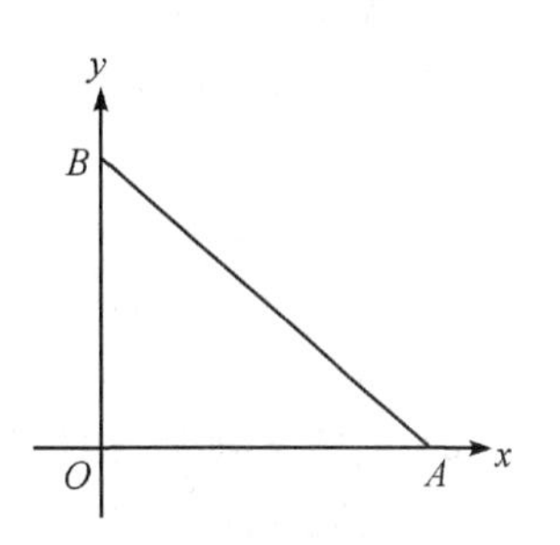

图 3-11

$$x^2 = (300t)^2 + (400t)^2 \Rightarrow x^2 = 90000t^2 + 160000t^2 \Rightarrow x = 500t \Rightarrow \frac{\mathrm{d}x}{\mathrm{d}t} = 500$$

当 $t=2$ 时，$\frac{\mathrm{d}x}{\mathrm{d}t}=500$.

所以，两架飞机在四点时相离的速率为 500 km/h.

例 3.14 将水以 24 cm^3/min 的速率倒入锥形槽中，当水的深度为 9 cm 时，水面上升得有多快？假设水槽的深度为 15 cm，顶半径为 5 cm.

解 如图 3-12 所示，圆锥形 ABC 为锥形水箱，在 t 时刻水面上升的高度为 $AE=h(t)$，半径为 $EF=r(t)$.

由于$\triangle ACD$ 和$\triangle AFE$ 是相似的，所以有

$$\frac{AE}{AD} = \frac{EF}{DC} \quad 或 \quad \frac{h}{15} = \frac{r}{5} \Rightarrow r = \frac{1}{3}h$$

图 3-12

设 V 为 t 时刻水槽内水的体积，则

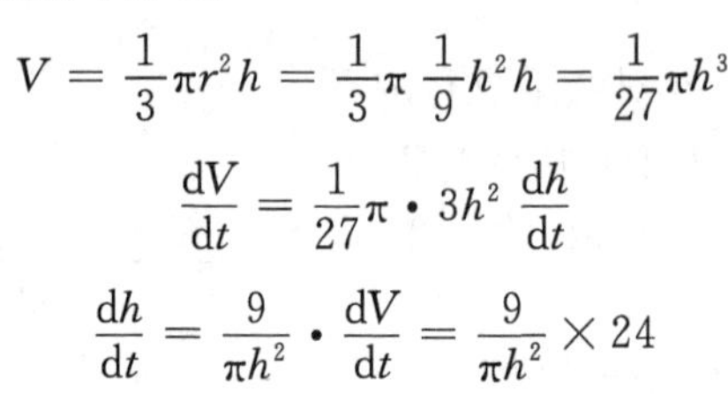

$$V = \frac{1}{3}\pi r^2 h = \frac{1}{3}\pi \frac{1}{9}h^2 h = \frac{1}{27}\pi h^3$$

$$\frac{\mathrm{d}V}{\mathrm{d}t} = \frac{1}{27}\pi \cdot 3h^2 \frac{\mathrm{d}h}{\mathrm{d}t}$$

$$\frac{\mathrm{d}h}{\mathrm{d}t} = \frac{9}{\pi h^2} \cdot \frac{\mathrm{d}V}{\mathrm{d}t} = \frac{9}{\pi h^2} \times 24$$

当 $h=9$ cm 时

$$\frac{\mathrm{d}h}{\mathrm{d}t} = \frac{9}{81\pi} \times 24 = \frac{8}{3\pi}\ \mathrm{cm/min}$$

习题 3

1. 求下列函数的单调增减区间.

(1) $f(x)=3x^2-6x+5$； (2) $f(x)=x^4-\frac{1}{3}x^3$；

(3) $f(x)=6+12x+3x^2-2x^3$； (4) $f(x)=x^3-12x$，$x\in[-3, 5]$.

2. 求下列函数在给定区间的最大值与最小值.

(1) $f(x)=x^3-6x^2+9x$，$x\in[0, 5]$；

(2) $f(x)=2x^3-15x^2+36x+10$，$x\in[1, 4]$.

3. 求函数的极大值、极小值及拐点.

(1) $f(x)=3x^2-6x+3$； (2) $f(x)=2x^3-9x^2-24x+3$；

(3) $f(x)=4x^3-15x^2+12x+7$； (4) $f(x)=x+\frac{100}{x}-5$.

4. 证明下列函数既没有极小值也没有极大值.

(1) $f(x)=x^3-6x^2+24x+4$； (2) $f(x)=x^3-6x^2+12x-3$.

5. 确定下列函数的凹凸区间.

(1) $f(x)=x^4-2x^3+5$； (2) $y=3x^5+10x^3+15x$.

6. 某人要用 144 m 长的栅栏围成一个长方形的花园，求能围起来的最大面积.

7. 证明周长相等的所有矩形中正方形面积最大.

8. 已知圆柱体的体积为 52 cm^3，如何设计可以使得表面积最小？

9. 求和为 10 而平方和最小的两个数.

10. 已知粒子沿直线运动，位移 s(m)与时间 t(s)的关系为

$$s = 2t^2 + 5t - 4$$

求粒子在 6 s 末的速度和加速度.

11. **(销售分析)**光盘的总数 S(以千张光盘计)与光盘发行至今的时间 t(月)的关系为

$$S(t) = \frac{90t^2}{t^2 + 50}$$

(1) 求 $S'(t)$.

(2) 求 $S(10)$和 $S'(10)$，并对这些结果做一个简短的解释.

(3) 使用(2)中的结果估计 11 个月后的总销售量.

12. **(医学)**药物通过病人的右臂血管注射到血液中，注射 t 小时后左臂血管血液中药物浓度(mg/cm^3)为

$$C(t) = \frac{0.14t}{t^2 + 1}$$

(1) 求 $C'(t)$.

(2) 求 $C'(0.5)$和 $C'(3)$，并分析结果.

13. 设生产某产品的固定成本为 2000 元，生产 x 个产品的可变成本为 $0.01x^2 + 10x$ 元，如果产品的销售价为 30 元，试求边际成本、边际利润及边际利润为零时的产量.

14. 设某产品需求量为 $x = 1000 - 10p$，其中 p 为价格，求边际收益函数以及 $x = 100$，200，500，600 时的边际收益，并分析当 $x = 500$ 时的结果.

15. (1) 一个装有油并保持垂直的圆柱形桶正在泄漏，油位以 2 cm/min 速度下降. 如果圆筒的半径和高度分别为 10.5 cm 和 40 cm，求出油体积减小的速率.

(2) 将水以 18 cm^3/min 速度倒入一个半径为 8 cm 的圆柱形桶里，求圆柱体中水平面上升的速度.

(3) 汽油以 24 cm^3/min 的速度泵入垂直的圆柱形油箱，油箱的半径是 9 cm. 求油平面上升的速度.

16. (1) 气球以 18 cm^3/min 的速率充气，当半径是 8 cm 时，半径增大的速率是多少？

(2) 一个球形的盐球在水中以这样的方式溶解：体积在任何时刻的减小速度都与表面积成正比. 证明盐球半径减小的速率是常数.

17. 一个高为 1.5 m 的人以 20 cm/s 的速度从高为 4.5 m 的灯柱旁走开. 当这个人离灯柱 42 cm 时，影子拉长得有多快？

18. (1) 风筝的高度为 24 m，线长 25 m. 如果风筝离开拉线人水平移动的速度是 36 km/h，那么绳子拉出的速度有多快？

(2) 一个 2.5 m 长的梯子靠在垂直的墙上. 如果顶部以 12 cm/s 的速度向下滑动，求下端离开墙 2 m 时的速度.

第 4 章　不定积分

本章，我们将介绍不定积分的概念及计算，从而为下一章定积分的计算做准备.

4.1　原函数与不定积分

4.1.1　原函数与不定积分的概念

设 $f(x)$为定义在区间(a, b)上的连续函数，如果 $F(x)$的导数等于 $f(x)$，即若$\frac{\mathrm{d}F(x)}{\mathrm{d}x}=f(x)$，则称 $F(x)$为 $f(x)$的一个**原函数**.

由于常数 C 的导数为 0，所以 $F(x)+C$ 也是 $f(x)$的一个原函数. 反之，任意两个原函数相差一个常数.

令 $F(x)$和 $G(x)$是 $f(x)$的任意两个原函数，则

$$\frac{\mathrm{d}[F(x)-G(x)]}{\mathrm{d}x}=\frac{\mathrm{d}F(x)}{\mathrm{d}x}-\frac{\mathrm{d}G(x)}{\mathrm{d}x}=f(x)-f(x)=0$$

于是存在常数 C，使得

$$F(x)-G(x)=C$$

从而，如果 $F(x)$是 $f(x)$的一个原函数，$F(x)+C$ 就是 $f(x)$的所有原函数，其中 C 为任意实数.

现在我们希望得到所有原函数的一般形式，这个一般形式，称为 $f(x)$的**不定积分**，记作

$$\int f(x)\mathrm{d}x$$

其中，$\int$ 为积分号；$f(x)$ 称为被积函数；x 称为积分变量.

如果 $F(x)$是 $f(x)$的一个原函数，则有

$$\int f(x)\mathrm{d}x=F(x)+C$$

4.1.2　不定积分的性质

设 $f(x)$和 $g(x)$都是区间(a, b)上的连续函数，k_1 和 k_2 为常数，则

(1) $\int[k_1 f(x)+k_2 g(x)]\mathrm{d}x=k_1\int f(x)\mathrm{d}x+k_2\int g(x)\mathrm{d}x$；

(2) $\left[\int f(x)\mathrm{d}x\right]'=f(x)$　或　$\mathrm{d}\left[\int f(x)\mathrm{d}x\right]=f(x)$；

(3) $\int f'(x)\mathrm{d}x=f(x)+C$　或　$\int \mathrm{d}f(x)=f(x)+C$.

4.2　积分法

4.2.1　基本公式

利用第 2 章中基本初等函数的求导公式不难得到下面的基本积分公式.

基本初等函数的积分公式

$$\int x^\alpha \mathrm{d}x = \frac{x^{\alpha+1}}{\alpha+1} + C(\alpha \neq -1).$$

$$\int \frac{1}{x\ln a}\mathrm{d}x = \log_a x + C, \int \frac{1}{x}\mathrm{d}x = \ln|x| + C.$$

$$\int a^x \mathrm{d}x = \frac{a^x}{\ln a} + C, \int \mathrm{e}^x \mathrm{d}x = \mathrm{e}^x + C.$$

$$\int \cos x \mathrm{d}x = \sin x + C, \int \sin x \mathrm{d}x = -\cos x + C.$$

$$\int \sec^2 x \mathrm{d}x = \tan x + C, \int \csc^2 x \mathrm{d}x = -\cot x + C.$$

$$\int \sec x \tan x \mathrm{d}x = \sec x + C, \int \csc x \cot x \mathrm{d}x = -\csc x + C.$$

例 4.1　计算：

(1) $\int \left(4x^{1/3} + 5x^{2/3} + \frac{1}{x^2}\right)\mathrm{d}x$；　(2) $\int \frac{2x+\sqrt{x}+1}{x}\mathrm{d}x$

(3) $\int \frac{1+2x^2}{x^2(1+x^2)}\mathrm{d}x$；　(4) $\int \frac{1}{\sin^2 x\cos^2 x}\mathrm{d}x$.

解

(1)
$$\begin{aligned}\int \left(4x^{1/3} + 5x^{2/3} + \frac{1}{x^2}\right)\mathrm{d}x &= \int 4x^{1/3}\mathrm{d}x + \int 5x^{2/3}\mathrm{d}x + \int x^{-2}\mathrm{d}x \\ &= 4\cdot\frac{x^{4/3}}{\frac{4}{3}} + 5\cdot\frac{x^{5/3}}{\frac{5}{3}} + \frac{x^{-1}}{-1} + C \\ &= 3x^{4/3} + 3x^{5/3} - \frac{1}{x} + C\end{aligned}$$

(2) $\int \frac{2x+\sqrt{x}+1}{x}\mathrm{d}x = 2\int \mathrm{d}x + \int \frac{\mathrm{d}x}{\sqrt{x}} + \int \frac{1}{x}\mathrm{d}x = 2x + 2\sqrt{x} + \ln|x| + C.$

(3) $\int \frac{1+2x^2}{x^2(1+x^2)}\mathrm{d}x = \int \frac{1}{1+x^2}\mathrm{d}x + \int \frac{1}{x^2}\mathrm{d}x = \arctan x - \frac{1}{x} + C.$

(4) $\int \frac{1}{\sin^2 x\cos^2 x}\mathrm{d}x = \int \frac{\sin^2 x + \cos^2 x}{\sin^2 x\cos^2 x}\mathrm{d}x = \int \sec^2 x\mathrm{d}x + \int \csc^2 x\mathrm{d}x = \tan x - \cot x + C.$

4.2.2　换元积分法

上节我们看到，当被积函数能化为符合基本积分公式的形式时积分很容易被求出来，但被积函数未必总是能化成那些形式，这时需要用一种新的方法，即换元积分法来解决. 换元积分法有两种类型.

第一类换元积分法

设 f 和 φ' 均为连续函数，且 $F'(u)=f(u)$，则

$$\int f[\varphi(x)]\varphi'(x)\mathrm{d}x=\left(\int f(u)\mathrm{d}u\right)_{u=\varphi(x)}=F(u)_{u=\varphi(x)}+C=F[\varphi(x)]+C.$$

例 4.2 计算：

(1) $\int\cos(3x+1)\mathrm{d}x$；

(2) $\int(kx+b)^{\alpha}\mathrm{d}x\ (\alpha\neq0,\ \alpha\neq-1)$；

(3) $\int\frac{\mathrm{d}x}{\sqrt{a^2-x^2}}(a>0)$；

(4) $\int\frac{\mathrm{d}x}{a^2+x^2}(a>0)$；

(5) $\int\mathrm{e}^{\sin x}\cos x\mathrm{d}x$；

(6) $\int\frac{\mathrm{d}x}{a^2-x^2}(a>0)$；

(7) $\int(2x+3)(4x+5)^4\mathrm{d}x$；

(8) $\int\frac{1-\mathrm{e}^{3x}}{\mathrm{e}^{5x}}\mathrm{d}x$.

解

(1) $$\int\cos(3x+1)\mathrm{d}x=\frac{1}{3}\int\cos(3x+1)\mathrm{d}(3x+1)=\frac{1}{3}\int\cos u\mathrm{d}u\Big|_{u=3x+1}$$
$$=\frac{1}{3}\sin u\Big|_{u=3x+1}+C=\frac{1}{3}\sin(3x+1)+C$$

(2) $$\int(kx+b)^{\alpha}\mathrm{d}x=\frac{1}{k}\int(kx+b)^{\alpha}\mathrm{d}(kx+b)=\frac{1}{k}\int u^{\alpha}\mathrm{d}u\Big|_{u=kx+b}$$
$$=\frac{1}{k(\alpha+1)}u^{\alpha+1}\Big|_{u=kx+b}+C=\frac{1}{k(\alpha+1)}(kx+b)^{\alpha+1}+C$$

(3) $$\int\frac{\mathrm{d}x}{\sqrt{a^2-x^2}}=\frac{1}{a}\int\frac{\mathrm{d}x}{\sqrt{1-\left(\frac{x}{a}\right)^2}}=\int\frac{\mathrm{d}\frac{x}{a}}{\sqrt{1-\left(\frac{x}{a}\right)^2}}=\int\frac{\mathrm{d}u}{\sqrt{1-u^2}}\Big|_{u=\frac{x}{a}}$$
$$=\arcsin u\Big|_{u=\frac{x}{a}}+C=\arcsin\frac{x}{a}+C$$

(4) 当我们熟悉这个方法后，可以省略设中间变量 u 的过程.

$$\int\frac{\mathrm{d}x}{a^2+x^2}=\frac{1}{a^2}\int\frac{\mathrm{d}x}{1+\left(\frac{x}{a}\right)^2}=\frac{1}{a}\int\frac{\mathrm{d}\frac{x}{a}}{1+\left(\frac{x}{a}\right)^2}=\frac{1}{a}\arctan\frac{x}{a}+C$$

(5) $$\int\mathrm{e}^{\sin x}\cos x\mathrm{d}x=\int\mathrm{e}^{\sin x}\mathrm{d}\sin x=\mathrm{e}^{\sin x}+C.$$

(6) $$\int\frac{\mathrm{d}x}{a^2-x^2}=\frac{1}{2a}\int\left(\frac{1}{a-x}+\frac{1}{a+x}\right)\mathrm{d}x=\frac{1}{2a}\left[-\int\frac{1}{a-x}\mathrm{d}(a-x)+\int\frac{1}{a+x}\mathrm{d}(a+x)\right]$$
$$=\frac{1}{2a}[-\ln|a-x|+\ln|a+x|]+C=\frac{1}{2a}\ln\left|\frac{a+x}{a-x}\right|+C.$$

(7) $$\int(2x+3)(4x+5)^4\mathrm{d}x=\frac{1}{2}\int(4x+6)(4x+5)^4\mathrm{d}x=\frac{1}{2}\int(4x+5+1)(4x+5)^4\mathrm{d}x$$

$$= \frac{1}{2}\int[(4x+5)^5+(4x+5)^4]\mathrm{d}x$$

$$= \frac{1}{8}\left[\frac{(4x+5)^6}{6}+\frac{(4x+5)^5}{5}\right]+C$$

(8) $\int \frac{1-\mathrm{e}^{3x}}{\mathrm{e}^{5x}}\mathrm{d}x = \int(\mathrm{e}^{-5x}-\mathrm{e}^{-2x})\mathrm{d}x = -\frac{1}{5\mathrm{e}^{5x}}+\frac{1}{2\mathrm{e}^{2x}}+C.$

例 4.3 计算：

(1) $\int \tan x\mathrm{d}x$；

(2) $\int \cot x\mathrm{d}x$；

(3) $\int \sqrt{1-\sin 2x}\mathrm{d}x$，$x\in\left(\frac{\pi}{2},\pi\right)$；

(4) $\int \frac{\mathrm{d}x}{1-\cos x}$；

(5) $\int \sin^3 x\mathrm{d}x$；

(6) $\int \sin^2 x\mathrm{d}x$；

(7) $\int \frac{\mathrm{d}x}{1-\sin x}$；

(8) $\int \sin 6x\cdot\cos 3x\mathrm{d}x$；

(9) $\int \sec x\mathrm{d}x$；

(10) $\int \frac{\mathrm{d}x}{\sin x\cos x}$.

解

(1) $\int \tan x\mathrm{d}x = \int\frac{\sin x}{\cos x}\mathrm{d}x = -\int\frac{1}{\cos x}\mathrm{d}\cos x = -\ln|\cos x|+C = \ln|\sec x|+C.$

(2) 利用类似的方法，有

$$\int \cot x\mathrm{d}x = \ln|\sin x|+C$$

(3) $\int \sqrt{1-\sin 2x}\mathrm{d}x = \int\sqrt{(\sin^2 x+\cos^2 x-2\sin x\cos x)}\mathrm{d}x = \int\sqrt{(\sin x-\cos x)^2}\mathrm{d}x$

$$= \int|\sin x-\cos x|\mathrm{d}x = -\cos x-\sin x+C.$$

(4) $\int \frac{\mathrm{d}x}{1-\cos x} = \int\frac{\mathrm{d}x}{2\sin^2\frac{x}{2}} = \frac{1}{2}\int\csc^2\frac{x}{2}\mathrm{d}x = \frac{1}{2}\left(-\frac{\cot\frac{x}{2}}{\frac{1}{2}}\right)+C = -\cot\frac{x}{2}+C.$

(5) $\int \sin^3 x\mathrm{d}x = \int(1-\cos^2 x)\sin x\mathrm{d}x = -\int(1-\cos^2 x)\mathrm{d}\cos x = -\cos x+\frac{1}{3}\cos^3 x+C.$

(6) $\int \sin^2 x\mathrm{d}x = \frac{1}{2}\int(1-\cos 2x)\mathrm{d}x = \frac{1}{2}\left(\int 1\mathrm{d}x-\int\cos 2x\mathrm{d}x\right) = \frac{1}{2}\left(x-\frac{\sin 2x}{2}\right)+C.$

(7) $\int \frac{\mathrm{d}x}{1-\sin x} = \int\frac{1+\sin x}{1-\sin^2 x}\mathrm{d}x = \int\frac{1+\sin x}{\cos^2 x}\mathrm{d}x$

$$= \int\sec^2 x\mathrm{d}x+\int\tan x\cdot\sec x\mathrm{d}x = \tan x+\sec x+C.$$

(8) $\int \sin 6x\cdot\cos 3x\mathrm{d}x = \frac{1}{2}\int(\sin 9x+\sin 3x)\mathrm{d}x = -\frac{1}{18}(\cos 9x+3\cos 3x)+C.$

(9) $\int \sec x\mathrm{d}x = \int\frac{\sec x(\sec x+\tan x)}{\sec x+\tan x}\mathrm{d}x = \int\frac{\mathrm{d}(\sec x+\tan x)}{\sec x+\tan x} = \ln|\sec x+\tan x|+C.$

类似地，可以得到

$$\int \csc x \mathrm{d}x = -\ln|\csc x + \cot x| + C$$

(10) 方法一：

$$\int \frac{\mathrm{d}x}{\sin x \cos x} = \int \frac{2\mathrm{d}x}{\sin 2x} = 2\int \csc 2x \mathrm{d}x = -\ln|\csc 2x + \cot 2x| + C;$$

方法二：

$$\int \frac{\mathrm{d}x}{\sin x \cos x} = \int \frac{\cos x \mathrm{d}x}{\sin x \cos^2 x} = \int \frac{1}{\tan x}\sec^2 x \mathrm{d}x = \int \frac{\mathrm{d}\tan x}{\tan x} = \ln|\tan x| + C;$$

方法三：

$$\int \frac{\mathrm{d}x}{\sin x \cos x} = \int \frac{\sin^2 x + \cos^2 x \mathrm{d}x}{\sin x \cos x} = \int \frac{\sin x}{\cos x}\mathrm{d}x + \int \frac{\cos x}{\sin x}\mathrm{d}x$$

$$= -\ln|\cos x| + \ln|\sin x| + C = \ln|\tan x| + C$$

例 4.4 计算：

(1) $\int \frac{x\mathrm{d}x}{2x^2+3}$；

(2) $\int \frac{\sqrt{\arctan x}\mathrm{d}x}{1+x^2}$；

(3) $\int \frac{(2ax+b)}{(ax^2+bx+c)^{1/2}}\mathrm{d}x$；

(4) $\int \frac{\mathrm{d}x}{x(1+\ln x)}$；

(5) $\int x\cos(ax^2+b)\mathrm{d}x$；

(6) $\int \sin^3 x \cos^3 x \mathrm{d}x$；

(7) $\int \mathrm{e}^{\cos^2 x}\sin x \cos x \mathrm{d}x$；

(8) $\int \frac{1}{1+\mathrm{e}^x}\mathrm{d}x$.

解

(1) $\int \frac{x\mathrm{d}x}{(2x^2+3)} = \frac{1}{2}\int \frac{\mathrm{d}x^2}{(2x^2+3)} = \frac{1}{4}\int \frac{\mathrm{d}(2x^2+3)}{(2x^2+3)} = \frac{1}{4}\ln(2x^2+3) + C.$

(2) $\int \frac{\sqrt{\arctan x}\mathrm{d}x}{1+x^2} = \int \sqrt{\arctan x}\mathrm{d}\arctan x = \frac{2}{3}(\arctan x)^{\frac{3}{2}} + C.$

(3) $\int \frac{(2ax+b)}{(ax^2+bx+c)^{1/2}}\mathrm{d}x = \int \frac{\mathrm{d}(ax^2+bx+c)}{(ax^2+bx+c)^{1/2}} = 2\sqrt{ax^2+bx+c} + C.$

(4) $\int \frac{\mathrm{d}x}{x(1+\ln x)} = \int \frac{\mathrm{d}\ln x}{1+\ln x} = \int \frac{\mathrm{d}(1+\ln x)}{1+\ln x} = \ln|1+\ln x| + C.$

(5) $\int x\cos(ax^2+b)\mathrm{d}x = \frac{1}{2}\int \cos(ax^2+b)\mathrm{d}x^2$

$$= \frac{1}{2a}\int \cos(ax^2+b)\mathrm{d}(ax^2+b) = \frac{1}{2a}\sin(ax^2+b) + C.$$

(6) $\int \sin^3 x \cos^3 x \mathrm{d}x = \int \sin^3 x \cos^2 x \mathrm{d}\sin x$

$$= \int \sin^3 x(1-\sin^2 x)\mathrm{d}\sin x = \frac{1}{4}\sin^4 x - \frac{1}{6}\sin^6 x + C.$$

(7) $\int \mathrm{e}^{\cos^2 x}\sin x \cos x \mathrm{d}x = -\int \mathrm{e}^{\cos^2 x}\cos x \mathrm{d}\cos x = -\frac{1}{2}\int \mathrm{e}^{\cos^2 x}\mathrm{d}\cos^2 x = -\frac{1}{2}\mathrm{e}^{\cos^2 x} + C.$

(8) $\int \frac{1}{1+\mathrm{e}^x}\mathrm{d}x = \int \frac{\mathrm{e}^{-x}}{1+\mathrm{e}^{-x}}\mathrm{d}x = -\int \frac{\mathrm{d}\mathrm{e}^{-x}}{1+\mathrm{e}^x} = -\ln(1+\mathrm{e}^{-x}) + C.$

如果 $\int f(x)\mathrm{d}x$ 直接计算很困难，我们可以做变量替换，令 $x=\varphi(t)$，得到关于 t 的原函数，再换回 x 的表达式.

第二类换元积分法

设 $f(x)$和 $\varphi'(t)$均连续，且 $\varphi'(t)>0$（或<0），则

$$\int f(x)\mathrm{d}x=\int f(x)\mathrm{d}x\Big|_{x=\varphi(t)}=\left\{\int f[\varphi(t)]\varphi'(t)\mathrm{d}t\right\}\Big|_{t=\varphi^{-1}(x)}$$

例 4.5　计算($a>0$)：

(1) $\int x\sqrt{x-4}\mathrm{d}x$；　(2) $\int \sqrt{a^2-x^2}\mathrm{d}x$；

(3) $\int \dfrac{\mathrm{d}x}{x\sqrt{x^2-a^2}}$；　(4) $\int \dfrac{\mathrm{d}x}{(a^2+x^2)^2}$；

(5) $\int \sqrt{\dfrac{a-x}{x}}\mathrm{d}x$；　(6) $\int \dfrac{\mathrm{d}x}{1+\sqrt[3]{x+2}}$；

(7) $\int \dfrac{\mathrm{d}x}{\sqrt{x^2-a^2}}$.

解

(1) 令 $x=t^2+4$，则 $\mathrm{d}x=2t\mathrm{d}t$，有

$$\int x\sqrt{x-4}\mathrm{d}x=\int(t^2+4)t\cdot 2t\mathrm{d}t=2\int(t^4+4t^2)\mathrm{d}t$$

$$=\left(\frac{2}{5}t^5+\frac{8}{3}t^3\right)_{t=\sqrt{x-4}}+C=\frac{2}{5}(x-4)^{\frac{5}{2}}+\frac{8}{3}(x-4)^{\frac{3}{2}}+C$$

(2) 令 $x=a\sin t$，则 $\mathrm{d}x=a\cos t\mathrm{d}t$，于是

$$\int\sqrt{a^2-x^2}\mathrm{d}x=a^2\int\cos^2 t\mathrm{d}t=\frac{a^2}{2}\int(1+\cos 2t)\mathrm{d}t=\frac{a^2}{2}\left(t+\frac{\sin 2t}{2}\right)+C$$

因为 $x=a\sin t$，在直角三角形中(见图 4-1(a))，$AC=a$，$BC=x$，所以 $AB=\sqrt{a^2-x^2}$. 从而有，$\cos t=\dfrac{\sqrt{a^2-x^2}}{a}$，$\sin 2t=2\sin t\cos t=2\dfrac{x}{a}\dfrac{\sqrt{a^2-x^2}}{a}$，故

$$\int\sqrt{a^2-x^2}\mathrm{d}x=a^2\int\cos^2 t\mathrm{d}t=\frac{a^2}{2}\int(1+\cos 2t)\mathrm{d}t=\frac{a^2}{2}\arcsin\frac{x}{a}+\frac{1}{2}x\sqrt{a^2-x^2}+C$$

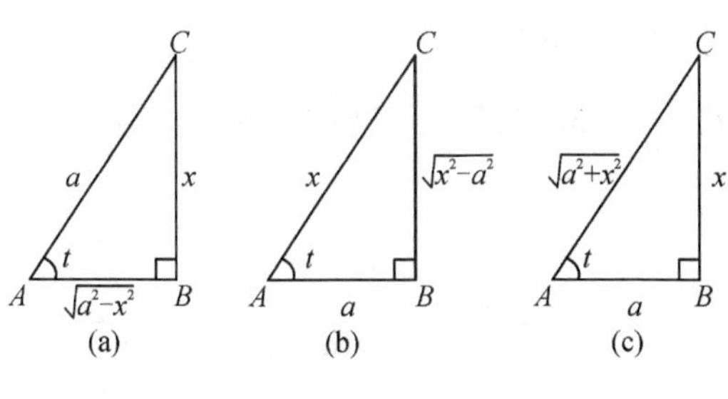

图 4-1

(3) 令 $x=a\sec t$，则 $\mathrm{d}x=a\sec t\tan t\mathrm{d}t$，于是

$$\int \frac{\mathrm{d}x}{x\sqrt{x^2-a^2}} = \int \frac{a\sec t\tan t\mathrm{d}t}{a\sec t\sqrt{a^2\sec^2 t-a^2}} = \int \frac{a\sec t\tan t\mathrm{d}t}{a\sec t\cdot a\tan t} = \frac{1}{a}\int \mathrm{d}t = \frac{1}{a}t + C$$

因为 $x=a\sec t$，$BC=\sqrt{x^2-a^2}$，由图 4 - 1(b)有 $\sec t=\frac{x}{a}$，$\cos t=\frac{a}{x}$，从而 $t=\arccos\frac{a}{x}$，故有

$$\int \frac{\mathrm{d}x}{x\sqrt{x^2-a^2}} = \frac{1}{a}\arccos\frac{a}{x} + C$$

(4) 令 $x=a\tan t$，则 $\mathrm{d}x=a\sec^2 t\mathrm{d}t$，$a^2+x^2=a^2+a^2\tan^2 t=a^2\sec^2 t$，于是有

$$\int \frac{\mathrm{d}x}{(a^2+x^2)^2} = \frac{1}{a^3}\int \cos^2 t\mathrm{d}t = \frac{1}{2a^3}\left(t+\frac{\sin 2t}{2}\right)+C$$

因为 $x=a\tan t$，$\tan t=\frac{x}{a}$，根据图 4 - 1(c)，我们知道 $AC=\sqrt{a^2+x^2}$，因此

$$\sin 2t = 2\sin t\cos t = 2\frac{x}{\sqrt{a^2+x^2}}\frac{a}{\sqrt{a^2+x^2}}$$

$$\int \frac{\mathrm{d}x}{(a^2+x^2)^2} = \frac{1}{2a^3}\left[\arctan\frac{x}{a}+\frac{x}{\sqrt{a^2+x^2}}\frac{a}{\sqrt{a^2+x^2}}\right]+C = \frac{1}{2a^3}\left[\arctan\frac{x}{a}+\frac{ax}{a^2+x^2}\right]+C$$

(5) 令 $x=a\sin^2 t$，则 $\mathrm{d}x=a\cdot 2\sin t\cos t\mathrm{d}t$，所以

$$\int\sqrt{\frac{a-x}{x}}\mathrm{d}x = \int\sqrt{\frac{a-a\sin^2 t}{a\sin^2 t}}\cdot 2a\sin t\cos t\mathrm{d}t = a\int 2\cos^2 t\mathrm{d}t = a\int(1+\cos 2t)\mathrm{d}t$$

$$= a\left(t+\frac{\sin 2t}{2}\right)+C = a(t+\sin t\cos t)+C$$

$$= a\left(\arcsin\sqrt{\frac{x}{a}}+\sqrt{\frac{x}{a}}\cdot\sqrt{1-\frac{x}{a}}\right)+C = a\arcsin\sqrt{\frac{x}{a}}+\sqrt{ax-x^2}+C$$

(6) 令 $x=t^3-2$，则 $\sqrt[3]{x+2}=t$，$\mathrm{d}x=3t^2\mathrm{d}t$，于是

$$\int \frac{\mathrm{d}x}{1+\sqrt[3]{x+2}} = \int \frac{3t^2}{1+t}\mathrm{d}t = 3\int\left(t-1+\frac{1}{t+1}\right)\mathrm{d}t = 3\left[\frac{t^2}{2}-t+\ln(1+t)\right]+C$$

$$= 3\left[\frac{1}{2}\sqrt[3]{(x+2)^2}-\sqrt[3]{x+2}+\ln(1+\sqrt[3]{x+2})\right]+C$$

(7) 令 $x=a\sec t$，则 $\mathrm{d}x=a\sec t\tan t\mathrm{d}t$，于是

$$\int \frac{\mathrm{d}x}{\sqrt{x^2-a^2}} = \int \frac{a\sec t\tan t\mathrm{d}t}{\sqrt{a^2\sec^2 t-a^2}} = \int \frac{a\sec t\tan t\mathrm{d}t}{a\tan t} = \int \sec t\mathrm{d}t \text{ (根据例 4.3(9))}$$

$$= \ln|\sec t+\tan t|+C$$

因为 $x=a\sec t$，$\sec t=\frac{x}{a}$，根据图 4 - 1(b)，我们知道 $BC=\sqrt{x^2-a^2}$，所以 $\tan t=\frac{\sqrt{x^2-a^2}}{a}$，所以

$$\int \frac{\mathrm{d}x}{\sqrt{x^2-a^2}} = \ln\left|\frac{x}{a}+\frac{\sqrt{x^2-a^2}}{a}\right|+C = \ln|x+\sqrt{x^2-a^2}|+C' \quad (C'=C-\ln a)$$

通过上面的例子可以归纳总结出，利用下面的变量替换可以去掉被积函数里的根号.

(1) $\sqrt{a^2-x^2}$，令 $x=a\sin t$；　(2) $\sqrt{x^2+a^2}$，令 $x=a\tan t$；

(3) $\sqrt{x^2-a^2}$，令 $x=a\sec t$；　(4) $\sqrt[n]{ax+b}$，令 $\sqrt[n]{ax+b}=t$.

4.2.3　分部积分

如果被积函数是乘积的形式，而且既无法化为基本初等函数求解也无法用换元积分法求解，可以考虑用下面的分部积分公式求解.

设 $u(x)$和 $v(x)$均可导，根据乘积的求导法则，有

$$(uv)' = u'v + uv'$$

或

$$uv' = (uv)' - vu'$$

两端对 x 积分，可以得到

分部积分公式

$$\int uv'\mathrm{d}x = \int (uv)'\mathrm{d}x - \int vu'\mathrm{d}x$$

或

$$\int u\mathrm{d}v = uv - \int v\mathrm{d}u$$

上述公式能否成功运用取决于对 $u(x)$和 $v(x)$的恰当选取，选取的标准是使第二个积分好求.

应当注意，在公式 $\int uv'\mathrm{d}x = \int (uv)'\mathrm{d}x - \int vu'\mathrm{d}x$ 中，右端第一项的常数可以合并到 $\int vu'\mathrm{d}x$ 里.

例 4.6　求下列不定积分.

(1) $\int x\mathrm{e}^x\mathrm{d}x$；　(2) $\int x\cos x\mathrm{d}x$；

(3) $\int \ln x\mathrm{d}x$；　(4) $\int x\arctan x\mathrm{d}x$；

(5) $\int x\ln x\mathrm{d}x$；　(6) $\int \mathrm{e}^x\sin x\mathrm{d}x$.

解

(1) 把 e^x 和 $\mathrm{d}x$ 结合写成 $\mathrm{d}\mathrm{e}^x$，根据分部积分公式

$$\int x\mathrm{e}^x\mathrm{d}x = \int x\mathrm{d}\mathrm{e}^x = x\mathrm{e}^x - \int \mathrm{e}^x\mathrm{d}x = x\mathrm{e}^x - \mathrm{e}^x + C$$

注意：如果把 x 和 $\mathrm{d}x$ 结合，则有

$$\int x\mathrm{e}^x\mathrm{d}x = \frac{1}{2}\int \mathrm{e}^x\mathrm{d}x^2 = \frac{1}{2}\left(x^2\mathrm{e}^x - \int x^2\mathrm{d}\mathrm{e}^x\right) = \frac{1}{2}\left(x^2\mathrm{e}^x - \int x^2\mathrm{e}^x\mathrm{d}x\right)$$

正如我们看到的，积分变得更复杂了，因此恰当地选择 $u(x)$和 $v(x)$是应用分部积分公式的关键.

(2) $\int x\cos x\mathrm{d}x = \int x\mathrm{d}\sin x = x\sin x - \int \sin x\mathrm{d}x = x\sin x + \cos x + C.$

(3) $\int \ln x\mathrm{d}x = x\ln x - \int x\mathrm{d}\ln x = x\ln x - \int x\frac{1}{x}\mathrm{d}x = x\ln x - x + C.$

(4) $\int x\arctan x\mathrm{d}x = \frac{1}{2}\int \arctan x\mathrm{d}x^2 = \frac{1}{2}\left(x^2\arctan x - \int x^2\mathrm{d}\arctan x\right)$

$$= \frac{1}{2}\left(x^2\arctan x - \int \frac{x^2}{1+x^2}\mathrm{d}x\right) = \frac{1}{2}(x^2\arctan x - x + \arctan x) + C$$

(5) $\int x\ln x\mathrm{d}x = \frac{1}{2}\int \ln x\mathrm{d}x^2 = \frac{1}{2}\left(x^2\ln x - \int x^2\frac{1}{x}\mathrm{d}x\right) = \frac{1}{2}\left(x^2\ln x - \frac{1}{2}x^2\right) + C.$

(6) 令 $I = \int \mathrm{e}^x\cos x\mathrm{d}x$，则

$$I = \int \mathrm{e}^x\cos x\mathrm{d}x = \int \cos x\mathrm{d}\mathrm{e}^x = \mathrm{e}^x\cos x - \int \mathrm{e}^x\mathrm{d}\cos x = \mathrm{e}^x\cos x + \int \mathrm{e}^x\sin x\mathrm{d}x$$

$$= \mathrm{e}^x\cos x + \int \sin x\mathrm{d}\mathrm{e}^x = \mathrm{e}^x\cos x + \mathrm{e}^x\sin x - \int \mathrm{e}^x\cos x\mathrm{d}x = \mathrm{e}^x\cos x + \mathrm{e}^x\sin x - I$$

然后把最后一个积分 I 移到左边，把常数移到右边，得

$$\int \mathrm{e}^x\cos x\mathrm{d}x = \frac{1}{2}\mathrm{e}^x(\sin x + \cos x) + C$$

习题 4

1. 计算下列不定积分.

(1) $\int (2x+1)(3x+2)\mathrm{d}x$；

(2) $\int \left(x^2 - \frac{1}{x^2}\right)\mathrm{d}x$；

(3) $\int \left(\sqrt{x} - \frac{1}{\sqrt{x}}\right)\mathrm{d}x$；

(4) $\int \frac{3x^2 - 5x + 2}{x}\mathrm{d}x$；

(5) $\int (x^2 + 3x + 5)x^{-1/3}\mathrm{d}x$；

(6) $\int (a - bx)^5\mathrm{d}x$；

(7) $\int \frac{\mathrm{d}x}{\sqrt{2x+7}}$；

(8) $\int \frac{3x-1}{x-2}\mathrm{d}x$；

(9) $\int \frac{\mathrm{d}x}{\sqrt{x+a} - \sqrt{x-a}}$；

(10) $\int \frac{3x+2}{\sqrt{5x+3}}\mathrm{d}x$；

(11) $\int \left[x + \frac{1}{(x+3)^2}\right]\mathrm{d}x$；

(12) $\int \frac{x^2+3x+3}{x+1}\mathrm{d}x$；

(13) $\int (\mathrm{e}^{px} + \mathrm{e}^{-qx})\mathrm{d}x$；

(14) $\int \mathrm{e}^x(\mathrm{e}^{2x} + 1)\mathrm{d}x$.

2. 计算下列不定积分.

(1) $\int \cos(a^2x + b)\mathrm{d}x$；

(2) $\int \sec^2(2x+3)\mathrm{d}x$；

(3) $\int \sin^2 ax\mathrm{d}x$；

(4) $\int \tan^2 ax\mathrm{d}x$；

(5) $\int \sin^4 x\mathrm{d}x$；

(6) $\int \frac{1}{\cos^2 x\sin^2 x}\mathrm{d}x$；

(7) $\int \frac{1}{\sec^2 x\tan^2 x}\mathrm{d}x$；

(8) $\int \sqrt{1 + \sin 2ax}\mathrm{d}x$；

(9) $\int \frac{dx}{1-\sin ax}$; (10) $\int \sin 7x \sin 5x dx$.

3. 计算下列不定积分.

(1) $\int 3x^2 (x^3+1)^3 dx$; (2) $\int \frac{2x+3}{(3x^2+9x+5)^3} dx$;

(3) $\int \frac{(x^2+1)dx}{\sqrt{x^3+3x+4}}$; (4) $\int \frac{1}{x}\ln x dx$;

(5) $\int \cos^5 x \sin^3 x dx$; (6) $\int (a\sin x - b)^3 \cos x dx$;

(7) $\int \cot x (\ln \sin x)^3 dx$; (8) $\int \tan^2\theta \sec^4\theta d\theta$;

(9) $\int \tan^3 x \sec^4 x dx$; (10) $\int \tan^3 x dx$;

(11) $\int e^{\sin x \cos x} \cos 2x dx$; (12) $\int \left(1-\frac{1}{x^2}\right) e^{x+1/x} dx$;

(13) $\int \frac{\sin\sqrt{x}}{\sqrt{x}} dx$; (14) $\int \frac{e^{2x}}{1+e^x} dx$;

(15) $\int \frac{e^x-1}{e^x+1} dx$.

4. 计算下列不定积分.

(1) $\int x\sqrt{x+1} dx$; (2) $\int (x+2)\sqrt{3x+2} dx$;

(3) $\int \frac{dx}{\sqrt{(a^2-x^2)^{3/2}}}$; (4) $\int \frac{x^2 dx}{\sqrt{a^2-x^2}}$;

(5) $\int \frac{dx}{\sqrt{x^2-4}}$; (6) $\int \frac{dx}{x^2\sqrt{x^2+1}}$;

(7) $\int \sqrt{\frac{a+x}{a-x}} dx$; (8) $\int \sqrt{\frac{x}{a-x}} dx$.

5. 计算下列不定积分.

(1) $\int x\ln x dx$; (2) $\int x e^{5x} dx$;

(3) $\int x\sec^2 x dx$; (4) $\int x\sin x dx$;

(5) $\int \sec^3 x dx$; (6) $\int \arcsin x dx$;

(7) $\int x\sec x \tan x dx$; (8) $\int x\sin^2 x dx$.

第5章　定积分及其应用

由第4章可知，不定积分的结果是一族函数，而本章介绍的定积分是一个常数. 定积分是微积分中非常重要的概念，可以用来度量面积、体积等.

5.1　定积分的概念

5.1.1　求抛物线下方的面积

曲线下面积(也称为曲边梯形的面积)的计算对于早期的数学家来说是一个很大的挑战，为说明此问题，我们先考虑一个例子. 求由曲线 $y=x^2$，x 轴，$x=0$ 和 $x=a$ 所围图形的面积.

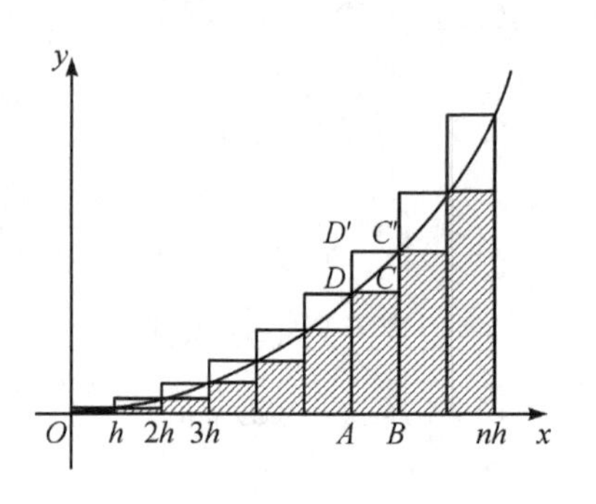

图5-1

如图5-1所示，把区间(0, a)划分成 n 个长度均为 $h=a/n$的子区间，在每个子区间上做一个矩形，所有形如 $ABCD$ 的阴影矩形都在曲线的下方，构成了一组矩形，我们来求这些矩形的面积(记作 s_n).

因为当 $x=h$ 时，$y=h^2$，所以有

$$\begin{aligned}s_n&=0+h\cdot h^2+h\cdot(2h)^2+h\cdot(3h)^2+\cdots+h\cdot[(n-1)h]^2\\&=h^3[1^2+2^2+\cdots+(n-1)^2]\\&=\frac{a^3}{n^3}\left[\frac{1}{6}(n-1)\cdot n\cdot(2n-1)\right]\\&=\frac{1}{6}a^3\left(1-\frac{1}{n}\right)\left(2-\frac{1}{n}\right)\end{aligned}$$

因此
$$\lim_{n\to\infty}s_n=\frac{1}{3}a^3$$

接下来我们考虑形如 $ABC'D'$的矩形. 这些矩形每一个都有一部分在抛物线的上方，记这些矩形的面积为 S_n. 因此有

$$\begin{aligned}S_n&=h\cdot h^2+h\cdot(2h)^2+h\cdot(3h)^2+\cdots+h\cdot(nh)^2\\&=h^3(1^2+2^2+\cdots+n^2)\\&=\frac{a^3}{n^3}\,\frac{1}{6}n(n+1)(2n+1)\\&=\frac{1}{6}a^3\left(1+\frac{1}{n}\right)\left(2+\frac{1}{n}\right)\end{aligned}$$

从而
$$\lim_{n\to\infty}S_n=\frac{1}{3}a^3$$

如果记所围部分的面积为 A，显然有

$$\lim_{n\to\infty} s_n \leqslant A \leqslant \lim_{n\to\infty} S_n$$

$$\frac{1}{3}a^3 \leqslant A \leqslant \frac{1}{3}a^3$$

根据夹逼定理得

$$A = \frac{1}{3}a^3$$

利用这种方法，很容易求出平面区域的面积.

5.1.2　求曲线下方的面积

设 $f(x)>0$ 为区间$[a, b]$上的连续函数，如果把区间$[a, b]$均分成长度为 $h=\frac{b-a}{n}$的 n 个小区间，则由曲线 $y=f(x)$，$x=a$，$x=b$ 和 x 轴所围的图形面积可以通过下面的式子来计算

$$\lim_{h\to 0} h[f(a)+f(a+h)+f(a+2h)+\cdots+f(a+(n-1)h)] \tag{5.1}$$

或

$$\lim_{h\to 0} h[f(a+h)+f(a+2h)+f(a+3h)+\cdots+f(a+nh)] \tag{5.2}$$

例 5.1　利用和式的极限求由曲线 $y=2x^2-3$，x 轴，$x=0$ 和 $x=a$ 所围图形的面积.

解　利用式(5.2)

$$A = \lim_{h\to 0} h[f(h)+f(2h)+f(3h)+\cdots+f(nh)]$$

我们有

$$\begin{aligned} A &= \lim_{h\to 0} h\{(2h^2-3)+[2(2h)^2-3]+[2(3h)^2-3]+\cdots+[2(nh)^2-3]\} \\ &= \lim_{h\to 0} h[2h^2(1^2+2^2+3^2+\cdots+n^2)-3n] \\ &= \lim_{h\to 0}\left[2h^3\cdot\frac{n(n+1)(2n+1)}{6}-3nh\right] \\ &= \lim_{h\to 0}\left[\frac{1}{3}\cdot nh(nh+h)(2nh+h)-3nh\right] \quad \left(h=\frac{a}{n}\right) \\ &= \frac{1}{3}\cdot a(a+0)(2a+0)-3a \\ &= \frac{2}{3}a^3-3a \end{aligned}$$

以上就是函数 $f(x)$关于 x 从 a 到 b 的**定积分**，记作 $\int_a^b f(x)\mathrm{d}x$.

因此有

$$\int_a^b f(x)\mathrm{d}x = \lim_{h\to 0} h[f(a)+f(a+h)+f(a+2h)+\cdots+f(a+(n-1)h)]$$

其中 $h=\frac{b-a}{n}$.

上面的定积分也可写成

$$\int_a^b f(x)\mathrm{d}x = \lim_{h\to 0} h[f(a+h)+f(a+2h)+f(a+3h)+\cdots+f(a+nh)]$$

特殊地，如果 $a=0$，则

$$\int_0^b f(x)\mathrm{d}x = \lim_{h\to 0} h[f(h)+f(2h)+f(3h)+\cdots+f(nh)]$$

其中 $h=\frac{b-0}{n}=\frac{b}{n}$.

5.1.3 黎曼和与定积分

现在我们来概括上一节的思想. 首先，假设函数 f 在 $[a, b]$上是连续的，如图 5-2 所示，在区间$[a, b]$内插入 $n+1$个分点 x_0，x_1，…，x_n，使得

$$a=x_0<x_1<\cdots<x_{n-1}<x_n=b$$

这些点将区间$[a, b]$分成 n 个子区间，记作

$$\Delta x_i=x_i-x_{i-1}, \ i=1, 2, 3, \cdots, n$$

子区间中最长的一个称为子区间的范数，记作 norm$=\max\Delta x_i$，$i=1, 2, \cdots, n$.

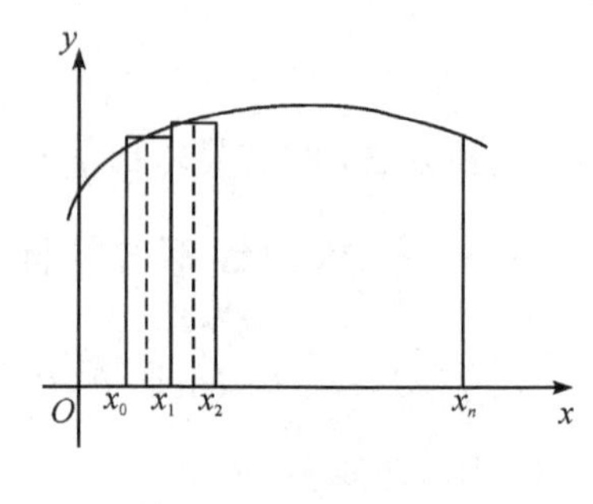

图 5-2

在区间$[x_{i-1}, x_i]$内任取一点 t_i，则和式 $\sum\limits_{i=1}^{n} f(t_i)\Delta x_i$ 称为函数 f 在区间$[a, b]$上的黎曼和. 当 n 趋于无穷大，且 $\Delta x_i\to 0$ 时，黎曼和的极限即为曲线下方的面积.

定义 5.1 如果无论 t_i 在区间$[x_{i-1}, x_i]$上如何选取，当范数(norm)趋于 0 时黎曼和 $\sum\limits_{i=1}^{n} f(t_i)\Delta x_i$ 都趋于 I，则称 I 为函数从 a 到 b 的**定积分**，记作

$$I=\int_a^b f(x)\mathrm{d}x=\lim_{\text{norm}\to 0}\sum_{i=1}^{n} f(t_i)\Delta x_i$$

这里 a 和 b 分别称为积分的下限和上限.

下面不加证明地给出定积分的存在定理.

定理 5.1 如果 f 在$[a, b]$上连续，则 $\int_a^b f(x)\mathrm{d}x$ 一定存在.

1. 定积分的几何意义

如果 $f(x)>0$ 在$[a, b]$上连续，则由曲线 $y=f(x)$，$x=a$，$x=b$ 和 x 轴所围图形的面积为 $\int_a^b f(x)\mathrm{d}x$.

利用定积分的几何意义，很容易计算 $\int_0^a \sqrt{x^2-a^2}\mathrm{d}x$ 的值，它等于$\frac{1}{4}$圆的面积$\frac{\pi}{4}a^2$.

2. 定积分的性质

(1) $\int_a^a f(x)\mathrm{d}x=0$;

(2) $\int_a^b f(x)\mathrm{d}x=-\int_b^a f(x)\mathrm{d}x$;

(3) $\int_a^b f(x)\mathrm{d}x=\int_a^c f(x)\mathrm{d}x+\int_c^b f(x)\mathrm{d}x$;

(4) $\int_a^b kf(x)\mathrm{d}x=k\int_a^b f(x)\mathrm{d}x$;

(5) $\int_a^b [\alpha f(x)+\beta g(x)]\mathrm{d}x=\alpha\int_a^b f(x)\mathrm{d}x+\beta\int_a^b g(x)\mathrm{d}x$;

(6) $f(x)\leqslant y(x)$，则$\int_a^b f(x)\mathrm{d}x\leqslant\int_a^b y(x)\mathrm{d}x$.

5.2　微积分的两个基本定理

令 $P(x, y)$和 $Q(x+\Delta x, y+\Delta y)$是曲线上两个相邻的点，$f(x)\geqslant 0$ 且在$[a, b]$上连续，如图 5-3 所示.

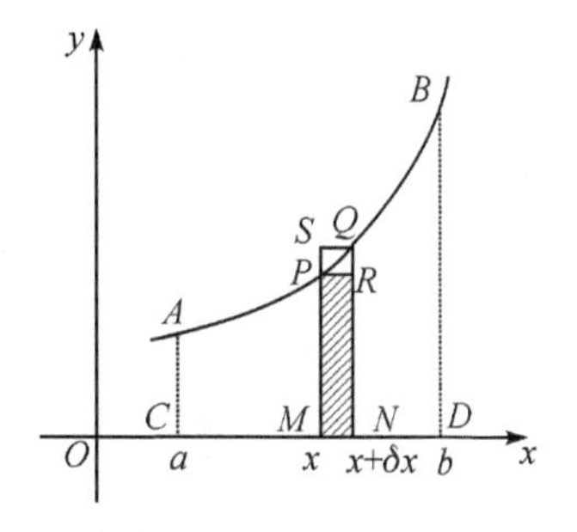

图 5-3

设 $A(x)=ACMP$ 的面积 $=\int_a^x f(x)\mathrm{d}x$，于是

$$A(x+\Delta x)=ACNQ\text{ 的面积 }=\int_a^{x+\Delta x} f(x)\mathrm{d}x$$

从而　　　$A(x+\Delta x)-A(x)=PMNQ$ 的面积

从图 5-3 可以看出

$$PMNR\text{ 的面积}<PMNQ\text{ 的面积}<SMNQ\text{ 的面积}$$

$$\Rightarrow f(x)\cdot\Delta x<A(x+\Delta x)-A(x)<f(x+\Delta x)\cdot\Delta x$$

$$\Rightarrow f(x)<\frac{A(x+\Delta x)-A(x)}{\Delta x}<f(x+\Delta x)$$

$$\Rightarrow \lim_{\Delta x\to 0}f(x)<\lim_{\Delta x\to 0}\frac{A(x+\Delta x)-A(x)}{\Delta x}<\lim_{\Delta x\to 0}f(x+\Delta x)$$

$$\Rightarrow f(x)<A'(x)<\lim_{\Delta x\to 0}f(x+\Delta x)$$

因为 $y=f(x)$是连续的，所以$\lim\limits_{\Delta x\to 0}f(x+\Delta x)=f(x)$

$$A'(x)=f(x)$$

注意：在上面的证明过程中，我们假设了 $f(x)\geqslant 0$（即曲线位于 x 轴的上方）.

根据上面的分析很容易得到微积分中一个非常重要的定理.

微积分第一基本定理

如果 f 是连续函数，且 $\Phi(x)=\int_a^x f(t)\mathrm{d}t$，则

$$\frac{\mathrm{d}}{\mathrm{d}x}\Phi(x)=f(x)$$

注意：

(1) $\Phi(x)$的自变量是定积分的上限，其变换范围为是从 a 到 x；

(2) 被积函数 $f(x)$的自变量与 $\Phi(x)$的自变量是不同的.

微积分第一基本定理建立了导数和定积分的关系，是一种互逆关系. 借助于这个基本定理，可以证明下面的定理.

微积分第二基本定理

如果 f 在$[a, b]$上连续，且 $F(x)$是 f 的一个原函数，则

$$\int_a^b f(x)\mathrm{d}x=F(x)\Big|_a^b=F(b)-F(a)$$

证明　令 $\Phi(x)=\int_a^x f(t)\mathrm{d}t$.

显然，有 $\Phi(a)=0$. 由于 Φ 和 F 都是 f 的原函数，它们仅仅相差了一个常数，所以必存

在某个常数 C，使得

$$\Phi(x)=F(x)+C$$

所以 $\Phi(a)=F(a)+C\Rightarrow 0=F(a)+C\Rightarrow F(a)=-C$，于是有

$$\Phi(x)=F(x)-F(a)$$

从而
$$\Phi(b)=F(b)-F(a)$$

又
$$\Phi(b)=\int_a^b f(t)\mathrm{d}t$$

所以

$$\int_a^b f(t)\mathrm{d}t=F(x)\Big|_a^b=F(b)-F(a) \tag{5.3}$$

这个定理称为微积分第二基本定理，公式(5.3)称为**牛顿-莱布尼茨公式(简记为 N－L 公式)**.

5.3 特殊情形下的面积

情形 1 如果曲线 $y=f(x)$位于 x 轴下方(即 $f(x)\leqslant 0$，见图 5－4)，则阴影部分的面积可表示为

$$\int_a^b -f(x)\mathrm{d}x=-\int_a^b f(x)\mathrm{d}x$$

情形 2 若曲线 $x=f(y)$位于 y 轴右侧(即 $f(y)\geqslant 0$，见图 5－5)，y 轴，$y=a$ 和 $y=b$ 与曲线 $x=f(y)$所围图形的面积为

$$\int_a^b x\mathrm{d}y$$

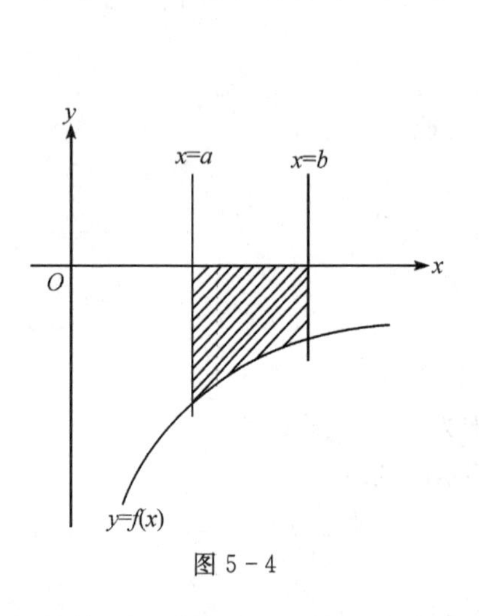

图 5－4

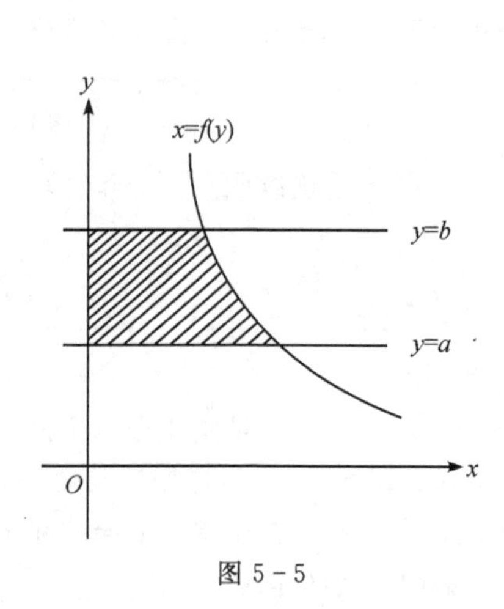

图 5－5

情形 3 若曲线 $x=f(y)$位于 y 轴的左侧(即 $f(y)\leqslant 0$，见图 5－6)，则由曲线 $x=f(y)$，y 轴，$y=a$ 和 $y=b$ 所围图形的面积为

$$\int_a^b(-x)\mathrm{d}y=-\int_a^b f(y)\mathrm{d}y$$

情形 4 (介于两曲线之间的面积)现在我们求由两个函数 f_1 和 f_2 所代表的两条曲线以及 $x=a$ 和 $x=b$ 所围图形的面积，见图 5－7.

$$PQRS\text{ 的面积}=\int_a^b f_1(x)\mathrm{d}x-\int_a^b f_2(x)\mathrm{d}x=\int_a^b[f_1(x)-f_2(x)]\mathrm{d}x$$

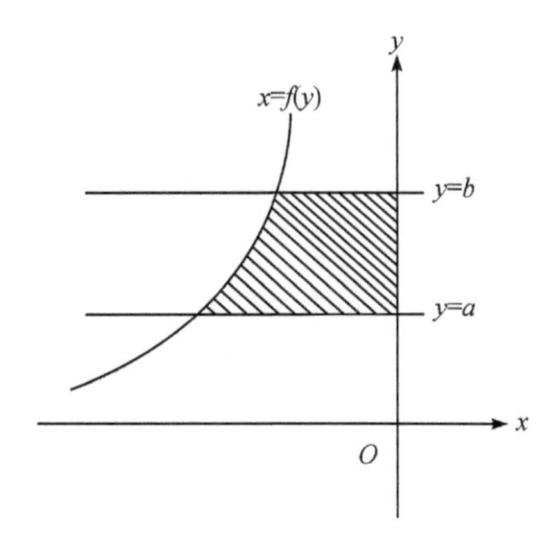

图 5 - 6

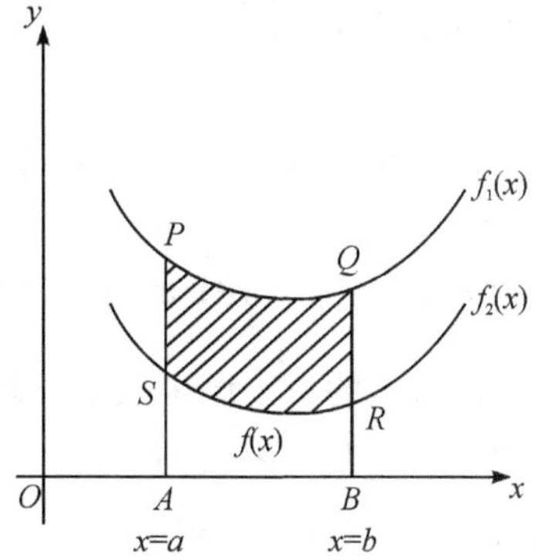

图 5 - 7

例 5.2　求由直线 $y=3x$，x 轴，$x=0$ 和 $x=4$ 所围图形的面积.

解　所求面积为 $=\int_0^4 3x\mathrm{d}x=\left.\dfrac{3x^2}{2}\right|_0^4=\dfrac{3\times4^2}{2}=24$.

例 5.3　求由曲线 $y^2=4ax$，x 轴及垂直于 x 轴且与曲线相交于点$(a,2a)$的直线所围图形的面积.

解　点$(0,0)$满足曲线方程，所以曲线过原点. 因此，所求面积为由曲线 $y^2=4ax$，x 轴，$x=0$ 和 $x=a$ 所围. 因此有

$$所求面积=\int_0^a 2\sqrt{ax}\,\mathrm{d}x=2\sqrt{a}\int_0^a x^{1/2}\mathrm{d}x=2\sqrt{a}\left.\frac{x^{3/2}}{3/2}\right|_0^a=\frac{4}{3}a^2$$

例 5.4　求由 x 轴和曲线 $y=x^2-4x+3$ 所围图形的面积.

解　首先求曲线 $y=x^2-4x+3$ 与 x 轴的交点. 令 $y=0$，于是有

$$(x-1)(x-3)=0\Rightarrow x=1\text{ 或 }x=3$$

$$所求面积=-\int_1^3(x^2-4x+3)\mathrm{d}x=-\left.\left(\frac{x^3}{3}-2x^2+3x\right)\right|_1^3=\frac{4}{3}$$

例 5.5　求由 y 轴，$y=11$ 和曲线 $x^2=4(y-2)$所围图形的面积.

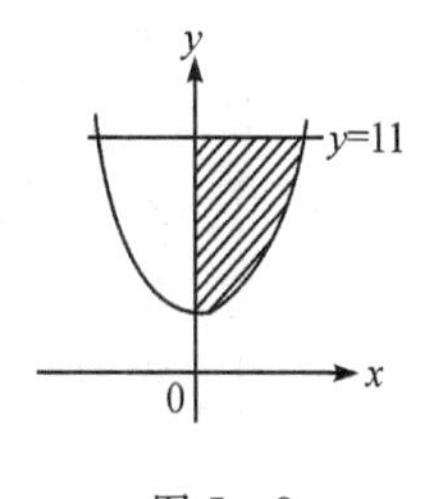

图 5 - 8

解　见图 5 - 8. 先求曲线 $x^2=4(y-2)$与 y 轴的交点，令 $x=0$，则 $y=2$，有

$$所求面积=\int_2^{11}2\sqrt{y-2}\,\mathrm{d}y=\left.\left[\frac{4}{3}(y-2)^{3/2}\right]\right|_2^{11}=36$$

例 5.6　计算由曲线 $y=x^2$ 和直线 $y=2x$ 所围图形的面积.

解　从给定方程中消去 y，得

$$x^2=2x\Rightarrow x(x-2)=0$$

所以，两曲线在 $x=0$ 和 $x=2$ 处相交，如图 5 - 9 所示。

因此所求面积为

$$\int_0^2(2x-x^2)\mathrm{d}x=\left.\left(x^2-\frac{1}{3}x^3\right)\right|_0^2=4-\frac{8}{3}=\frac{4}{3}$$

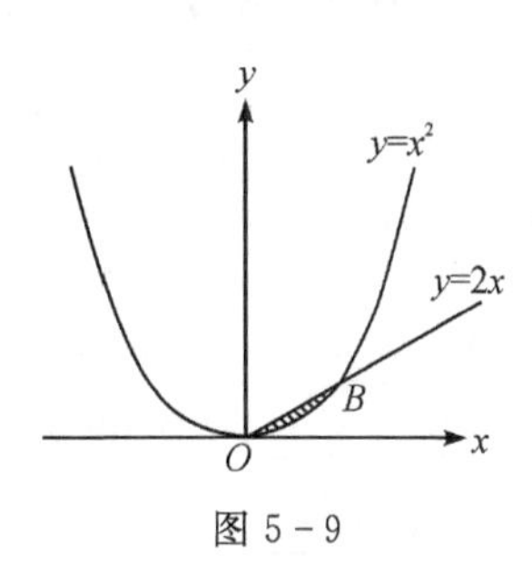

图 5 - 9

5.4 定积分的计算

5.4.1 应用 N-L 公式计算定积分

计算定积分 $\int_a^b f(x)\mathrm{d}x$ 时分以下几个步骤：

(1) 求 $f(x)$的一个原函数 $F(x)$；

(2) 把 $x=b$ 代入 $F(x)$；

(3) 把 $x=a$ 代入 $F(x)$；

(4) 计算 $F(b)-F(a)$ 即得定积分$\int_a^b f(x)\mathrm{d}x$ 的值.

例 5.7 计算下列定积分.

(1) $\int_0^3 x^5\mathrm{d}x$；　　(2) $\int_{-1}^2 (x^2+x+1)\mathrm{d}x$；

(3) $\int_0^1\left(\mathrm{e}^{2x}+\frac{3}{x+1}\right)\mathrm{d}x$；　　(4) $\int_{-2}^0 (x-\mathrm{e}^{-x})\mathrm{d}x$；

(5) $\int_{\frac{\pi}{6}}^{\frac{\pi}{2}}\sin3x\mathrm{d}x$；　　(6) $\int_5^2 (3x-4)^4\mathrm{d}x$.

解 (1) $\int_0^3 x^5\mathrm{d}x=\left(\frac{1}{6}x^6\right)\Big|_0^3=\frac{1}{6}(3^6-0)=\frac{243}{2}$.

(2) $\int_{-1}^2 (x^2+x+1)\mathrm{d}x=\left(\frac{1}{3}x^3+\frac{1}{2}x^2+x\right)\Big|_{-1}^2=\frac{15}{2}$.

(3) $\int_0^1\left(\mathrm{e}^{2x}+\frac{3}{x+1}\right)\mathrm{d}x=\left(\frac{1}{2}\mathrm{e}^{2x}+3\ln(x+1)\right)\Big|_0^1=\frac{1}{2}\mathrm{e}^2+3\ln2-\frac{1}{2}$.

(4) $\int_{-2}^0 (x-\mathrm{e}^{-x})\mathrm{d}x=\left(\frac{1}{2}x^2+\mathrm{e}^{-x}\right)\Big|_{-2}^0=-1-\mathrm{e}^2$.

(5) $\int_{\frac{\pi}{6}}^{\frac{\pi}{2}}\sin3x\mathrm{d}x=-\frac{1}{3}\cos3x\Big|_{\frac{\pi}{6}}^{\frac{\pi}{2}}=0$.

(6) $\int_5^2 (3x-4)^4\mathrm{d}x=\frac{1}{15}(3x-4)^5\Big|_5^2=-\frac{161019}{15}$.

5.4.2 利用换元法计算定积分

设函数 f 连续，$x=\varphi(t)$在区间$[\alpha,\beta]$上有连续的导数，且 $\varphi(\alpha)=a$，$\varphi(\beta)=b$，则

$$\int_a^b f(x)\mathrm{d}x=\int_\alpha^\beta f[\varphi(t)]\varphi'(t)\mathrm{d}t$$

例 5.8 求下列定积分.

(1) $\int_0^4\frac{\mathrm{d}x}{1+\sqrt{x}}$；　　(2) $\int_0^{\pi/6}\frac{\cos\theta\mathrm{d}\theta}{\sqrt{1-\sin\theta}}$；

(3) $\int_0^1 \sqrt{1-x^2}\,\mathrm{d}x$；　　(4) $\int_{\sqrt{2}}^2 \dfrac{\mathrm{d}x}{\sqrt{x^2-1}}$.

解　(1) 令 $x=t^2$，则 $\mathrm{d}x=2t\mathrm{d}t$. 当 $x=0$ 时，$t=0$；当 $x=4$ 时，$t=2$，所以有

$$\int_0^4 \frac{\mathrm{d}x}{1+\sqrt{x}} = \int_0^2 \frac{2t\mathrm{d}t}{1+t} = 2\left[t-\ln(1+t)\right]\Big|_0^2 = 4-2\ln 3$$

(2) 令 $t=\sin\theta$，则 $\mathrm{d}t=\cos\theta\mathrm{d}\theta$. 当 $\theta=0$ 时 $t=0$；$\theta=\dfrac{\pi}{6}$时，$t=\dfrac{1}{2}$，所以有

$$\int_0^{\pi/6} \frac{\cos\theta\mathrm{d}\theta}{\sqrt{1-\sin\theta}} = \int_0^{\frac{1}{2}} \frac{\mathrm{d}t}{\sqrt{1-t}} = -2\sqrt{1-t}\Big|_0^{\frac{1}{2}} = 2-\sqrt{2}$$

(3) 令 $x=\sin t$，则 $\mathrm{d}x=\cos t\mathrm{d}t$. 当 $x=0$ 时，$t=0$；当 $x=1$ 时，$t=\dfrac{\pi}{2}$，所以有

$$\int_0^1 \sqrt{1-x^2}\,\mathrm{d}x = \int_0^{\frac{\pi}{2}} \sqrt{1-\sin^2 t}\,\mathrm{d}\sin t = \int_0^{\frac{\pi}{2}} \cos^2 t\mathrm{d}t = \frac{1}{2}\left(t+\frac{1}{2}\sin 2t\right)\Big|_0^{\frac{\pi}{2}} = \frac{\pi}{4}$$

本题也可利用定积分的几何意义来求解.

(4) 令 $x=\sec t$，则 $\mathrm{d}x=\sec t\tan t\mathrm{d}t$. 当 $x=\sqrt{2}$时，$t=\dfrac{\pi}{4}$；当 $x=2$ 时，$t=\dfrac{\pi}{3}$，所以有

$$\int_{\sqrt{2}}^2 \frac{\mathrm{d}x}{\sqrt{x^2-1}} = \int_{\frac{\pi}{4}}^{\frac{\pi}{3}} \sec t\mathrm{d}t = \ln(\sec t+\tan t)\Big|_{\frac{\pi}{4}}^{\frac{\pi}{3}} = \ln\frac{2+\frac{\sqrt{3}}{3}}{\sqrt{2}+1}$$

5.4.3　利用分部积分法计算定积分

定积分的分部积分公式

设 $u(x)$，$v(x)$在区间$[a, b]$上均有连续的导数，则 $\int_a^b u\mathrm{d}v = uv\Big|_a^b - \int_a^b v\mathrm{d}u$.

例 5.9　计算下列定积分.

(1) $\int_1^{\mathrm{e}} \ln x\mathrm{d}x$；　(2) $\int_0^4 \mathrm{e}^{\sqrt{x}}\mathrm{d}x$；　(3) $\int_0^{\pi/2} x\sin x\mathrm{d}x$.

解　(1) $\int_1^{\mathrm{e}} \ln x\mathrm{d}x = x\ln x\Big|_1^{\mathrm{e}} - \int_1^{\mathrm{e}} x\dfrac{1}{x}\mathrm{d}x = 1$.

(2) 令$\sqrt{x}=t$，则 $x=t^2$，$\mathrm{d}x=2t\mathrm{d}t$，因此

$$\int_0^4 \mathrm{e}^{\sqrt{x}}\mathrm{d}x = \int_0^2 \mathrm{e}^t\cdot 2t\mathrm{d}t = 2\int_0^2 t\mathrm{d}\mathrm{e}^t = 2\left(t\mathrm{e}^t\Big|_0^2 - \int_0^2 \mathrm{e}^t\mathrm{d}t\right) = 2(\mathrm{e}^2+1)$$

(3) $\int_0^{\pi/2} x\sin x\mathrm{d}x = -x\cos x\Big|_0^{\pi/2} + \int_0^{\pi/2}\cos x\mathrm{d}x = -x\cos x\Big|_0^{\pi/2} + \sin x\Big|_0^{\pi/2} = 1$.

例 5.10　求椭圆$\dfrac{x^2}{9}+\dfrac{y^2}{16}=1$ 的面积.

解　如图 5-10 所示，图形关于 x 轴和 y 轴对称，所以要求整个椭圆的面积可以先求第一象限的面积，然后乘以 4 即可.

这里 $OA=3$, $OB=4$. 第一象限部分是由曲线 $y=\frac{4}{3}\sqrt{9-x^2}$ $(x>0)$和 $x=0$, $x=3$ 所围，所以它的面积为

$$A=\int_0^3 \frac{4}{3}\sqrt{9-x^2}\,\mathrm{d}x$$

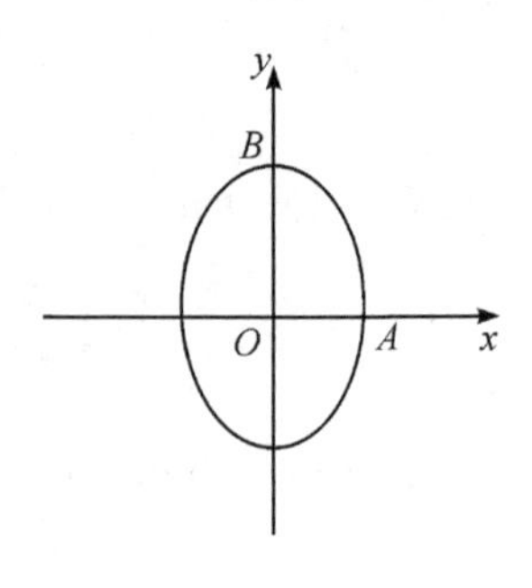

图 5 - 10

令 $x=3\sin\theta$, 则 $\mathrm{d}x=3\cos\theta\mathrm{d}\theta$, 有

$$\sqrt{9-x^2}=\sqrt{9-9\sin^2\theta}=3\cos\theta$$

当 $x=0$ 时, $\theta=0$;当 $x=3$ 时, $\theta=\frac{\pi}{2}$, 故有

$$A=\int_0^{\pi/2}\frac{4}{3}\cdot 3\cos\theta\cdot 3\cos\theta\mathrm{d}\theta=12\int_0^{\pi/2}\cos^2\theta\mathrm{d}\theta$$

$$=12\int_0^{\pi/2}\frac{1+\cos 2\theta}{2}\mathrm{d}\theta=6\left(\theta+\frac{\sin 2\theta}{2}\right)\Big|_0^{\frac{\pi}{2}}=3\pi$$

因此，所求面积为 $4A=4\times 3\pi=12\pi$.

5.5 求旋转体的体积

旋转体是由平面图形绕直线(也叫旋转轴)旋转而成. 本节我们仅仅讨论绕 x 轴和 y 轴旋转的情况.

5.5.1 圆盘法

求旋转体的体积与求由曲线 $y=f(x)$, x 轴，直线 $x=a$ 和 $x=b$ 所围图形的面积的思想方法相同.

如图 5 - 11 所示，首先，想象把这个旋转体用垂直于 x 轴的平面切成许多薄片(即圆盘)，每个圆盘都有厚度 Δx_i 和半径 $f(x_i)$，所求旋转体的体积就是这些圆盘的体积之和，即

$$V\approx\sum_{i=1}^{n-1}\pi[f(x_i)]^2\Delta x_i,\ \Delta x_i=\frac{b-a}{n}$$

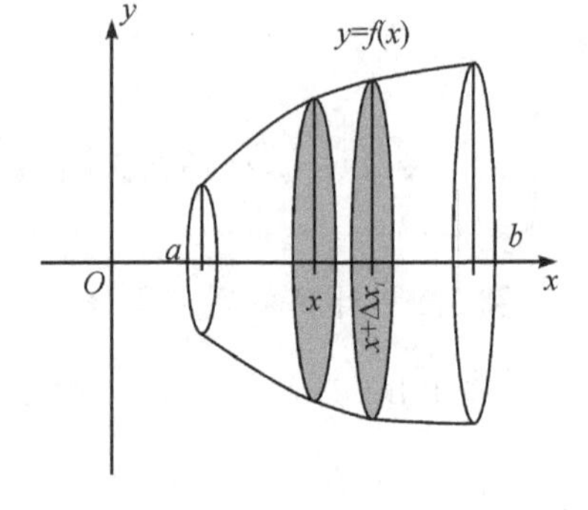

图 5 - 11

所以，当 $n\to\infty$时, $\Delta x_i\to 0$, 从而有

$$V=\lim_{\Delta x\to 0}\sum_{i=1}^{n-1}\pi[f(x_i)]^2\Delta x_i=\int_a^b\pi[f(x)]^2\mathrm{d}x$$

这个过程可以简化为两步.

第一步：近似. 选出一个薄片为代表，计算出薄片的体积，得到**体积元素**为

$$\Delta V\approx\pi[f(x)]^2\Delta x$$

第二步：精确. 对体积元素进行积分，则可得到整体体积为

$$V=\int_a^b\pi[f(x)]^2\mathrm{d}x$$

这两步称为定积分的**元素法**.

因此，我们得到如下公式.

旋转体的体积公式

当平面图形由曲线 $y=f(x)$，$x=a$ 和 $x=b$ 围成，且绕 x 轴旋转时，所得旋转体的体积为

$$V=\int_a^b \pi[f(x)]^2\mathrm{d}x$$

当平面图形由曲线 $x=f^{-1}(y)$，$y=c$ 和 $y=d$ 围成，且绕 y 轴旋转时，所得旋转体的体积为

$$V=\int_c^d \pi[f^{-1}(y)]^2\mathrm{d}y$$

例 5.11　曲线 $y=\sqrt{x-1}$ $(1\leqslant x\leqslant 5)$绕 x 轴旋转，画出这个旋转体，并求它的体积. 如果这条曲线绕 y 轴旋转，画出这个旋转体并求其体积.

解　在限制范围内曲线绕 x 轴旋转后如图 5－12 所示.

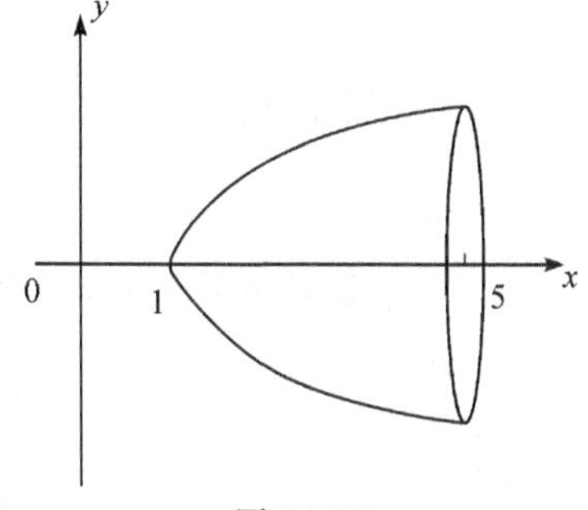

图 5－12

通过近似找体积元素，有

$$\Delta V\approx\pi\left(\sqrt{x-1}\right)^2\Delta x$$

所以，旋转体的体积为

$$V=\int_1^5\pi\left(\sqrt{x-1}\right)^2\mathrm{d}x=\pi\int_1^5(x-1)\mathrm{d}x=8\pi$$

如果这条曲线绕 y 轴旋转，所得旋转体如图5－13所示.

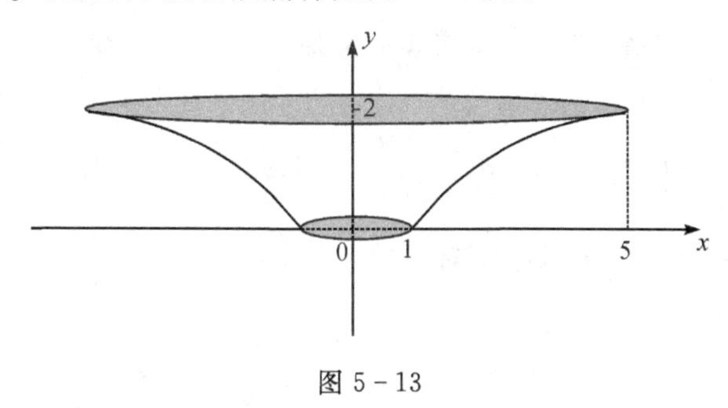

图 5－13

这里要清楚，y 是积分变量且它的变化范围是0～2，需要把 x 表示成 y 的函数.

$$y=\sqrt{x-1}\Rightarrow y^2=x-1\Rightarrow x=y^2+1\ (y>0)$$

当 $x=1$ 时，$y=0$；当 $x=5$ 时，$y=2$，从而有

$$V=\pi\int_0^2(y^2+1)^2\mathrm{d}y=13\frac{11}{15}\pi$$

例 5.12　设圆锥体的半顶角为 α，高为 h，求该圆锥体的体积.

解 如图 5-14 所示. 圆锥由介于 $y=0$ 和 $y=h$ 之间的直角三角形的斜边绕 y 轴旋转而成.

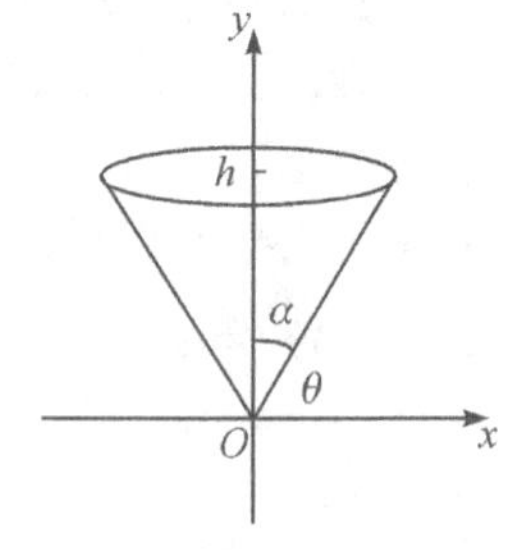

图 5-14

首先设直线方程为

$$y = kx$$

由于 $k=\tan\theta$，因此 $\theta=\frac{\pi}{2}-\alpha$，从而有 $k=\tan\left(\frac{\pi}{2}-\alpha\right)=\cot\alpha$. 故所求直线方程为

$$y = \cot\alpha \cdot x$$

接下来，绕 y 轴旋转，根据元素法有

$$\Delta V \approx \pi x^2 \Delta y$$

因 $a=0$，$b=h$ 并且 $y=x\cot\alpha$，有 $x=y\tan\alpha$，即

$$V = \pi\int_0^h (y\tan\alpha)^2 \mathrm{d}y = \frac{1}{3}\pi \tan^2\alpha \cdot h^3$$

熟练后，可以跳过第一步. 于是，所求旋转体体积为

$$V = \pi\int_0^h x^2 \mathrm{d}y$$

注意： 由于是对 y 的积分，所以需要把 x 表示成 y 的函数 $x=y\tan\alpha$，故有

$$V = \pi\int_0^h (y\tan\alpha)^2 \mathrm{d}y = \frac{1}{3}\pi \tan^2\alpha \cdot h^3$$

例 5.13 求由方程 $f(x)=\sqrt{25-x^2}$ 和直线 $g(x)=3$ 所围区域绕 x 轴旋转的旋转体的体积.

解 所围图形如图 5-15 所示，首先确定交点坐标. 令 $f(x)=g(x)$，有 $\sqrt{25-x^2}=3 \Rightarrow x=\pm 4$，旋转体内部是空心的.

然后，求生成的两个旋转体的体积之差，类似于求两条曲线之间的面积.

$$V= \pi\int_{-4}^4 [f(x)]^2 \mathrm{d}x - \pi\int_{-4}^4 [g(x)]^2 \mathrm{d}x = \pi\int_{-4}^4 [f^2(x) - g^2(x)]\mathrm{d}x$$

$$= 2\pi\int_0^4 \left[\left(\sqrt{25-x^2}\right)^2 - 3^2\right]\mathrm{d}x \text{（根据对称性）}$$

$$= \frac{256}{3}\pi$$

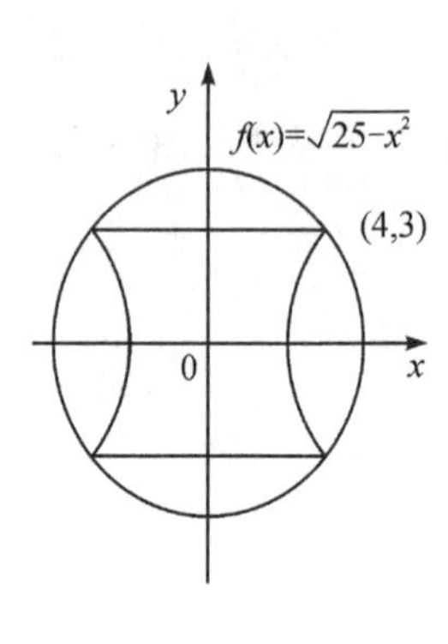

图 5-15

5.5.2 柱壳法

求旋转体的体积还有一种方法，称为柱壳法。对于许多求旋转体体积的问题，用柱壳法比用圆盘法要容易.

现在，考虑平面区域，如图 5-16(a)所示，对其进行分割，然后绕 y 轴旋转，每一个窄带绕 y 轴旋转后都会形成一个柱壳，如图 5-16(b)所示. 为了得到整个体积，可以先计算其中一个柱壳的体积. 因为把柱壳剪开后是一个长方体，故体积为

$$\Delta V \approx 2\pi x f(x)\Delta x$$

于是，这个旋转体的体积为

$$V=\int_a^b 2\pi x f(x)\mathrm{d}x$$

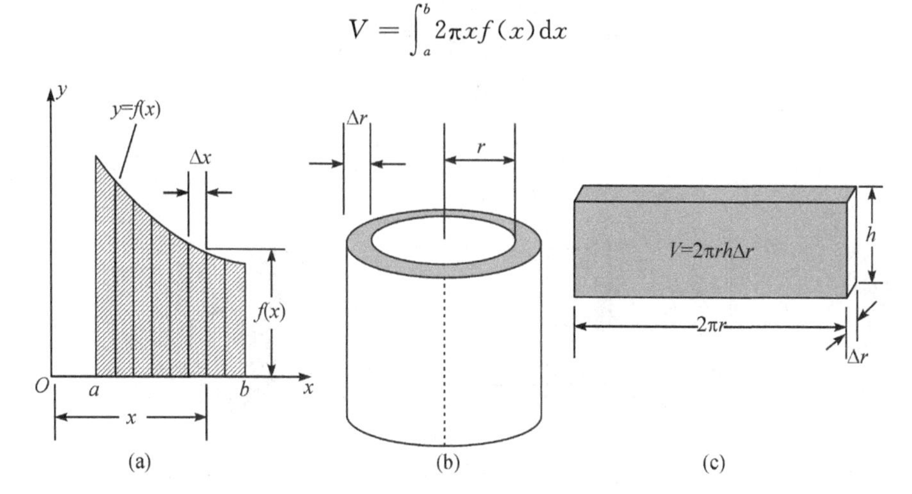

图 5－16

例 5.14　设平面区域是由 $y=\dfrac{1}{\sqrt{x}}$，x 轴和直线 $x=1$、$x=9$ 所围，求该平面图形绕 y 轴旋转所得旋转体的体积.

解　由图 5－17 可以看出，柱壳的体积为

$$\Delta V\approx 2\pi x f(x)\Delta x=2\pi x\frac{1}{\sqrt{x}}\Delta x$$

于是对体积元素积分，得旋转体的体积

$$V=2\pi\int_1^9 x\frac{1}{\sqrt{x}}\mathrm{d}x=2\pi\int_1^9\sqrt{x}\mathrm{d}x=2\pi\left(\frac{2}{3}x^{3/2}\right)\bigg|_1^9=\frac{104}{3}\pi$$

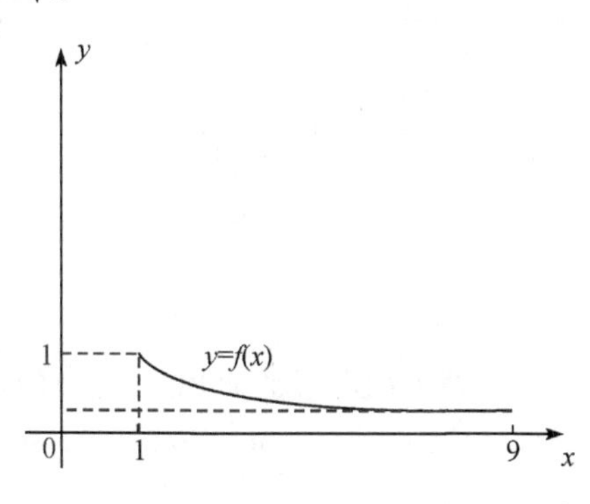

图 5－17

例 5.15　设一平面区域由直线 $y=\dfrac{r}{h}x$，$x=h$ 和 x 轴所围，其绕 x 轴旋转后为一个圆锥体（假设 $xr>0$，$h>0$）. 分别利用圆盘法和柱壳法求该圆锥体的体积.

解　圆盘法：见图 5－18.

先求体积元素

$$\Delta V \approx \pi\left[f(x)\right]^2\Delta x = \pi\left(\frac{r}{h}x\right)^2\Delta x$$

于是有

$$V = \int_0^h \pi\left(\frac{r}{h}x\right)^2 \mathrm{d}x = \frac{1}{3}\pi r^2 h$$

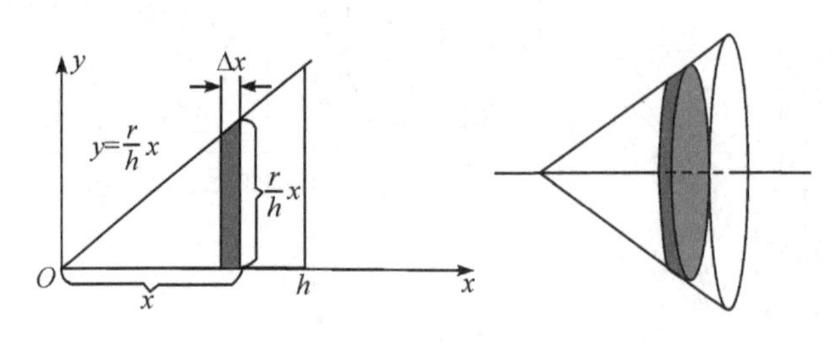

图 5 - 18

柱壳法：见图 5 - 19.

$$\Delta V \approx 2\pi y\left(h-\frac{h}{r}y\right)^2\Delta y$$

于是有

$$V = \int_0^r 2\pi y\left(h-\frac{h}{r}y\right)^2 \mathrm{d}y = \frac{1}{3}\pi r^2 h$$

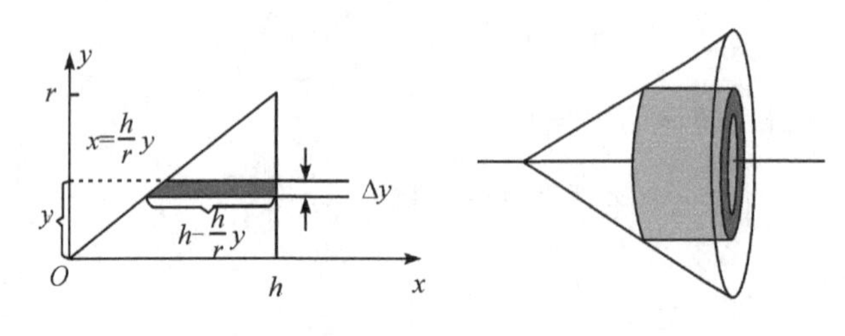

图 5 - 19

习题 5

1. 利用求和式的极限方法求下列平面图形的面积.（每个题目中都包括 x 轴）

(1) $y=1-x$，$x=0$，$x=b(0<b<1)$；　　(2) $y=4x^2$，$x=0$，$x=c(c>0)$；

(3) $y=\mathrm{e}^x$，$x=0$，$x=a(a>0)$.

2. 求由 x 轴和下列曲线及直线所围区域的面积.

(1) $x^2=4by$，$x=a$，$x=b(b>a>0)$；　　(2) $y=4x^3$，$x=2$，$x=4$；

(3) $y=3x^2-2$，$x=1$，$x=4$；　　(4) $y^2-x-4=0$，$x=2$，$x=5$；

(5) $y=\mathrm{e}^{ax}$，$x=b$，$x=c(c>b>0)$；　　(6) $y=\ln(1+x)$，$x=0$，$x=1$.

3. 求下列以 x 轴、所给曲线和点的纵坐标为边界的图形在第一象限区域的面积.

(1) $y^2=8ax(a>0)$，点$(4a, 0)$；

(2) $y^2=4a(x-a)$，点$(h, 0)(h>a>0)$；

(3) $x^2=4by(b>0)$，点$(b, 0)$.

4. 求下列曲线和直线所围平面图形的面积.

(1) 曲线 $y^2=16x$，直线 $y=2x$；

(2) 曲线 $y=x^3$，直线 $x=y$，位于第一象限；

(3) 曲线 $y^2=x^3$，直线 $x=4$；

(4) 曲线 $y^2=4ax$，曲线 $x^2=4ay$；

(5) 曲线 $y=x^2-8x+15$，x 轴.

5. 计算下面的定积分.

(1) $\int_1^2 (2x^2+3x+4)\mathrm{d}x$；　　(2) $\int_0^{-1} \frac{\mathrm{d}x}{x+2}$；

(3) $\int_0^1 x^3\sqrt{1+2x^4}\mathrm{d}x$；　　(4) $\int_1^2 \mathrm{e}^{2x^2-1}x\mathrm{d}x$；

(5) $\int_0^a \frac{x\mathrm{d}x}{(a^2+x^2)^{3/2}}$；　　(6) $\int_0^1 \frac{2x\mathrm{d}x}{x^2+3}$；

(7) $\int_0^{\pi/4} \tan^2\theta\mathrm{d}\theta$；　　(8) $\int_0^1 \cos^2\pi x\mathrm{d}x$；

(9) $\int_0^{\pi/2} \sin^3 x\mathrm{d}x$；　　(10) $\int_0^{\pi/4} \frac{\mathrm{d}x}{1-\sin x}$；

(11) $\int_0^{\pi/2} \sqrt{1+\sin x}\mathrm{d}x$；　　(12) $\int_0^{\pi/2} \cos 3x\cos 2x\mathrm{d}x$；

(13) $\int_0^{\pi/4} \cos^3 x\sin^2 x\mathrm{d}x$；　　(14) $\int_0^{\pi/4} \tan^3 x\mathrm{d}x$

(15) $\int_0^{\pi/4} \tan^2 x\sec^4 x\mathrm{d}x$；　　(16) $\int_0^{-1} \frac{\mathrm{d}x}{\sqrt{4-x^2}}$；

(17) $\int_{\pi}^{\pi/2} x\cos x\mathrm{d}x$；　　(18) $\int_1^{\mathrm{e}} x\ln x\mathrm{d}x$.

6. 求由曲线 $y=\sin x(0\leqslant x\leqslant\pi)$和 x 轴所围图形分别绕 x 轴，y 轴旋转所形成的旋转体的体积.

7. 设曲线 $y=\frac{1}{x}\left(\frac{1}{5}\leqslant x\leqslant 1\right)$绕 y 轴旋转，求旋转体体积.

8. 求过原点和点(r, h)的直线方程，并利用微积分证明：以 r 为底面半径，高为 h 的直圆锥的体积为 $V=\frac{1}{3}\pi r^2 h$.

9. 求以$(0, 0)$为圆心，半径为 r 的圆的方程，并利用定积分证明以 r 为半径的球体的体积为 $V=\frac{4}{3}\pi r^3$.

10. 求由曲线 $y=\sqrt{x}$和 $y=\sqrt{x^3}$所围区域绕坐标轴旋转而形成的旋转体的体积. (1) 绕 y 轴旋转；(2) 绕 x 轴旋转.

11. 求由曲线 $y=\frac{1}{x}$，直线 $x=1$，$x=2$ 绕直线 $x=1$ 旋转所形成的旋转体的体积.

第 6 章　多元函数微分及其应用

一元函数的图像是平面上的一条曲线，在本章介绍完多元函数的概念后，我们将看到二元函数的图像是一个曲面．二元及二元以上的函数经常出现在科学和工程技术领域，熟练掌握这些函数是很重要的．本章中我们将学习如何绘制简单曲面，并研究如何确定 $f(x, y)$ 关于变量 x 和 y 的变化率，以及如何求多元函数的最值．

6.1　多元函数

6.1.1　认识多元函数

我们知道 $f(x)$ 表示带有一个自变量的函数：输入 x，输出 $y=f(x)$，这里 x 叫作**自变量**，$y=f(x)$ 叫作**因变量**．

带有两个自变量的函数 $z=f(x, y)$ 的定义域是二维平面上的区域，由点的所有坐标 (x, y) 组成，把坐标代入函数得到一个实数．我们考虑带有两个自变量 x 和 y 的函数，例如

$$f(x, y)=x+2y+3$$

它的定义域是 xOy 平面．如果我们指定 x 和 y 值，就得到唯一的 $f(x, y)$ 值．如，令 $x=3$，$y=1$，则 $f(x, y)=3+2+3=8$，写作 $f(3, 1)=8$．

例 6.1　确定下列函数的定义域，并画出定义域的草图．

(1) $f(x, y)=\sqrt{x+y}$；(2) $f(x, y)=\sqrt{x}+\sqrt{y}$；(3) $f(x, y)=\ln(9-x^2-9y^2)$．

解　(1) 由于负数不能开平方，所以这里 $x+y\geqslant 0$；

(2) 类似地，还必须要求 $x\geqslant 0$ 且 $y\geqslant 0$；

(3) 由于负数不能取对数，所以，$9-x^2-9y^2>0 \Rightarrow x^2+9y^2<9$．

图 6-1 所示分别为三个函数的定义域图形．

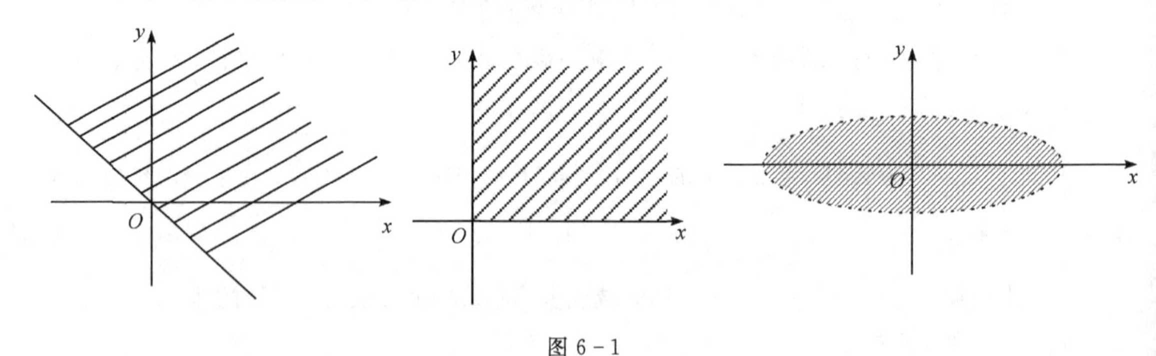

图 6-1

例 6.2　对于下列函数求值 $f(2, 1)$，$f(-1, -3)$ 和 $f(0, 0)$．

(1) $f(x, y)=x^2+y^2+1$；(2) $f(x, y)=2x+xy+y^3$．

解　(1) $f(2, 1)=2^2+1^2+1=6$;

$f(-1, -3)=(-1)^2+(-3)^2+1=11$;

$f(0, 0)=1$.

(2) $f(2, 1)=4+2+1=7$;

$f(-1, -3)=-2+3-27=-26$;

$f(0, 0)=0$.

类似地，可以定义含有三个自变量 x, y, z 的函数，$u=f(x, y, z)$.

6.1.2　二元函数的几何意义

读者可以回顾最简单的情形——平面，平面是曲面的特殊情况.

平面的一般方程为

$$Ax+By+Cz=D$$

其中：A, B, C, D 为常数. 如果 $C\neq 0$，我们可以把方程写成 $z=\frac{D}{C}-\frac{A}{C}x-\frac{B}{C}y$.

一般地，曲面可以由 $z=f(x, y)$来表示，该表达式中涉及两个自变量 x 和 y.

6.1.3　绘制曲面图像

平面相对容易画，因为它是平的，我们只需要知道它与三个坐标轴的交点. 对于更一般的曲面，我们将绘制曲面上的曲线，如果画出的曲线足够多，自然会得到曲面的形状.

例如绘制 $z=x^2+y^2$ 的图像.

第一步　把 x 固定在 x_0，从而表达式变为

$$z=x_0^2+y^2$$

由于此时 z 是单变量 y 的函数，$z=x_0^2+y^2$ 定义了一条位于平面 $x=x_0$ 上的抛物线. 图 6-2中已画了这条抛物线，选择不同的 x_0，可以得到一系列抛物线，每条抛物线都在不同的平面上，所有的抛物线都是曲面的一部分. 用平行于 yOz 的平面去截这个曲面，每个截面与曲面的交线是一条曲线. 这时，我们还没有画出足够多的曲线，还不能够看出曲面的全貌，还需要画其他一些曲线.

第二步　把 y 固定在 y_0.

这里 $y=y_0$(该方程表示平面平行于 xOz 平面). 此时，曲面方程变为

$$z=x^2+y_0^2$$

由于固定了 $y=y_0$，所以 z 是单变量 x 的函数，也由一条抛物线表示，但该抛物线在平面 $y=y_0$ 上，如图 6-3 所示. 通过选取不同的 y_0 可以得到不同的抛物线，每条抛物线均在曲面 $z=x^2+y^2$ 上.

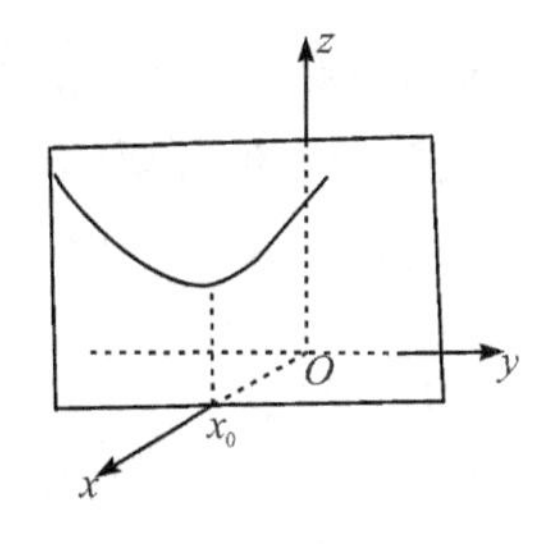

图 6-2

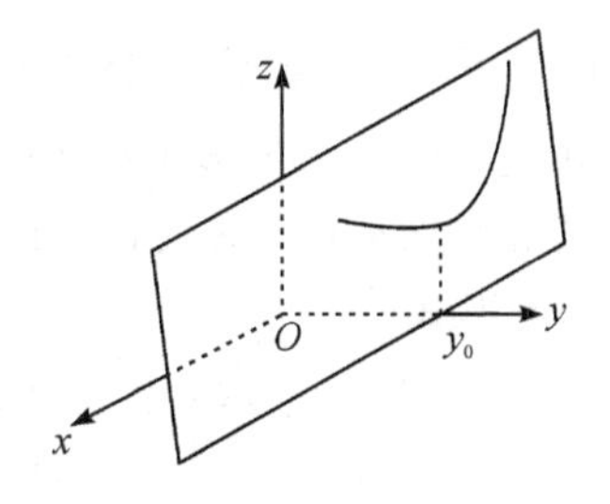

图 6-3

第三步　把 z 固定在 z_0.

有 $z=z_0$(该方程表示平行于 xOy 面的平面)，方程变为 $z_0=x^2+y^2$. 这是一个以 $x=0$，$y=0$ 为圆心，半径为 $\sqrt{z_0}$ 的圆的方程，如图 6 - 4 所示.（显然，我们必须选择 $z_0\geqslant 0$). 当 z_0 变化时，每个圆都在不同的 $z=z_0$ 平面上.

我们把图 6 - 4 中所有的圆和之前得到的曲线组合在一起，就可以看出曲面 $z=x^2+y^2$ 的全貌了，如图 6 - 5 所示.

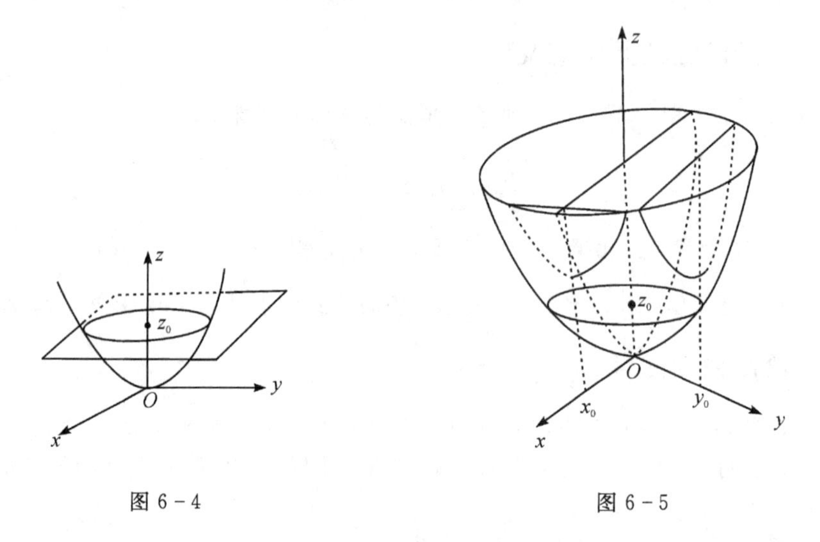

图 6 - 4　　　　图 6 - 5

这个曲面叫作**抛物面**，可以由一条抛物线绕 z 轴旋转得到.

借助于画图软件我们可以绘制更复杂的函数图像，所以对绘制曲面的要求降低了，然而我们要求初学者应会画简单的二元函数图像，这对理解二元函数是有帮助的.

6.2　二元函数的极限

本节将介绍函数的极限. 在介绍多元函数的极限之前，先来回顾一元函数的极限

$$\lim_{x\to x_0} f(x)=a$$

等价于

$$\lim_{x\to x_0^-} f(x)=\lim_{x\to x_0^+} f(x)=a$$

换句话说，$\lim\limits_{x\to x_0} f(x)=a$ 是指当自变量 x 从 x_0 的两侧趋近于 x_0 时，所对应的函数值 $f(x)$ 趋近于 a.

注意：这种情形下，x 趋近于 x_0 的路径只有两条，即从左侧和从右侧. 一元函数的极限存在，要求自变量 x 无论是从左侧还是从右侧趋近于 x_0 时，函数值都要趋近于 a.

二元函数的情况与之类似. 我们讨论的是，当 x 趋于 x_0，y 趋于 y_0 时，函数 $f(x, y)$ 的极限. 可以写成如下形式

$$\lim_{\substack{x\to x_0\\ y\to y_0}} f(x, y)\quad 或\quad \lim_{(x, y)\to(x_0, y_0)} f(x, y)$$

像一元函数的极限一样，二元函数的自变量 (x, y) 无论以什么路径趋近于点 (x_0, y_0)，所对

应的函数值都必须趋于同一个值．由于(x_0, y_0)是平面上的点，这种路径有无穷多个．

换句话说，如果极限存在，我们需要检验无穷多条路径，并且无论在哪个路径上，自变量趋于点(x_0, y_0)时，函数都趋于同一个值．反之，如果能找到两条路径，在这两条路径上函数趋于不同的值，则可知函数的极限不存在．

例 6.3　判定下列函数的极限是否存在．

(1) $\lim\limits_{\substack{x\to 0\\y\to 0}}\dfrac{x-y}{x+y}$；　　(2) $\lim\limits_{\substack{x\to 0\\y\to 0}}\dfrac{x^3y}{x^6+y^2}$．

解　(1) 当点(x, y)沿着直线 $y=kx$ 趋向于$(0, 0)$点时，有

$$\lim_{\substack{x\to 0\\y\to 0}}\frac{x-y}{x+y}=\lim_{\substack{x\to 0\\y=kx}}\frac{x-kx}{x+kx}=\frac{1-k}{1+k}$$

上式说明，若 k 不同，即当点(x, y)沿着不同的直线 $y=kx$ 趋向于$(0, 0)$时，$f(x, y)$趋于不同的数值，因此$\lim\limits_{\substack{x\to 0\\y\to 0}}\dfrac{x-y}{x+y}$不存在．

(2) 当点(x, y)沿着直线 $y=x$ 趋向于$(0, 0)$点时，有

$$\lim_{\substack{x\to 0\\y\to 0}}\frac{x^3y}{x^6+y^2}=\lim_{\substack{x\to 0\\y=x}}\frac{x^4}{x^6+x^2}=\lim_{\substack{x\to 0\\y=x}}\frac{x^2}{x^4+1}=0$$

当点(x, y)沿着曲线 $y=x^3$ 趋向于$(0, 0)$点时，有

$$\lim_{\substack{x\to 0\\y\to 0}}\frac{x^3y}{x^6+y^2}=\lim_{\substack{x\to 0\\y=x^3}}\frac{x^6}{x^6+x^6}=\frac{1}{2}$$

由于点(x, y)沿不同路径趋向于$(0, 0)$点时 $f(x, y)$趋于不同的数值，因此$\lim\limits_{\substack{x\to 0\\y\to 0}}\dfrac{x^3y}{x^6+y^2}$不存在．

那么如何求二元函数的极限呢？可以采用一元函数求极限的思想进行求解，即利用函数的连续性．

6.3　二元函数的连续性

从几何角度看，二元函数的连续性与第 1 章中一元函数的连续性是一样的，即如果图像在某个点处没有空洞或断裂，则函数在这个点是连续的．

定义 6.1　如果$\lim\limits_{\substack{x\to x_0\\y\to y_0}}f(x, y)=f(x_0, y_0)$，则称函数 $f(x, y)$ 在点(x_0, y_0)处**连续**．

这个定义提供了一种利用连续函数求极限的方法．

例 6.4　判断下列极限是否存在，如果存在，求出其极限值．

(1) $\lim\limits_{\substack{x\to 5\\y\to 1}}\dfrac{xy}{x+y}$；　　(2) $\lim\limits_{\substack{x\to 5\\y\to 0}}\dfrac{e^x+e^y}{\cos x-\sin y}$．

解　因为所给函数在给定点都是连续的，所以我们只需要把点代入求函数值即可．

(1) $\lim\limits_{\substack{x\to 5\\y\to 1}}\dfrac{xy}{x+y}=\dfrac{5}{6}$；

(2) $\lim\limits_{\substack{x\to 5\\y\to 0}}\dfrac{e^x+e^y}{\cos x-\sin y}=\dfrac{e^5+1}{\cos 5}$．

6.4 偏导数

当多元函数的一个或几个自变量改变而导致多元函数发生改变时，计算函数本身的变化是很重要的. 可以通过保持除一个变量外的所有变量不变并求出函数相对于剩下的一个变量的变化率来研究，这个过程叫作偏微分. 在本节中，我们将对此进行探讨.

6.4.1 一阶偏导数

1. 关于 x 的偏导数

对于一元函数 $y=f(x)$，自变量 x 的变化会导致因变量 y 的变化，y 关于 x 的变化率可以由导数给出，记作$\frac{\mathrm{d}f}{\mathrm{d}x}$. 多个变量的函数也会出现类似的情况. 为了清楚起见，我们将集中讨论只有两个自变量的函数.

在对应关系 $z=f(x, y)$ 中，x 和 y 是自变量，z 是因变量. 在 6.1 节中我们看到，随着 x 和 y 的变化，z 的轨迹是一个曲面. 现在这两个自变量可以同时变化，从而使 z 发生变化. 与其考虑这种一般情况，不如先把其中一个自变量固定下来，仅考虑另一自变量发生变化，这等价于函数值沿着一条曲线移动，这条曲线是由平行于坐标面的平面与曲面相交而得到的.

例如，对函数 $z=x^3+2x^2y+y^2+2x+1$，如果保持 y 不变，仅让 x 变化，则函数 f 的变化率是多少?

假设 y 取固定值 3，则

$$f(x, 3)=x^3+6x^2+9+2x+1=x^3+6x^2+2x+10$$

实际上，这是一个只与 x 有关的函数，如果对 x 求导，则有

$$f'(x, 3)=3x^2+12x+2$$

即 f 关于 x 存在偏导数，记作$\frac{\partial f}{\partial x}$（读作偏 f 除以偏 x）. 此例中，当 $y=3$ 时

$$\frac{\partial f}{\partial x}=3x^2+12x+2$$

如果将原函数 $f(x, y)=x^3+2x^2y+y^2+2x+1$ 中的 y 看作常数，则函数关于 x 的偏导数为

$$\begin{aligned}\frac{\partial f}{\partial x}&=3x^2+4xy+0+2+0\\&=3x^2+4xy+2\end{aligned}$$

f 关于 x 的偏导数

对于一个二元函数 $z=f(x, y)$，f 对 x 的偏导数记作$\frac{\partial f}{\partial x}$，可以通过把 y 看作常数按关于 x 的一元函数求导的方法得到，也记作 $f_x(x, y)$，f_x 或$\frac{\partial z}{\partial x}$.

2. 关于 y 的偏导数

对于二元函数 $f(x, y)$，变量 x 和 y 是同等关系，对 x 实施的运算，对 y 也可以实施. 因此，可以保持 x 不变，确定 f 关于 y 的变化率，记作$\frac{\partial f}{\partial y}$.

f 关于 y 的偏导数

对于一个二元函数 $z=f(x, y)$，f 对 y 的偏导数记作$\frac{\partial f}{\partial y}$，可以通过把 x 看作常数按关于 y 的一元函数求导的方法得到. 也记作 $f_y(x, y)$或 f_y 或$\frac{\partial z}{\partial y}$.

再回到 $f(x, y)=x^3+2x^2y+y^2+2x+1$，有

$$\frac{\partial f}{\partial y}=0+2x^2+2y+0+0=2x^2+2y$$

例 6.5　求$\frac{\partial f}{\partial y}$. 设 (1) $f(x, y)=x+\sqrt{y}$；(2) $f(x, y)=\arccos\frac{x}{y}$.

解　(1) $\frac{\partial f}{\partial y}=0+\frac{1}{2\sqrt{y}}=\frac{1}{2\sqrt{y}}$；

(2) $\frac{\partial f}{\partial y}=-\frac{1}{\sqrt{1-\frac{x^2}{y^2}}}\left(-\frac{x}{y^2}\right)=\frac{x}{y\sqrt{y^2-x^2}}$.

还可以计算函数 f 在某一点的偏导数$\frac{\partial f}{\partial x}$和$\frac{\partial f}{\partial y}$，如 $x=1$，$y=-2$.

例 6.6　设 $f(x, y)=x^2+y^3+2xy$，求 $f_x(1, -2)$和 $f_y(-3, 2)$.

解　方法 1：

$$f_x(x, y)=2x+2y \Rightarrow f_x(1, -2)=2-4=-2$$

$$f_y(x, y)=3y^2+2x \Rightarrow f_y(-3, 2)=12-6=6$$

方法 2：

$$f(x, -2)=x^2-8-4x \Rightarrow f_x(x, -2)=(x^2-8-4x)'_x=2x-4$$

$$\Rightarrow f_x(1, -2)=(2x-4)_{x=1}=-2$$

类似地

$$f(-3, y)=9+y^3-6y \Rightarrow f_y(-3, y)=(9+y^3-6y)'_y=3y^2-6$$

$$\Rightarrow f_y(-3, 2)=(3y^2-6)_{y=2}=12-6=6$$

3. 偏导数的几何意义

根据偏导数的定义，函数 $z=f(x, y)$在(x_0, y_0)点关于 x 的偏导数，就是先固定 y，即做平面 $y=y_0$，与曲面 $z=f(x, y)$相交后得到一条交线RPQ，如图 6-6 所示. 该交线在点(x_0, y_0)处的切线 l 的斜率就是函数 $z=f(x, y)$在点(x_0, y_0)处关于 x 的偏导数 $f_x(x_0, y_0)$. 类似地，做平面 $x=x_0$，与曲面相交后的曲线在点(x_0, y_0)处的切线的斜率就是 $f_y(x_0, y_0)$.

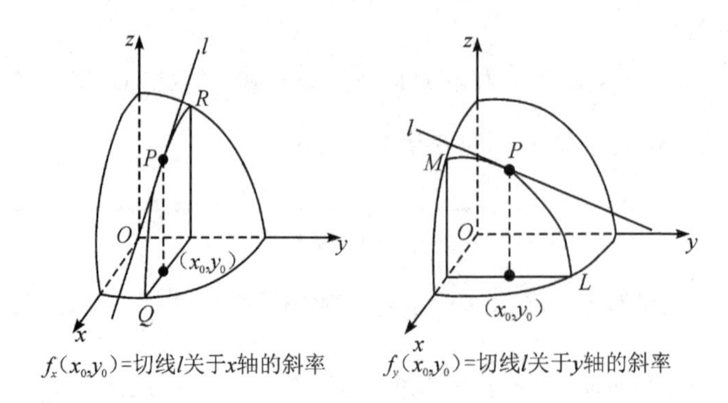

图 6－6

4. 含两个以上自变量的函数的偏导数

二元函数 $f(x,y)$有两个偏导数，$\frac{\partial f}{\partial x}$和$\frac{\partial f}{\partial y}$. 类似地，一个有三个变量的函数$f(x,y,z)$有三个偏导数，$\frac{\partial f}{\partial x}$,$\frac{\partial f}{\partial y}$和$\frac{\partial f}{\partial z}$，对于含三个以上变量的函数，依此类推，每个偏导数的计算方法与前面类似.

例 6.7　设 (1) $f=x^2+yz+z^3$；(2) $f=e^z+\ln(x^2+y^2-1)$. 求$\frac{\partial f}{\partial x}$和$\frac{\partial f}{\partial z}$.

解　(1) $\frac{\partial f}{\partial x}=2x,\frac{\partial f}{\partial z}=y+3z^2$；(2) $\frac{\partial f}{\partial x}=\frac{2x}{x^2+y^2-1}$，$\frac{\partial f}{\partial z}=e^z$.

6.4.2　二阶偏导数

$f(x,y)$连续两次对 x 求导(保持 y 不变)，即$\frac{\partial^2 f}{\partial x^2}=\frac{\partial}{\partial x}\left(\frac{\partial f}{\partial x}\right)$，记作$\frac{\partial^2 f}{\partial x^2}$（或 $f_{xx}(x,y)$).
类似地，可以得到其他二阶偏导数.

$$\frac{\partial^2 f}{\partial x^2}=\frac{\partial}{\partial x}\left(\frac{\partial f}{\partial x}\right)=f_{xx}(x,y),\quad \frac{\partial^2 f}{\partial x\partial y}=\frac{\partial}{\partial y}\left(\frac{\partial f}{\partial x}\right)=f_{xy}(x,y)$$

$$\frac{\partial^2 f}{\partial y^2}=\frac{\partial}{\partial y}\left(\frac{\partial f}{\partial y}\right)=f_{yy}(x,y),\quad \frac{\partial^2 f}{\partial y\partial x}=\frac{\partial}{\partial x}\left(\frac{\partial f}{\partial y}\right)=f_{yx}(x,y)$$

其中，$\frac{\partial^2 f}{\partial x\partial y}$和$\frac{\partial^2 f}{\partial y\partial x}$称为**二阶混合偏导数**. $\frac{\partial^2 f}{\partial x\partial y}$表示先对 x 再对 y 求导，$\frac{\partial^2 f}{\partial y\partial x}$表示先对 y 再对 x 求导.

例 6.8　设 $f(x,y)=x^3+x^2y^2+2y^3+2x+y$，求$\frac{\partial^2 f}{\partial x^2}$和$\frac{\partial^2 f}{\partial y^2}$.

解　$\frac{\partial f}{\partial x}=3x^2+2xy^2+0+2+0=3x^2+2xy^2+2\Rightarrow\frac{\partial^2 f}{\partial x^2}=6x+2y^2$.

$\frac{\partial f}{\partial y}=0+2x^2y+6y^2+0+1=2x^2y+6y^2+1\Rightarrow\frac{\partial^2 f}{\partial y^2}=2x^2+12y$.

例 6.9　设 $f(x, y)=x^3+x^2y^2+2y^3+2x+y$，求 $f_{xx}(-1, 1)$和 $f_{yy}(2, -2)$.

解

$$f_{xx}(-1, 1)=(6x+2y^2)\Big|_{(-1, 1)}=-4$$

$$f_{yy}(2, -2)=(2x^2+12y)\Big|_{(2, -2)}=-16$$

例 6.10　设 $f(x, y)=x^3+2x^2y^2+y^3$，求$\frac{\partial^2 f}{\partial x\partial y}$.

解　$\frac{\partial f}{\partial x}=3x^2+4xy^2+0$，$\frac{\partial^2 f}{\partial x\partial y}=0+8xy=8xy$.

另一个混合偏导数是先对 y 再对 x 求导，即$\frac{\partial}{\partial x}\left(\frac{\partial f}{\partial y}\right)$. 对于例 6.10 中的函数，$\frac{\partial f}{\partial y}=4x^2y+3y^2$，则$\frac{\partial^2 f}{\partial y\partial x}=8xy$.

注意：这里$\frac{\partial^2 f}{\partial x\partial y}=\frac{\partial^2 f}{\partial y\partial x}$. 此等式适用于本教材中的所有二元函数.

例 6.11　设 $f(x, y)=x^3+2x^2y^2+y^3$，求 $f_{yx}(1, 2)$.

解　$\frac{\partial f}{\partial y}=4x^2y+3y^2$，$f_{yx}=8xy$，所以 $f_{yx}(1, 2)=16$.

6.5　驻点

许多工程技术领域都需要计算二元函数的最优值，如热力学领域. 与一元函数不同，确定驻点的性质的方法更为复杂.

6.5.1　二元函数的驻点

图 6-7 是由计算机画出的二次曲面的图像，其中 x 和 y 都取值于区间[−1.8, 1.8]. 曲面上有四个有趣的特征点，在点 A 处有一个**极大值**，点 B 处有一个**极小值**，点 C 和 D 是所谓的**鞍点**.

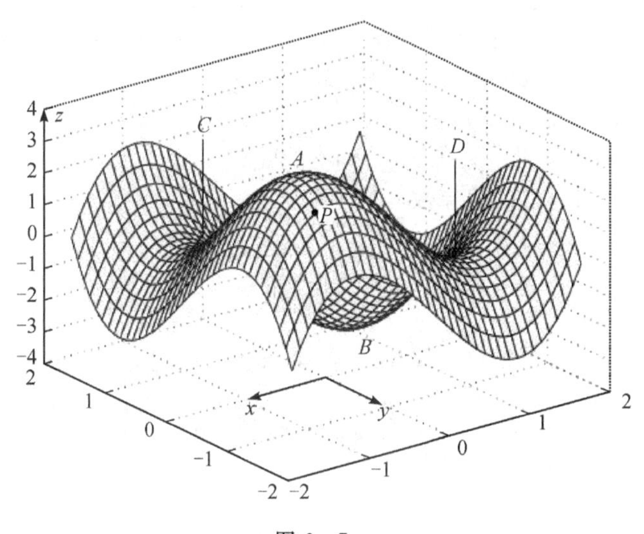

图 6-7

曲面上 A 点邻近处是最高的，如果在曲面上从 A 点开始，无论我们朝哪个方向移动，高度都会立即下降. 在点 B 处，曲面在邻近区域内的高度是最低的，如果我们从 B 点开始，无论朝哪个方向移动，高度都会立刻增加.

C 点和 D 点的特征完全不同，在某些方向上，当我们沿着曲面远离这两个点时，高度会降低，而在另一些方向上，高度会增加，形状与马鞍相似.

在光滑曲面上的任意一点 P 处，都可以画出一个与曲面接触的唯一平面，这个平面称为点 P 的切平面(切平面是切线的自然泛化，可以在光滑曲线的每一点上画出来). 在图 6－7 中，在点 A，B，C，D 处，曲面的切平面是水平的，这样的点称为函数的**驻点**. 在下一节中，我们将讨论如何寻找驻点，以及如何使用函数 $f(x, y)$ 的偏导数来确定它们的性质.

例 6.12 图 6－8 中，点 A，B 和 O 分别有什么特征？

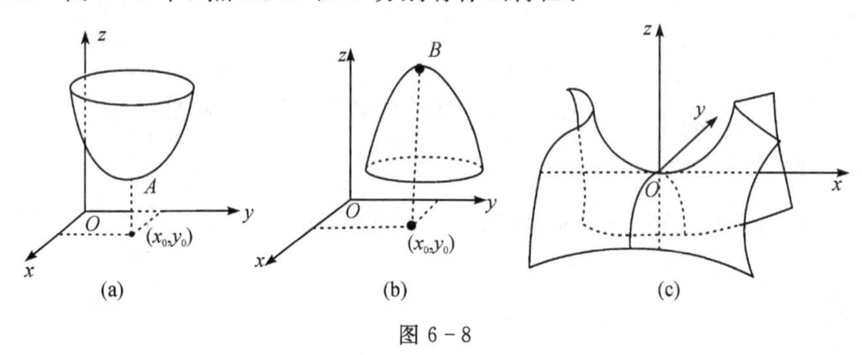

图 6－8

解 在图 6－8(a)中函数在点 A 处取得极小值，在图 6－8(b)中函数在点 B 处取得极大值，在图 6－8(c)中点 O 是鞍点.

6.5.2 确定驻点

前面我们指出，曲面 $z=f(x, y)$ 在驻点处的切平面是水平的. 一个保证点 (x_0, y_0) 是函数 $f(x, y)$ 的驻点的条件是 $f_x=0$，同时 $f_y=0$.

例 6.13 求函数 $f(x, y)=8x^2+6y^2-2y^3+5$ 的驻点.

解 $f_x=16x$，$f_y=6y(2-y)$. 令 $f_x=16x=0$，$f_y=6y(2-y)=0$，解方程组，得

$$x=0, y=0 \quad \text{或} \quad x=0, y=2$$

因此，驻点为 $(0, 0)$ 和 $(0, 2)$.

例 6.14 求函数 $f(x, y)=x^4+y^4-36xy$ 的驻点.

解 首先求函数 $f(x, y)$ 的偏导数，

$$\frac{\partial f}{\partial x}=4x^3-36y, \frac{\partial f}{\partial y}=4y^3-36x$$

然后解方程组

$$\begin{cases}\dfrac{\partial f}{\partial x}=4x^3-36y=0 & (6.1)\\[2ex] \dfrac{\partial f}{\partial y}=4y^3-36x=0 & (6.2)\end{cases}$$

由式(6.2)得

$$x=\frac{y^3}{9} \tag{6.3}$$

把式(6.3)代入式(6.1)得

$$\frac{y^9}{9^3}-9y=0 \Rightarrow y(y^4-3^4)(y^4+3^4)=0 \Rightarrow y=0 \text{ 或 } y=\pm 3$$

由式(6.3)可得，当 $y=0$ 时，$x=0$；当 $y=3$ 时，当 $x=3$；当 $y=-3$ 时，$x=-3$. 所以，驻点为 (0，0)，(3，3) 和 (−3，−3).

6.5.3 驻点的特征

我们不加证明地给出一个相对简单的方法来确定驻点的性质. 这个方法依赖于二阶偏导数的值 f_{xx}，f_{yy}，f_{xy}的值及 $AC-B^2$ 的值，其中

$$A=f_{xx},\ B=f_{xy},\ C=f_{yy}$$

具体方法如下：

定理 6.1　确定驻点性质的方法

(1) 在每个驻点处求出二阶偏导数；

(2) 计算每个驻点处的 $AC-B^2$；

(3) 对每个驻点依次检验：

如果 $A>0$，且 $AC-B^2>0$，则函数在该驻点处取得极小值；

如果 $A<0$，且 $AC-B^2>0$，则函数在该驻点处取得极大值；

如果 $AC-B^2<0$，则该驻点是鞍点，函数在该点处不取得极值；

如果 $AC-B^2=0$，则该方法失效，需要用其他方法判定.

例 6.15　函数 $f(x,y)=x^4+y^4-36xy$ 有三个驻点(0，0)，(−3，−3)，(3，3). 确定每个驻点的性质.

解　我们有$\dfrac{\partial f}{\partial x}=4x^3-36y$，$\dfrac{\partial f}{\partial y}=4y^3-36x$，则

$$\frac{\partial^2 f}{\partial x^2}=12x^2,\ \frac{\partial^2 f}{\partial y^2}=12y^2,\ \frac{\partial^2 f}{\partial x\partial y}=-36$$

$AC-B^2$ 的计算如下表所示.

检验项	驻点处各项取值		
	(0，0)	(−3，−3)	(3，3)
A	0	108	108
C	0	108	108
B	−36	−36	−36
$AC-B^2$	<0	>0	>0

故，(0，0) 点是鞍点；(−3，−3) 和 (3，3) 都是极小值点.

对于大多数函数，上述方法可以区分各种类型的驻点，但要注意下面的示例，这个示例中，上述方式失效.

设函数 $f(x,y)=x^4+y^4+2x^2y^2$，则有$\dfrac{\partial f}{\partial x}=4x^3+4xy^2$，$\dfrac{\partial f}{\partial y}=4y^3+4x^2y$. 于是

$$\frac{\partial^2 f}{\partial x^2}=12x^2+4y^2,\ \frac{\partial^2 f}{\partial y^2}=12y^2+4x^2,\ \frac{\partial^2 f}{\partial x\partial y}=8xy$$

令$\frac{\partial f}{\partial x}=\frac{\partial f}{\partial y}=0$，解方程组

$$\begin{cases}4x^3+4xy^2=0\\4y^3+4x^2y=0\end{cases}\Rightarrow x=y=0$$

从而函数有唯一的驻点(0，0). 不幸的是，在(0，0)点的所有二阶偏导数均为0，$AC-B^2=0$，因此由定理6.1不能得出结论. 然而，我们很容易看出，此例中的(0，0)点实际上是极小值点. 通过观察，$f(x,y)=x^4+y^4+2x^2y^2=(x^2+y^2)^2\geqslant 0$，使得(0，0)点恰好是驻点，而且(0，0)是极小值点也是最小值点.

例 6.16 (利润最大问题)假设冲浪板公司的年利润为

$$P(x,y)=-22x^2+22xy-11y^2+110x-44y-23$$

其中，x是每年生产的标准冲浪板的数量(以千为单位)；y是每年生产的比赛冲浪板的数量(以千为单位)；P是利润(以千元为单位). 为了实现最大利润，每种类型的冲浪板应该生产多少？最大利润是多少？

解　第一步　求驻点：　$P_x(x,y)=-44x+22y+110=0$

$$P_y(x,y)=22x-22y-44=0$$

解方程组得点(3，1)是函数的驻点.

第二步　计算 $A=P_{xx}(3,1)$，$B=P_{xy}(3,1)$和$C=P_{yy}(3,1)$：

$$A=P_{xx}(3,1)=-44,\ B=P_{xy}(3,1)=22\ 和\ C=P_{yy}(3,1)=-22$$

第三步　计算 $AC-B^2$ 并对驻点(3，1)进行分类：

$$AC-B^2=484>0\quad 且\quad A=-44<0$$

由于驻点是唯一的，因此$P(3,1)=120$是最大值，即当每年生产3000个标准冲浪板和1000个比赛冲浪板时利润达到最大值，为12万元.

习题 6

1. 确定下列函数的定义域并画出定义域的草图.

(1) $f(x,y)=\sqrt{2x+4y-1}$；　　(2) $f(x,y)=\ln\frac{1}{x-y}$；

(3) $f(x,y)=\sqrt{\frac{1}{x^2}-\frac{1}{y^2}}$；　　(4) $f(x,y)=\sqrt{x+y}-\sqrt{x-3}$.

2. 画出下列二次曲面的草图.

(1) $z=1+x^2+y^2$；　　(2) $z=\sqrt{x^2+y^2}$；

(3) $z=2x^2+3y^2$；　　(4) $4x^2+9y^2+z^2=1$.

3. 计算下列极限，如果极限不存在，请说明原因.

(1) $\lim\limits_{\substack{x\to\pi\\y\to 0}}\frac{x\sin y}{x-y}$；　　(2) $\lim\limits_{\substack{x\to 3\\y\to -7}}\frac{6x-y+xy}{2x^3+y^3}$；

(3) $\lim\limits_{\substack{x\to 0\\y\to 0}}\frac{2x^2+7y^2}{4y^2+x^2}$；　　(4) $\lim\limits_{\substack{x\to 0\\y\to 0}}\frac{2x^4y}{x^8+6y^2}$.

4. 求下列函数的一阶偏导数.

(1) $f(x,y)=xy+\frac{x}{y}$；　　(2) $f(x,y)=\arctan(xy)$；

(3) $f(x, y)=(1+y^2)^x$；　(4) $f(x, y, z)=\ln\sqrt{x^2+y^2+z^2}$.

5. 对第 4 题中的函数(1)～(3)，求 $f_x(1, 1)$，$f_x(-1, -1)$，$f_y(1, 2)$，$f_y(2, 1)$.

6. 求曲面 $36z=4x^2+9y^2$ 与平面 $x=3$ 的交线在点(3，2，2)处切线的斜率.

7. 求曲面 $2z=\sqrt{9x^2+9y^2-36}$ 与平面 $y=1$ 的交线在点 $\left(2, 1, \frac{3}{2}\right)$ 处切线的斜率.

8. 对于下列函数，求 $\frac{\partial^2 f}{\partial x^2}, \frac{\partial^2 f}{\partial y^2}, \frac{\partial^2 f}{\partial x \partial y}, \frac{\partial^2 f}{\partial y \partial x}$.

(1) $f(x, y)=x+2y+3$；　(2) $f(x, y)=x^2+y^2$；

(3) $f(x, y)=x^3+xy+y^3$；　(4) $f(x, y)=x^4+xy^3+2x^3y^2$；

(5) $f(x, y, z)=xy+yz$.

9. 对于第 8 题中的函数(1)～(4)，求 $f_{xx}(1, -3)$，$f_{yy}(-2, -2)$，$f_{xy}(-1, 1)$.

10. 已知 $f(x, y, z)=3x^2y-xyz+y^2z^2$，求下列各值.

(1) $f_x(x,y,z)$；　(2) $f_y(0,1,1)$；　(3) $f_z(x,y,z)$.

11. 对于下列函数，求 $\frac{\partial f}{\partial x}$ 和 $\frac{\partial^2 f}{\partial x \partial t}$.

(1) $f(x, t)=x\sin(xt)+x^2t$；　(2) $f(x, t, z)=zxt-e^{xt}$；

(3) $f(x, t)=3\cos(t+x^2)$.

12. 确定每种情况下函数驻点的性质.

(1) $f(x, y)=8x^2+6y^2-2y^3+5$；　(2) $f(x, y)=x^3+15x^2-20y^2+10$；

(3) $f(x, y)=4-x^2-xy-y^2$；　(4) $f(x, y)=2x^2+y^2+3xy-3y-5x+8$；

(5) $f(x, y)=(x^2+y^2)^2-2(x^2-y^2)+1$；

(6) $f(x, y)=x^4+y^4+2x^2y^2+2x^2+2y^2+1$.

13.（自动化-劳动力组合为最低成本问题）一个公司生产电视机每年需要支付人工和设备的成本(以百万元计)为

$$C(x, y)=2x^2+2xy+3y^2-16x-18y+54$$

其中，x 是每年花费在劳动力上的金额；y 是每年花费在自动化设备上的金额(均以百万美元计). 确定每年应如何投入使得成本最小，最小成本是多少？

14.(包装设计)设计一个没有顶部的矩形盒子，中间有一个隔板，如图所示. 这个盒子的体积必须是 48 cm³. 问如何设计使用料最省？

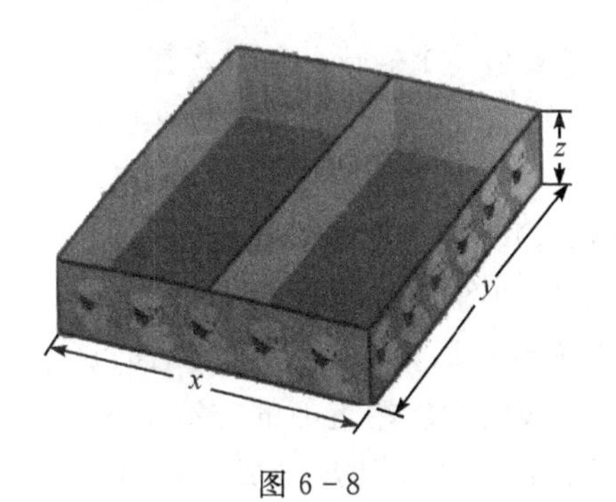

图 6 - 8

第 7 章　重积分及其应用

上一章我们讨论了多元函数的偏导数，接下来我们要讨论多元函数的积分. 本章涉及的函数有两个或三个自变量，多元函数的积分情况会更复杂.

7.1　二重积分

7.1.1　矩形域上的重积分

在讨论二重积分之前，让我们先来回顾一元函数的定积分的定义. 首先，当计算定积分 $\int_a^b f(x)\mathrm{d}x$ 时，要把 x 看作是取自区间 $[a, b]$ 上的量，即定积分可认为是对区间的积分.

注意： 上面我们假定了 $a<b$，如果 $b<a$，则可以用 $[b, a]$.

当给定积分下定义时，我们首先把它看成面积问题. 为了回答曲线下方的面积问题，我们把区间 $[a, b]$ 分成 n 个宽度为 Δx 的子区间，从每个子区间里任取 t_i，如图 7－1所示.

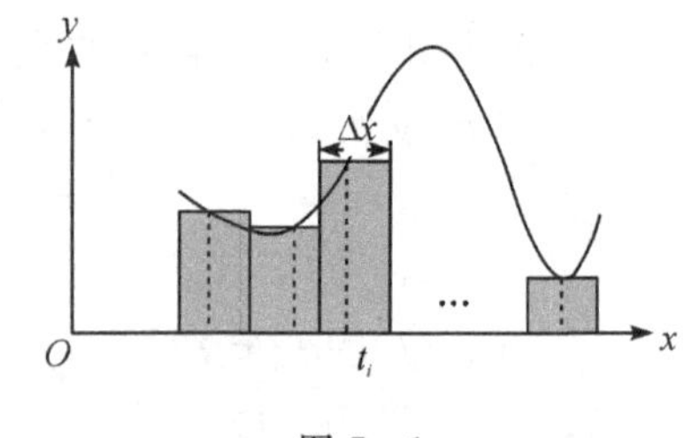

图 7－1

图 7－1 中每个矩形的高为 $f(t_i)$，我们可以用这些矩形的面积做如下的近似计算：

$$A \approx f(t_1)\Delta x_1 + f(t_2)\Delta x_2 + \cdots + f(t_n)\Delta x_n$$

为了得到面积的精确值，我们对上述和式当 n 趋于无穷时取极限，这也是定积分的定义.

$$\int_a^b f(x)\mathrm{d}x = \lim_{n\to\infty}\sum_{i=1}^{n} f(t_i)\Delta x_i$$

在本节中，我们要对一个包含两个变量的函数 $z=f(x, y)$ 进行积分. 对于单变量函数，是在区间(即一维空间)上积分；不难理解，对于一个双变量函数，是在 $\mathbf{R}^2$ 上(即二维空间)积分的.

首先，我们假设 $\mathbf{R}^2$ 中的区域是一个矩形，我们将其表示为

$$D = [a, b]\times[c, d]$$

其中，$a\leqslant x\leqslant b$，$c\leqslant y\leqslant d$.

同样，我们首先假设 $f(x, y)\geqslant 0$(这不是必须的). 先画出由函数 $f(x, y)$ 所表示的定义在矩形域 D 上的曲面 S 的图像，如图 7－2 所示.

像一元函数一样，我们先不考虑积分而是先来求曲面 S 下(xOy 平面以上)区域的体积.

我们用近似的方法求体积(如同之前用近似的方法求面积一样). 先把 $[a, b]$ 分成 n 个子区间，再把 $[c, d]$ 分成 m 个子区间. 这将把 D 分成一系列更小的矩形域，在每一个矩形域上都选一个点 (t_i, τ_j). 在每一个小矩形上我们要构造一个"盒子"，它的高度是 $f(t_i, \tau_j)$，草图如图 7－3 所示.

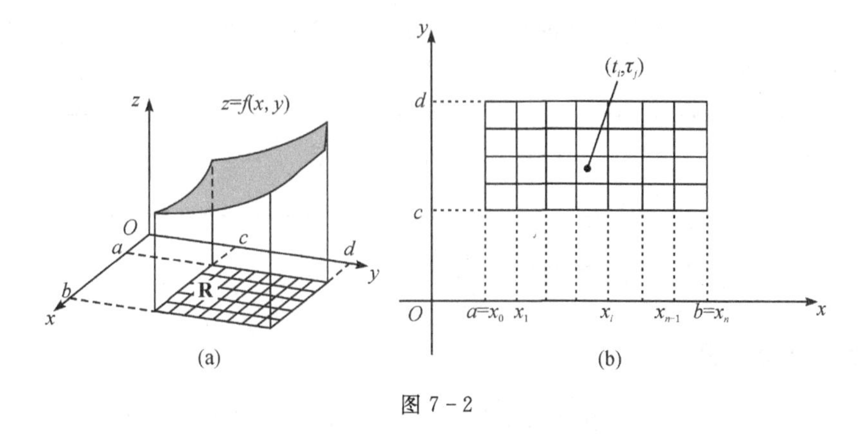

图 7－2

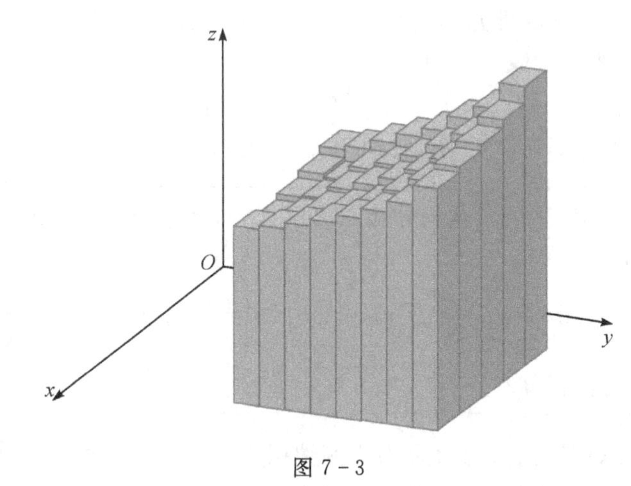

图 7－3

每个小盒子的底都是面积为 ΔA 的矩形，高为 $f(t_i, \tau_j)$，所以，每个小盒子的体积为 $f(t_i, \tau_j)\Delta A$. 曲面 S 下的体积的近似值为

$$V \approx \sum_{i=1}^{n}\sum_{j=1}^{m} f(t_i, \tau_j)\Delta A$$

因为需要沿着 x 方向和 y 方向把体积相加，所以我们得到两个累加和.

为了更好地估计体积，我们将 n 和 m 逐渐增大，而为了得到体积的精确值，我们需要取 n 和 m 趋于无穷时的极限，即

$$V = \lim_{n, m\to\infty}\sum_{i=1}^{n}\sum_{j=1}^{m} f(t_i, \tau_j)\Delta A$$

这看起来很像一元函数积分的定义. 实际上，这也是二重积分的定义，更准确地说，是一个二元函数在一个矩形上的积分，记作

$$\iint_D f(x, y)\mathrm{d}A = \lim_{n, m\to\infty}\sum_{i=1}^{n}\sum_{j=1}^{m} f(t_i, \tau_j)\Delta A$$

注意二重积分符号与单积分符号的不同，我们用两个积分号来表示处理的是一个二维区域上的积分. 这里也有一个微分，注意微分是 $\mathrm{d}A$ 而不是我们习惯看到的 $\mathrm{d}x$ 和 $\mathrm{d}y$. 还要注意，在这个符号中积分没有上下限. 我们把 D 写在这两个积分号的下面，表示积分的区域.

如上所述，矩形区域 D 上的二重积分的几何意义是介于曲面 $z=f(x, y)$ 下方和平面 xOy 上方的体积，即

$$\text{体积} = \iint_D f(x, y)\mathrm{d}A$$

特殊的，当 $f(x, y)=1$ 时，投影域面积 $= \iint_D \mathrm{d}A$.

7.1.2 累次积分

在上一节中，我们给出了二重积分的定义. 然而，就像单积分的定义一样，该定义在实际中很难应用，所以我们需要研究如何计算二重积分. 我们仍然假设积分是在矩形域 D 上(下节我们将看到更一般的区域)，其中

$$D=[a, b]\times[c, d]$$

下面的定理告诉我们如何在矩形域上计算二重积分.

定理 7.1 矩形域上的二重积分化累次积分. 如果 $f(x, y)$ 在区域 $D=[a, b]\times[c, d]$ 上连续，则

$$\iint_D f(x, y)\mathrm{d}A = \int_a^b\left[\int_c^d f(x, y)\mathrm{d}y\right]\mathrm{d}x = \int_c^d\left[\int_a^b f(x, y)\mathrm{d}x\right]\mathrm{d}y$$

这两个积分称为累次积分.

注意，实际上计算矩形域上的二重积分有两种积分次序，先 y 后 x，或先 x 后 y. 在某种程度上这只是一种表示方法，并没有告诉我们如何计算二重积分. 我们先考虑先 y 后 x 的情况.

$$\iint_D f(x, y)\mathrm{d}A = \int_a^b\left[\int_c^d f(x, y)\mathrm{d}y\right]\mathrm{d}x$$

我们通过保持 x 恒定先对 y 积分来计算

$$\int_c^d f(x, y)\mathrm{d}y$$

就好像这是一个单变量的积分，将得到一个只包含 x 的函数，我们再对这个函数关于 x 积分.

我们对偏导也做过类似的处理，一个二元函数对 y 求导，就是把 x 看成常数，好像它是一个一元函数一样. 类似地，二重积分的计算可以把 y 看成常数，先对 x 积分再对 y 积分，或者把 x 看成常数，先对 y 积分再对 x 积分.

7.1.3 二重积分的性质

下面是二重积分的一些性质，注意，所有这三个性质实际上都是单积分的性质的扩展.

(1) $\iint_D[f(x, y)+g(x, y)]\mathrm{d}A = \iint_D f(x, y)\mathrm{d}A + \iint_D g(x, y)\mathrm{d}A$.

(2) $\iint_D cf(x, y)\mathrm{d}A = c\iint_D f(x, y)\mathrm{d}A$，其中 c 为任意常数.

(3) 如果区域 D 可以分成两个区域 D_1 和 D_2，则积分可以写成

$$\iint\limits_D f(x, y)\mathrm{d}A = \iint\limits_{D_1} f(x, y)\mathrm{d}A + \iint\limits_{D_2} f(x, y)\mathrm{d}A$$

例 7.1　计算下列矩形域上的二重积分.

(1) $\iint\limits_D 6xy^2\mathrm{d}A$，$D = [2, 4]\times[1, 2]$；

(2) $\iint\limits_D (2x - 4y^3)\mathrm{d}A$，$D = [-5, 4]\times[0, 3]$；

(3) $\iint\limits_D [x^2y^2 + \cos(\pi x) + \sin(\pi y)]\mathrm{d}A$，$D = [-2, -1]\times[0, 1]$；

(4) $\iint\limits_D \dfrac{1}{(2x+3y)^2}\mathrm{d}A$，$D = [0, 1]\times[1, 2]$；

(5) $\iint\limits_D x\mathrm{e}^{xy}\mathrm{d}A$，$D = [-1, 2]\times[0, 1]$.

解　(1) $\iint\limits_D 6xy^2\mathrm{d}A$，$D = [2, 4]\times[1, 2]$.

先对哪个变量积分并不重要，不管积分的顺序如何，结果都是一样的. 为了说明这一点，我们对两种顺序都进行计算来验证会得到相同的答案.

方法 1　我们先对 y 积分，需要计算的累次积分是

$$\iint\limits_D 6xy^2\mathrm{d}A = \int_2^4\int_1^2 6xy^2\mathrm{d}y\mathrm{d}x$$

由于先对 y 积分，所以需要确定 y 的积分限，把 x 看成常数(后求的积分保持不变)

$$\iint\limits_D 6xy^2\mathrm{d}A = \int_2^4 (2xy^3)\Big|_1^2\mathrm{d}x = \int_2^4 (16x - 2x)\mathrm{d}x = \int_2^4 14x\mathrm{d}x$$

接下来，再求关于 x 的积分

$$\iint\limits_D 6xy^2\mathrm{d}A = 7x^2\Big|_2^4 = 84$$

方法 2　先对 x 再对 y 积分

$$\iint\limits_D 6xy^2\mathrm{d}A = \int_1^2 (3x^2y^2)\Big|_2^4\mathrm{d}y = \int_1^2 (48y^2 - 12y^2)\mathrm{d}y = \int_1^2 36y^2\mathrm{d}y = 84$$

和方法 1 的结果一样. 所以，可以按任意顺序积分.

(2) $\iint\limits_D (2x - 4y^3)\mathrm{d}A$，$D = [-5, 4]\times[0, 3]$.

先对 y 积分后对 x 积分有

$$\begin{aligned}\iint\limits_D (2x - 4y^3)\mathrm{d}A &= \int_{-5}^4\int_0^3 (2x - 4y^3)\mathrm{d}y\mathrm{d}x \\ &= \int_{-5}^4 (2xy - y^4)\Big|_0^3\mathrm{d}x = \int_{-5}^4 (6x - 81)\mathrm{d}x \\ &= (3x^2 - 81x)\Big|_{-5}^4 = -756\end{aligned}$$

(3) $\iint\limits_D [x^2y^2 + \cos(\pi x) + \sin(\pi y)]\mathrm{d}A$，$D = [-2, -1]\times[0, 1]$.

先对 x 积分后对 y 积分，有

$$\iint_D [x^2y^2+\cos(\pi x)+\sin(\pi y)]\mathrm{d}A=\int_0^1\int_{-2}^{-1}\left[\frac{1}{3}x^3y^2+\frac{1}{\pi}\sin(\pi x)+x\sin(\pi y)\right]\mathrm{d}x\mathrm{d}y$$

$$=\int_0^1\left[\frac{7}{3}y^2+\sin(\pi y)\right]\mathrm{d}y=\left[\frac{7}{9}y^3-\frac{1}{\pi}\cos(\pi y)\right]\Big|_0^1=\frac{7}{9}+\frac{2}{\pi}$$

(4) $\iint_D \frac{1}{(2x+3y)^2}\mathrm{d}A$，$D=[0，1]\times[1，2]$.

在这个问题中，因为 x 的上下限比较简单，所以我们先对 x 积分.

$$\iint_D \frac{1}{(2x+3y)^2}\mathrm{d}A=\int_1^2\left[-\frac{1}{2}(2x+3y)^{-1}\right]\Big|_0^1\mathrm{d}y=-\frac{1}{2}\int_1^2\left(\frac{1}{2+3y}-\frac{1}{3y}\right)\mathrm{d}y$$

$$=-\frac{1}{2}\left(\frac{1}{3}\ln|2+3y|-\frac{1}{3}\ln y\right)\Big|_1^2=-\frac{1}{6}(\ln 8-\ln 2-\ln 5)$$

(5) $\iint_D x\mathrm{e}^{xy}\mathrm{d}A$，$D=[-1，2]\times[0，1]$.

虽然我们可以选择任意一种积分次序，但对于这个问题，先对 y 积分要容易得多.

$$\iint_D x\mathrm{e}^{xy}\mathrm{d}A=\int_{-1}^2\left(x\frac{1}{x}\mathrm{e}^{xy}\right)\Big|_0^1\mathrm{d}x=\int_{-1}^2(\mathrm{e}^x-1)\mathrm{d}x=\mathrm{e}^2-2-(\mathrm{e}^{-1}+1)=\mathrm{e}^2-\frac{1}{\mathrm{e}}-3$$

现在我们看看如果先对 x 积分会出现什么问题.

$$\iint_D x\mathrm{e}^{xy}\mathrm{d}A=\int_0^1\int_{-1}^2 x\mathrm{e}^{xy}\mathrm{d}x\mathrm{d}y=\int_0^1\left[\frac{x}{y}\mathrm{e}^{xy}-\frac{1}{y^2}\mathrm{e}^{xy}\right]_{-1}^2\mathrm{d}y$$

$$=\int_0^1\left[\left(\frac{2}{y}\mathrm{e}^{2y}-\frac{1}{y^2}\mathrm{e}^{2y}\right)-\left(-\frac{1}{y}\mathrm{e}^{-y}-\frac{1}{y^2}\mathrm{e}^{-y}\right)\right]\mathrm{d}y$$

我们看到计算变得更复杂. 所以，恰当地选择积分次序，可以使问题得以化简.

当被积函数为下列情形时，二重积分可以直接化成两个定积分的乘积.

定理 7.2 如果 $f(x，y)=g(x)h(y)$，则在矩形域 $D=[a，b]\times[c，d]$上的积分为

$$\iint_D f(x，y)\mathrm{d}A=\iint_D g(x)h(y)\mathrm{d}A=\left(\int_a^b g(x)\mathrm{d}x\right)\left(\int_c^d h(y)\mathrm{d}y\right)$$

例 7.2 计算 $\iint_D x\cos^2 y\mathrm{d}A$，$D=[-2，3]\times\left[0，\frac{\pi}{2}\right]$.

解 因为被积函数是关于 x 的函数与关于 y 的函数的乘积，所以根据上述结论，有

$$\iint_D x\cos^2 y\mathrm{d}A=\int_{-2}^3 x\mathrm{d}x\int_0^{\frac{\pi}{2}}\cos^2 y\mathrm{d}y=\left[\frac{1}{2}x^2\right]_{-2}^3\int_0^{\frac{\pi}{2}}\frac{1+\cos 2y}{2}\mathrm{d}y$$

$$=\frac{5}{2}\times\frac{1}{2}\left[y+\frac{1}{2}\sin 2y\right]_0^{\frac{\pi}{2}}=\frac{5}{8}\pi$$

7.2 一般域上的二重积分

在上一节中，我们研究了矩形区域上的二重积分，但大部分积分区域都不是矩形的，所

以我们需要考虑下面的二重积分，

$$\iint_D f(x, y)\mathrm{d}A$$

其中，D 是任意的区域.

7.2.1　两种类型积分域上二重积分的计算

我们将讨论的两种类型的积分域的草图如图 7-4 所示。

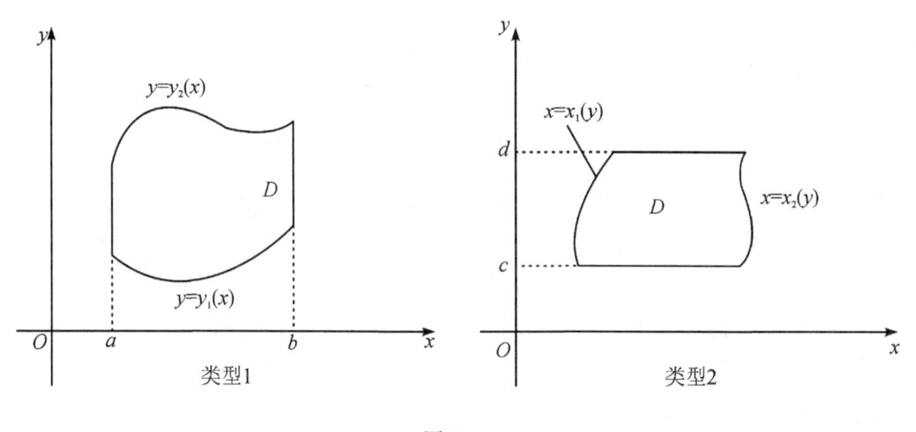

图 7-4

我们经常用集合记号来描述这些区域，也称为区域的边界不等式.

类型 1 中区域的定义如下，它被称为 x 型域.

$$D = \{(x, y) \mid a \leqslant x \leqslant b, y_1(x) \leqslant y \leqslant y_2(x)\}$$

类型 2 中区域的定义如下，它被称为 y 型域.

$$D = \{(x, y) \mid x_1(y) \leqslant x \leqslant x_2(y), c \leqslant y \leqslant d\}$$

接下来，我们利用二重积分的几何意义来探讨如何在 x 型域上计算二重积分，y 型域上二重积分的计算方法是类似的. 如图 7-5 所示，假定$(x, y) \in D$ 时 $f(x, y) \geqslant 0$，根据二重积分的几何意义，$\iint\limits_D f(x, y)\mathrm{d}x\mathrm{d}y = V$ 是曲顶柱体的体积. 下面应用平行截面面积已知的立体的体积公式来计算这个曲顶柱体的体积 V.

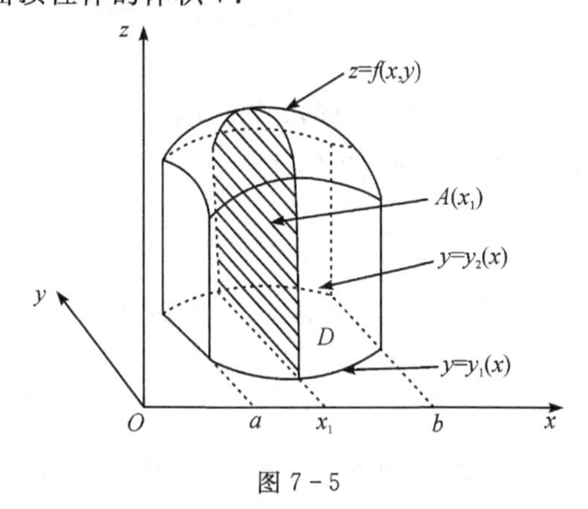

图 7-5

首先在区间$[a, b]$上任取一点x_1，过x_1点做平行于yOz面的平面. 用这个平面去截曲顶柱体，所得截面是一个以区间$[y_1(x_1), y_2(x_1)]$为底、$z=f(x_1, y)$(x固定) 为曲边的曲边梯形(图7-5中阴影部分)，故截面的面积为

$$A(x_1)=\int_{y_1(x_1)}^{y_2(x_1)} f(x_1, y)\mathrm{d}y$$

根据已知平行截面面积求立体的体积公式，可得曲顶柱体体积的体积为

$$V=\iint_D f(x, y)\mathrm{d}x\mathrm{d}y=\int_a^b\left[\int_{y_1(x)}^{y_2(x)} f(x, y)\mathrm{d}y\right]\mathrm{d}x$$

这个体积就是二重积分$\iint\limits_D f(x, y)\mathrm{d}x\mathrm{d}y$的值，从而有

$$\iint_D f(x, y)\mathrm{d}x\mathrm{d}y=\int_a^b\left[\int_{y_1(x)}^{y_2(x)} f(x, y)\mathrm{d}y\right]\mathrm{d}x=\int_a^b\mathrm{d}x\int_{y_1(x)}^{y_2(x)} f(x, y)\mathrm{d}y$$

上式右端的积分叫作先对y后对x积分的**二次积分或累次积分**. 也就是说，先把x看成常数，把$f(x, y)$只看作y的函数，对y从$y_1(x)$到$y_2(x)$计算定积分；然后再把计算的结果(是x的函数)对x在区间$[a, b]$上计算定积分.

在上述讨论中，我们假定了当$(x, y)\in D$时$f(x, y)\geqslant 0$，实际上，对于$f(x, y)$可正可负的情形，结论同样成立.

根据以上分析，可得如下定理.

定理 7.3 一般域上的二重积分化累次积分.

(1) 当积分域为x型域时，二重积分可化为如下累次积分

$$\iint_D f(x, y)\mathrm{d}A=\int_a^b\left[\int_{y_1(x)}^{y_2(x)} f(x, y)\mathrm{d}y\right]\mathrm{d}x=\int_a^b\mathrm{d}x\int_{y_1(x)}^{y_2(x)} f(x, y)\mathrm{d}y$$

(2) 当积分域为y型域时，二重积分可化为如下累次积分

$$\iint_D f(x, y)\mathrm{d}A=\int_c^d\left[\int_{x_1(y)}^{x_2(y)} f(x, y)\mathrm{d}x\right]\mathrm{d}y=\int_c^d\mathrm{d}y\int_{x_1(y)}^{x_2(y)} f(x, y)\mathrm{d}x$$

根据上面的结论，在x型域中定y的上下限时，可以通过射线穿线法. 即在x的取值范围内任取一点做射线从下到上穿过积分域，先相交的交点的y值是y的下限，后相交的交点的y值是y的上限. 类似地，在y型域中定x的上下限时，也可以通过射线穿线法. 即在y的取值范围内任取一点做射线从左到右穿过积分域，先相交的交点的x值是x的下限，后相交的交点的x值是x的上限.

让我们来看一些一般积分域上二重积分的计算例题.

例 7.3 在所给区域D上计算下列二重积分.

(1) $\iint\limits_D \mathrm{e}^{\frac{x}{y}}\mathrm{d}A$, $D=\{(x, y)\mid 1\leqslant y\leqslant 2, y\leqslant x\leqslant y^3\}$.

(2) $\iint\limits_D (4xy-y^3)\mathrm{d}A$, D由曲线$y=\sqrt{x}$和$y=x^3$围成.

(3) $\iint\limits_D (6x^2-40y)\mathrm{d}A$, D是以为$A(0, 3)$, $B(1, 1)$和$C(5, 3)$为顶点的三角形区域.

解 (1) 积分域如图7-6所示. 由于所给区域为y型域，所以选择先x后y的积分次序.

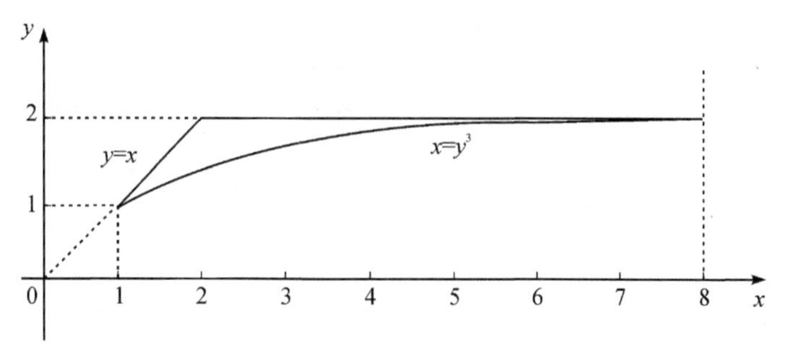

图 7 - 6

$$\iint_D e^{\frac{x}{y}} dA = \int_1^2 \left[\int_y^{y^3} e^{\frac{x}{y}} dx\right] dy = \int_1^2 y e^{\frac{x}{y}} \Big|_y^{y^3} dy$$

$$= \int_1^2 y e^{y^2} - y e dy$$

$$= \left(\frac{1}{2} e^{y^2} - \frac{1}{2} y^2 e\right) \Big|_1^2 = \frac{1}{2} e^4 - 2e$$

(2) 积分域如图 7 - 7 所示. 由于所给区域既是 x 型域又是 y 型域，所以两种积分次序均可. 这里，我们把积分域看成 x 型，即 $D = \{(x, y) \mid 0 \leqslant x \leqslant 1, x^3 \leqslant y \leqslant \sqrt{x}\}$.

$$\iint_D (4xy - y^3) dA = \int_0^1 \int_{x^3}^{\sqrt{x}} [(4xy - y^3) dy] dx = \int_0^1 \left(2xy^2 - \frac{1}{4} y^4\right) \Big|_{x^3}^{\sqrt{x}} dx$$

$$= \int_0^1 \left(\frac{7}{4} x^2 - 2x^7 + \frac{1}{4} x^{12}\right) dx = \left(\frac{7}{12} x^3 - \frac{1}{4} x^8 + \frac{1}{52} x^{13}\right) \Big|_0^1 = \frac{55}{156}$$

读者可尝试在另一种积分次序下计算此二重积分.

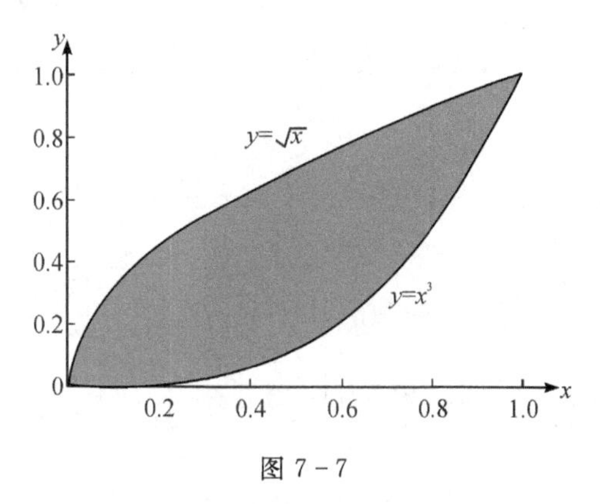

图 7 - 7

(3) 我们先画出积分域的草图，如图 7 - 8 所示.

方法 1　可以把积分域看成 y 型域，如图 7 - 8(a)所示. 由于两点可以确定直线方程，所以首先求出三角形各边的方程，并把 x 表示成 y 的函数.

直线 AB 的方程为 $x = -\frac{1}{2}(y-3)$，直线 BC 的方程为 $x = 2(y - \frac{1}{2})$，从而积分区域可表示为

$$D = \left\{(x, y) \mid 1 \leqslant y \leqslant 3, -\frac{1}{2}(y-3) \leqslant x \leqslant 2(y - \frac{1}{2})\right\}$$

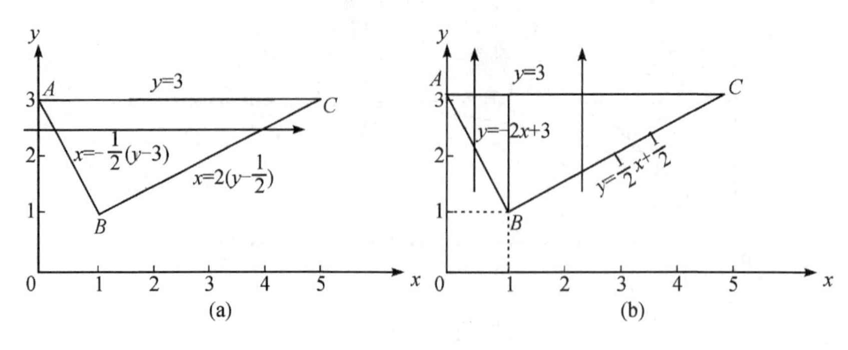

图 7-8

于是二重积分化为先 x 后 y 的累次积分

$$\iint_D (6x^2-40y)\,\mathrm{d}A=\int_1^3\left[\int_{-\frac{1}{2}y+\frac{3}{2}}^{2y-1}(6x^2-40y)\,\mathrm{d}x\right]\mathrm{d}y=\int_1^3(2x^3-40xy)\Big|_{-\frac{1}{2}y+\frac{3}{2}}^{2y-1}\mathrm{d}y$$

$$=\int_1^3\left[100y-100y^2+2\,(2y-1)^3-2\left(-\frac{1}{2}y+\frac{3}{2}\right)^3\right]\mathrm{d}y$$

$$=\left[50y^2-\frac{100}{3}y^3+\frac{1}{4}\,(2y-1)^4+\left(-\frac{1}{2}y+\frac{3}{2}\right)^4\right]\Bigg|_1^3=-\frac{935}{3}$$

方法 2　如果把这个积分区域看成 x 型域，就需要对它进行分块，如图 7-8(b)所示. 边界方程中，要把 y 表示成 x 的函数.

直线 AB 的方程为 $y=-2x+3$，直线 BC 的方程为 $y=\frac{1}{2}x+\frac{1}{2}$. 把积分域 D 看成 $D=D_1\cup D_2$，其中

$$D_1=\{(x,\ y)\,|\,0\leqslant x\leqslant 1,\ -2x+3\leqslant y\leqslant 3\}$$

$$D_2=\left\{(x,\ y)\,|\,1\leqslant x\leqslant 5,\frac{1}{2}x+\frac{1}{2}\leqslant y\leqslant 3\right\}$$

于是二重积分为

$$\iint_D(6x^2-40y)\,\mathrm{d}A=\iint_{D_1}(6x^2-40y)\,\mathrm{d}A+\iint_{D_2}(6x^2-40y)\,\mathrm{d}A$$

$$=\int_0^1\left[\int_{-2x+3}^3(6x^2-40y)\,\mathrm{d}y\right]\mathrm{d}x+\int_1^5\left[\int_{\frac{1}{2}x+\frac{1}{2}}^3(6x^2-40y)\,\mathrm{d}y\right]\mathrm{d}x$$

$$=\int_0^1(6x^2y-20y^2)\Big|_{-2x+3}^3\mathrm{d}x+\int_1^5(6x^2y-20y^2)\Big|_{\frac{1}{2}x+\frac{1}{2}}^3\mathrm{d}x$$

$$=\int_0^1\left[12x^3-180+20\,(3-2x)^2\right]\mathrm{d}x$$

$$+\int_1^5\left[-3x^3+15x^2-180+20\left(\frac{1}{2}x+\frac{1}{2}\right)^2\right]\mathrm{d}x$$

$$=\left[3x^4-180x-\frac{10}{3}\,(3-2x)^3\right]_0^1$$

$$+\left[-\frac{3}{4}x^4+5x^3-180x+\frac{40}{3}\left(\frac{1}{2}x+\frac{1}{2}\right)^3\right]_1^5$$

$$=-\frac{935}{3}$$

显然，方法 1 要比方法 2 简单. 所以，我们在选择积分次序时，应尽量避免把积分区域分块.

7.2.2　交换积分次序

事实上，有些时候，一种积分次序下是无法得到积分结果的，这时我们需要交换积分次序，来看下面几个例子.

例 7.4　计算下列二重积分.

(1) $\int_0^3\int_{x^2}^9 x^3 e^{y^3}\,dy dx$；　　　(2) $\int_0^8\int_{\sqrt[3]{y}}^2 \sqrt{x^4+1}\,dx dy$.

解　(1) 由于在所给积分次序下，被积函数的原函数无法用初等函数来表示(称之为积不出来)，所以考虑交换积分次序.

由题设条件知，区域的边界不等式为

$$D=\{(x,\ y)\mid 0\leqslant x\leqslant 3,\ x^2\leqslant y\leqslant 9\}$$

根据边界不等式画出积分区域的草图，如图 7-9 所示.

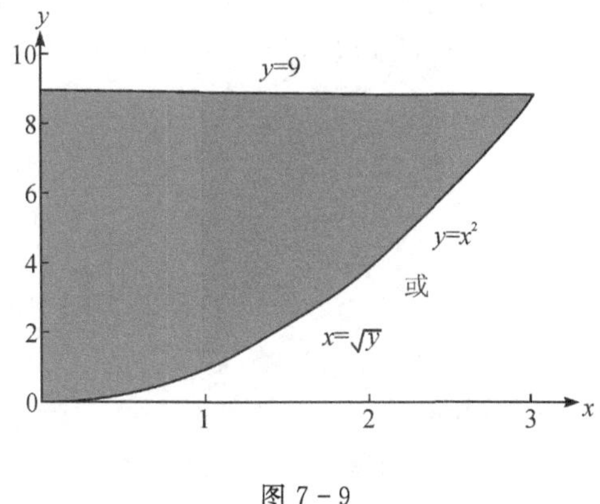

图 7-9

根据草图，写出积分区域的边界不等式

$$D=\left\{(x,\ y)\mid 0\leqslant y\leqslant 9,\ 0\leqslant x\leqslant \sqrt{y}\right\}$$

从而二重积分可以化为先 x 后 y 的累次积分

$$\int_0^3\int_{x^2}^9 x^3 e^{y^3}\,dy dx=\int_0^9\int_0^{\sqrt{y}} x^3 e^{y^3}\,dx dy=\int_0^9 \frac{1}{4}x^4 e^{y^3}\Big|_0^{\sqrt{y}}\,dy$$

$$=\int_0^9 \frac{1}{4}y^2 e^{y^3}\,dy=\frac{1}{12}e^{y^3}\Big|_0^9=\frac{1}{12}(e^{729}-1)$$

注意：交换积分次序时，不是保持积分限不变仅仅交换次序，而是要根据积分域的图形，确定出新的积分次序下的上下限.

(2) $\int_0^8\int_{\sqrt[3]{y}}^2 \sqrt{x^4+1}\,dx dy$ 和第一个积分一样，在原积分次序下，无法找到原函数，需要交换积分次序. 由题设条件知，区域的边界不等式为

$$D=\left\{(x,\ y)\mid 0\leqslant y\leqslant 8,\ \sqrt[3]{y}\leqslant x\leqslant 2\right\}$$

根据边界不等式画出积分区域的草图，如图 7-10 所示.

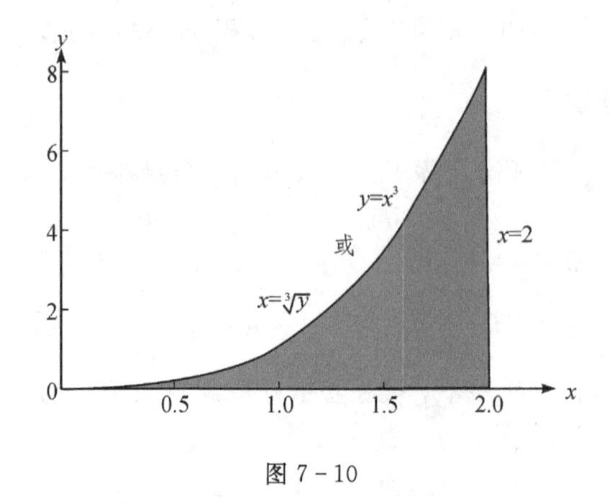

图 7-10

根据草图，写出积分区域的边界不等式.

$$D = \{(x, y) \mid 0 \leqslant x \leqslant 2, 0 \leqslant y \leqslant x^3\}$$

从而二重积分可以化为先 y 后 x 的累次积分

$$\int_0^8\int_{\sqrt[3]{y}}^2\sqrt{x^4+1}\,\mathrm{d}x\mathrm{d}y = \int_0^2\int_0^{x^3}\sqrt{x^4+1}\,\mathrm{d}y\mathrm{d}x$$

$$= \int_0^2 y\sqrt{x^4+1}\Big|_0^{x^3}\mathrm{d}x$$

$$= \int_0^2 x^3\sqrt{x^4+1}\,\mathrm{d}x = \frac{1}{6}(17^{\frac{3}{2}}-1)$$

7.3 极坐标系下的二重积分

7.3.1 极坐标系

在二维空间中，点 P 的位置可以由笛卡儿坐标(x, y)来确定，还可以由极坐标来确定. 在极坐标系中，点 P 的位置是由点 P 到原点的距离 r 和原点与点 P 的连线与极轴正向的夹角 θ 来确定的. 点 P 的极坐标(r, θ)见图 7-11.

由于 r 的变化范围为 0 到正无穷，θ 的变化范围为 0 到 2π，所以极坐标(r, θ)覆盖了笛卡儿坐标平面上的所有点，即笛卡儿坐标平面上的每一个点(x, y)都可以和(r, θ)一一对应.

从图 7-12 中很容易得到直角坐标和极坐标的相互变换关系.

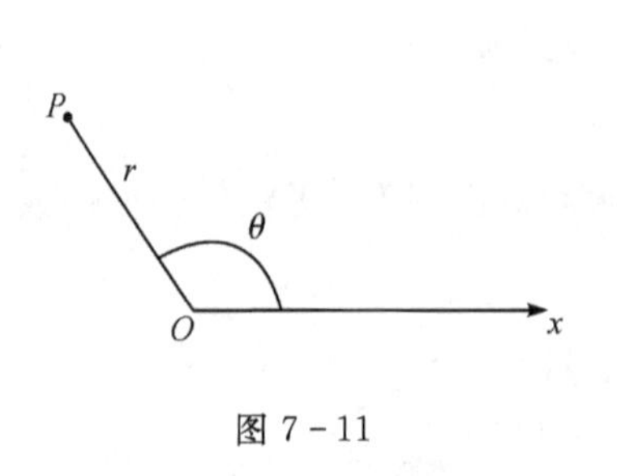

图 7-11

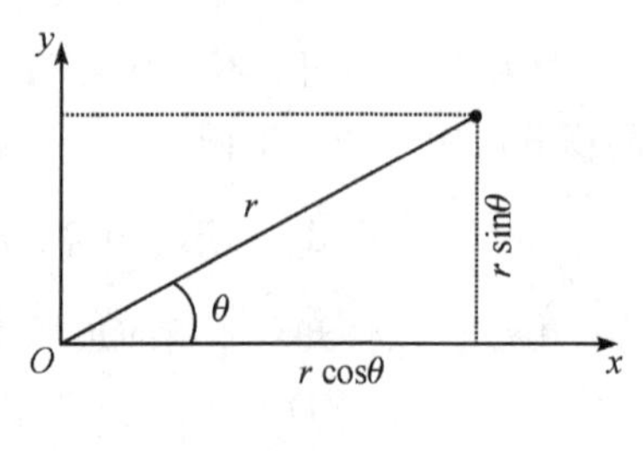

图 7-12

变换公式

$$\begin{cases} x=r\cos\theta \\ y=r\sin\theta \end{cases} \qquad \begin{cases} r=\sqrt{x^2+y^2} \\ \theta=\arctan\dfrac{y}{x} \end{cases}$$

7.3.2　极坐标系下二重积分的计算

我们已经学过很多二重积分了. 但在我们所见过的每一种情况下积分区域都可以很容易地用笛卡儿坐标系下的简单函数来描述. 在本节中，我们要考虑一些更容易用极坐标来描述的区域. 例如，圆域、环域、圆域或环域的一部分. 在这些情况下，使用笛卡儿坐标可能有些麻烦. 例如，下面的积分，

$$\iint\limits_D f(x,\ y)\mathrm{d}A\ (D\text{ 是以 }O\text{ 为圆心、半径为 2 的圆域})$$

我们需要用一组不等式来描述这个区域

$$-2\leqslant x\leqslant 2,\ -\sqrt{4-x^2}\leqslant y\leqslant\sqrt{4-x^2}$$

从而二重积分为

$$\iint\limits_D f(x,\ y)\mathrm{d}A=\int_{-2}^{2}\int_{-\sqrt{4-x^2}}^{\sqrt{4-x^2}}f(x,\ y)\mathrm{d}y\mathrm{d}x$$

由于积分的上下限比较复杂，所以这个二重积分的计算可能会很困难. 但在极坐标系下，这个积分域却可以用一组简单的不等式来刻画

$$0\leqslant\theta\leqslant 2\pi,\ 0\leqslant r\leqslant 2$$

由于上下限是常数，有可能使积分变得简单. 所以，如果我们能把二重积分公式转换成极坐标的形式将更容易得到结果. 但我们不能直接把 $\mathrm{d}x$ 和 $\mathrm{d}y$ 换成 $\mathrm{d}r$ 和 $\mathrm{d}\theta$. 在直角坐标系下计算二重积分时，$\mathrm{d}A=\mathrm{d}x\mathrm{d}y$，但在极坐标系下，$\mathrm{d}A\neq\mathrm{d}x\mathrm{d}y$. 所以，我们需要确定极坐标下的 $\mathrm{d}A$.

假设积分域 D 如图 7－13 所示. 此积分域在极坐标系下可以用以下不等式来刻画

$$\alpha\leqslant\theta\leqslant\beta,\ r_1(\theta)\leqslant r\leqslant r_2(\theta)$$

为了得到极坐标系下的 $\mathrm{d}A$，类似于直角坐标的方法，我们用一族射线和一族同心圆划分 D，如图 7－14 所示.

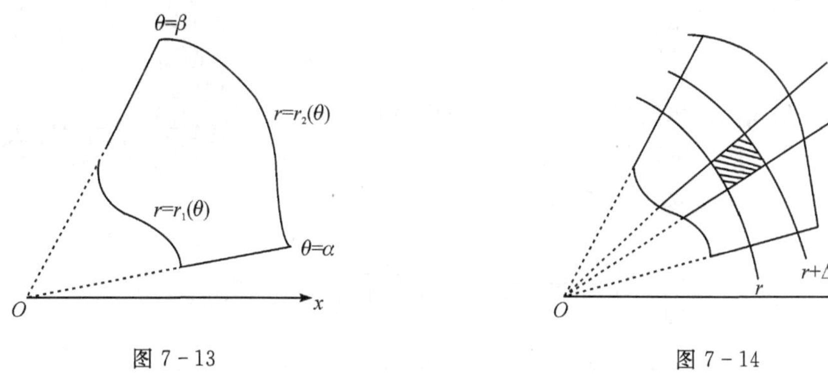

图 7－13　　　图 7－14

由于划分得很小，可以把每个子区域近似地看作是一个矩形. 于是阴影部分的面积

$\Delta A\approx r\Delta r\Delta\theta$. 分割的子区域必须足够小，此假设才是合理的，因为二重积分的定义就是由两个极限构成的，说明分得越细密，精确程度越高. 所以当无限细分时，有 $dA=rdrd\theta$.

现在，我们可以利用转换公式把直角坐标系下的二重积分转化为极坐标下的二重积分

$$\iint_D f(x, y)dA=\int_\alpha^\beta\int_{r_1(\theta)}^{r_2(\theta)} f(r\cos\theta, r\sin\theta)rdrd\theta$$

下面来看几个例子.

例 7.5 在极坐标系下计算下列二重积分.

(1) $\iint_D 2xy\,dA$，D 是第一象限介于以原点为中心、半径为 2 和 5 的圆之间的部分.

(2) $\iint_D e^{x^2+y^2}dA$，D 是以原点为圆心的单位圆.

解 (1) 由题设条件知，所给区域的边界可由下面的不等式来刻画

$$2\leqslant r\leqslant 5,\ 0\leqslant\theta\leqslant\frac{\pi}{2}$$

于是在极坐标系下的二重积分可化为先对 r 再对 θ 的累次积分

$$\iint_D 2xy\,dA=\int_0^{\frac{\pi}{2}}\left[\int_2^5 2(r\cos\theta)(r\sin\theta)r\,dr\right]d\theta=\int_0^{\frac{\pi}{2}}\int_2^5 r^3\sin(2\theta)drd\theta$$

$$=\int_0^{\frac{\pi}{2}}\frac{1}{4}r^4\sin(2\theta)\Big|_2^5 d\theta=\int_0^{\frac{\pi}{2}}\frac{609}{4}\sin(2\theta)d\theta$$

$$=-\frac{609}{8}\cos(2\theta)\Big|_0^{\frac{\pi}{2}}=\frac{609}{4}$$

(2) 由题设条件知，所给区域的边界可由下面的不等式来刻画.

$$0\leqslant\theta\leqslant 2\pi,\ 0\leqslant r\leqslant 1$$

于是在极坐标系下的二重积分可化为先对 r 再对 θ 的累次积分.

$$\iint_D e^{x^2+y^2}dA=\int_0^{2\pi}\left[\int_0^1 re^{r^2}dr\right]d\theta=\int_0^{2\pi}\frac{1}{2}e^{r^2}\Big|_0^1 d\theta=\int_0^{2\pi}\frac{1}{2}(e-1)d\theta=\pi(e-1)$$

例 7.6 求由球面 $x^2+y^2+z^2=9$ 和柱面 $x^2+y^2=5$ 所围且位于平面 $z=0$ 之上的立体的体积.

解 所围立体如图 7-15 所示.

根据二重积分的几何意义，顶为曲面 $z=f(x, y)=\sqrt{9-x^2-y^2}$，底为 $x^2+y^2\leqslant 5$. 所求立体的体积为

$$V=\iint_D f(x, y)dA$$

根据被积函数和积分域的特征，适合用极坐标系求解. 积分域在极坐标系下的不等式为

$$0\leqslant\theta\leqslant 2\pi,\ 0\leqslant r\leqslant\sqrt{5}$$

所以有

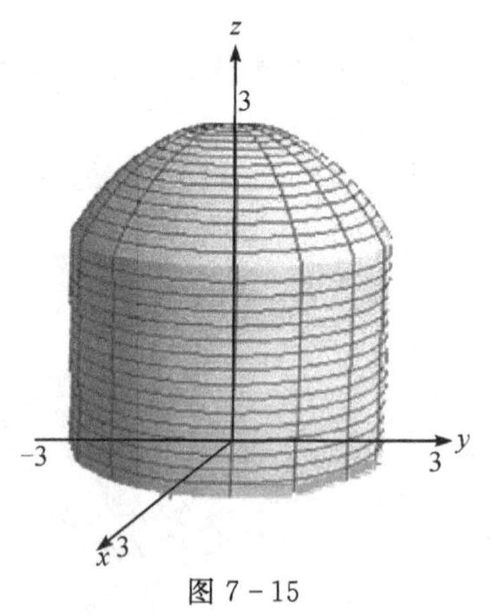

图 7-15

$$V=\iint_D\sqrt{9-x^2-y^2}dA=\int_0^{2\pi}\int_0^{\sqrt{5}}r\sqrt{9-r^2}drd\theta$$

$$=\int_0^{2\pi}-\frac{1}{3}(9-r^2)^{\frac{3}{2}}\Big|_0^{\sqrt{5}}d\theta=\int_0^{2\pi}\frac{19}{3}d\theta=\frac{38\pi}{3}$$

例 7.7　求由曲面 $z=x^2+y^2$ 和平面 $z=16$ 所围立体的体积.

解　所围立体如图 7-16 所示.

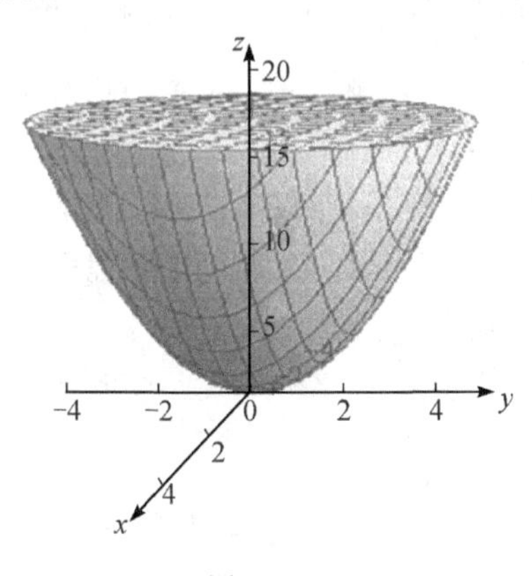

图 7-16

从图 7-16 可以看出，底面的投影域为 D：$x^2+y^2\leqslant 16$. 所求立体的体积为以 $z=16$ 为顶、以 D 为底的柱体体积与以曲面 $z=x^2+y^2$ 为顶、以 D 为底的柱体体积之差.

$$V=\iint_D 16\mathrm{d}A-\iint_D (x^2+y^2)\mathrm{d}A=\iint_D [16-(x^2+y^2)]\mathrm{d}A$$

根据被积函数与积分域的特征，该问题适合用极坐标系来求解. 在极坐标系下，积分域和被积函数可表示为

$$0\leqslant\theta\leqslant 2\pi,\ 0\leqslant r\leqslant 4,\ z=16-r^2$$

故所求立体的体积为

$$V=\iint_D [16-(x^2+y^2)]\mathrm{d}A=\int_0^{2\pi}\int_0^4 r(16-r^2)\mathrm{d}r\mathrm{d}\theta$$

$$=\int_0^{2\pi}\left(8r^2-\frac{1}{4}r^4\right)\bigg|_0^4\mathrm{d}\theta=\int_0^{2\pi}64\mathrm{d}\theta=128\pi$$

例 7.8　计算二重积分 $\int_{-1}^1\left[\int_{-\sqrt{1-x^2}}^0\cos(x^2+y^2)\mathrm{d}y\right]\mathrm{d}x$.

解　根据重积分的上下限，先画出积分域的草图如图 7-17 所示.

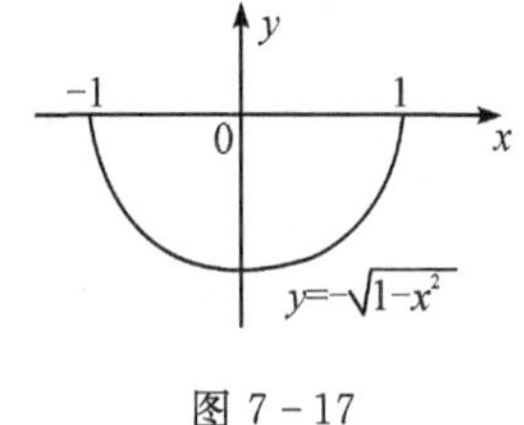

图 7-17

根据被积函数与积分域的特征，本问题适合用极坐标求解. 在极坐标系下，积分域和被积函数可表示为

$$\pi\leqslant\theta\leqslant 2\pi,\ 0\leqslant r\leqslant 1,\ \cos(x^2+y^2)=\cos r^2$$

$$\int_{-1}^1\left[\int_{-\sqrt{1-x^2}}^0\cos(x^2+y^2)\mathrm{d}y\right]\mathrm{d}x=\int_\pi^{2\pi}\int_0^1 r\cos(r^2)\mathrm{d}r\mathrm{d}\theta=\int_\pi^{2\pi}\frac{1}{2}\sin(1)\mathrm{d}\theta=\frac{\pi}{2}\sin(1)$$

7.4　二重积分的应用

7.4.1　立体的体积

我们知道，当被积函数 $f(x,y)\geqslant 0$ 时，二重积分 $\iint_D f(x,y)\mathrm{d}A$ 的几何意义是以 S 为顶，

以 D 为底的立体的体积. 其中曲面 S 的方程为 $z = f(x, y)$.

特殊地，当 $f(x, y)=1$ 时，$\iint\limits_D f(x, y)\mathrm{d}A = \iint\limits_D \mathrm{d}A$ 表示的是平面区域 D 的面积.

例 7.9 求在 xOy 平面上以 $y=x^2$ 和 $y=8-x^2$ 为边界的区域内，位于曲面 $z=16xy+200$ 下方和平面 xOy 上方的立体的体积.

解 图 7-18 是曲面和其在 xOy 平面上投影的图像. 图 7-19 是平面上投影域即底的图形.

根据积分域的形状，可以看成 x 型域. 联立边界方程可以求出交点的坐标为 $(-2, 4)$ 和 $(2, 4)$. 所以积分域边界的不等式为

$$-2 \leqslant x \leqslant 2,\ x^2 \leqslant y \leqslant 8-x^2$$

故所求立体的体积为

$$\begin{aligned} V &= \iint\limits_D (16xy+200)\mathrm{d}A = \int_{-2}^{2}\int_{x^2}^{8-x^2} (16xy+200)\mathrm{d}y\mathrm{d}x \\ &= \int_{-2}^{2} (8xy^2+200y)\Big|_{x^2}^{8-x^2} \mathrm{d}x = \int_{-2}^{2} (-128x^3-400x^2+512x+1600)\mathrm{d}x \\ &= \left(-32x^4-\frac{400}{3}x^3+256x^2+1600x\right)\Big|_{-2}^{2} = \frac{12800}{3} \end{aligned}$$

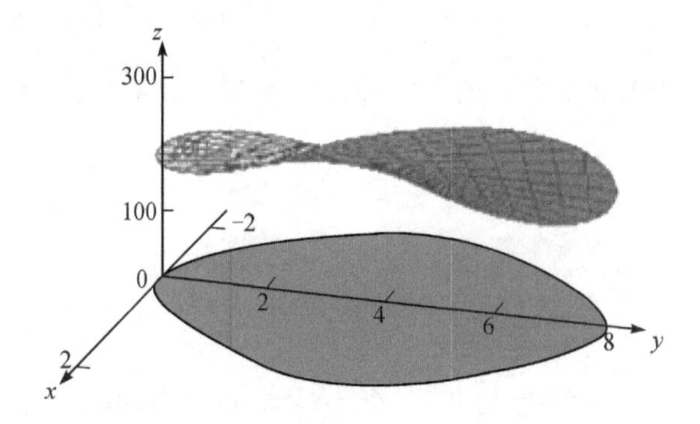

图 7-18

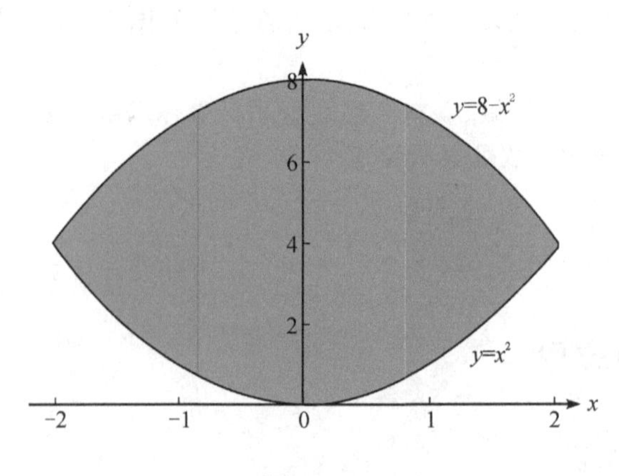

图 7-19

例 7.10 求由平面 $4x+2y+z=10$，$y=3x$，$z=0$ 和 $x=0$ 所围立体的体积.

解 本例与例 7.3 不同，这里没有明确给出积分区域，所以，我们首先要找到它.

第一个平面，$4x+2y+z=10$，是立体的顶，我们实际上是在平面 $z=10-4x-2y$ 下和 xOy 平面内的区域 D 上求体积，如图 7－20 所示.

第二个平面，$y=3x$，给出了立体的一个侧面. 底面区域 D 由直线 $y=3x$，$x=0$ 及平面 $z=10-4x-2y$ 与 xOy 面的交线所围. 令 $z=0$，联立方程 $z+4x+2y=10$ 和 $y=3x$，可求出交点为(1，3)，如图 7－21 所示.

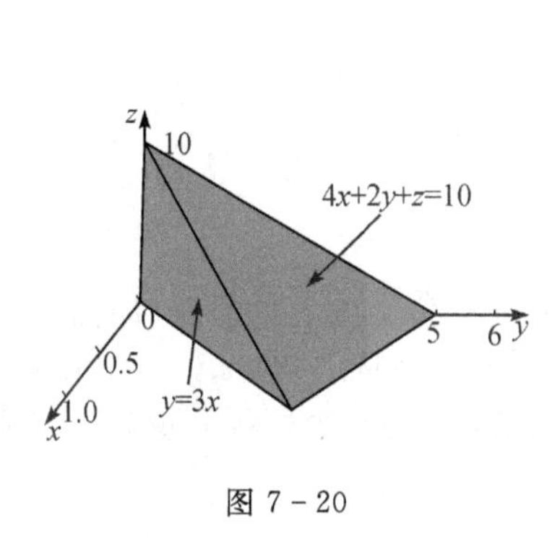

图 7－20

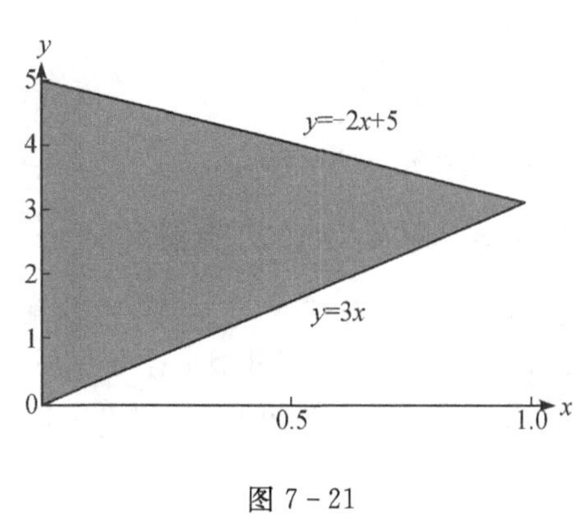

图 7－21

由图 7－21 得到，区域 D 的边界不等式为

$$0 \leqslant x \leqslant 1,\ 3x \leqslant y \leqslant -2x+5$$

故所求立体的体积为

$$\begin{aligned} V &= \iint_D (10-4x-2y)\mathrm{d}A = \int_0^1 \int_{3x}^{-2x+5} (10-4x-2y)\mathrm{d}y\mathrm{d}x \\ &= \int_0^1 (10y-4xy-y^2)\Big|_{3x}^{-2x+5} \mathrm{d}x = \int_0^1 (25x^2-50x+25)\mathrm{d}x \\ &= \left(\frac{25}{3}x^3-25x^2+25x\right)\Big|_0^1 = \frac{25}{3} \end{aligned}$$

7.4.2 平面图形的面积

在定积分中，我们知道位于曲线 $y=f(x)$下方和 x 轴上的区间$[a, b]$上方的平面图形面积可由下面的定积分得到，

$$A = \int_a^b f(x)\mathrm{d}x$$

事实上，我们也可以利用二重积分计算平面图形的面积. 因为当 $f(x, y)=1$ 时，$\iint_D f(x, y)\mathrm{d}A = \iint_D \mathrm{d}A$ 表示的也是平面区域 D 的面积.

求图 7－22 中由曲线 $y=y_1(x)$和 $y=y_2(x)$以及直线 $x=a$ 和 $x=b$ 所围的区域 D 的面积.

由定积分我们知道，该区域的面积可由下式计算

$$A = \int_a^b [y_2(x)-y_1(x)]\mathrm{d}x$$

根据二重积分，区域 D 的面积 $= \iint_D \mathrm{d}A = \int_a^b \int_{y_1(x)}^{y_2(x)} \mathrm{d}y\mathrm{d}x = \int_a^b y\Big|_{y_1(x)}^{y_2(x)} \mathrm{d}x = \int_a^b [y_2(x)-y_1(x)]\mathrm{d}x$

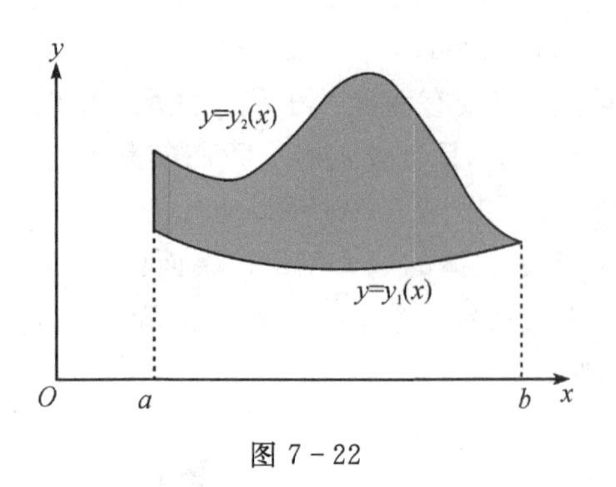

图 7-22

这两个结果完全一样.

7.4.3 平面薄板的质量和质心

设一个薄板占据了平面上的一块区域 D，且在点(x, y)处单位面积上的质量(即面密度)为 $\mu(x, y)$。把该区域划分成 n 个子区域 ΔA_i，在第 i 个子区域上任取一点(x_i, y_i)，则该子区域的质量的近似值为 $\mu(x_i, y_i)\Delta A_i$，把所有的近似值加起来，得到整个薄板质量的近似值，

$$m \approx \sum_{i=1}^{n} \mu(x_i, y_i)\Delta A_i$$

通过对上式取极限可以得到质量的精确值，即

$$m = \iint_D \mu(x, y)\mathrm{d}A$$

现在研究平面薄板 D 的质心(或重心).

(1) 设 xOy 平面上有 n 个质点，质量分别为 $m_1, m_2, \cdots, m_n$，它们分别位于点(x_1, y_1), (x_2, y_2), $\cdots$, (x_n, y_n)处，构成一个离散质点系. 设该质点系的质心坐标为$(\bar{x}, \bar{y})$，则由物理知识知道，该质点系的总质量 $M = \sum_{i=1}^{n} m_i$，质心坐标为

$$\bar{x} = \frac{M_y}{M} = \frac{\sum_{i=1}^{n} m_i x_i}{\sum_{i=1}^{n} m_i}, \quad \bar{y} = \frac{M_x}{M} = \frac{\sum_{i=1}^{n} m_i y_i}{\sum_{i=1}^{n} m_i}$$

其中，$M_y = \sum_{i=1}^{n} m_i x_i$，$M_x = \sum_{i=1}^{n} m_i y_i$ 分别称为该质点系关于 y 轴和 x 轴的**静力矩**.

(2) 设有一个非均匀分布的平面薄板，在 xOy 平面上占据区域 D，点(x, y)处的面密度为 $\mu(x, y)$(假定 $\mu(x, y)$在 D 上连续)，则该薄板的质心坐标为

$$\bar{x} = \frac{M_y}{M} = \frac{\iint_D x\mu(x, y)\mathrm{d}A}{\iint_D \mu(x, y)\mathrm{d}A}, \quad \bar{y} = \frac{M_x}{M} = \frac{\iint_D y\mu(x, y)\mathrm{d}A}{\iint_D \mu(x, y)\mathrm{d}A}$$

其中，$M_y=\iint\limits_D x\mu(x,\ y)\mathrm{d}A$，$M_x=\iint\limits_D y\mu(x,\ y)\mathrm{d}A$ 分别为薄板关于 y 轴和 x 轴的**静力矩**.

特别地，如果平面薄板质量均匀分布，即面密度 $\mu(x,\ y)=k$（常量），则质心坐标为

$$\bar{x}=\frac{1}{A}\iint\limits_D x\,\mathrm{d}A,\ \bar{y}=\frac{1}{A}\iint\limits_D y\,\mathrm{d}A$$

其中，$A=\iint\limits_D \mathrm{d}\sigma$ 为区域 D 的面积. 这时薄板的质心完全由区域 D 的形状所决定，并把均匀薄板的质心叫作这个平面图形的**形心**. 因此，上式也是平面图形 D 的**形心坐标公式**.

例 7.11　设一个平面薄板由 x 轴、直线 $x=2$ 和曲线 $y=x^2$ 围成，其面密度函数为 $\mu=xy$，求它的质量.

解　$m=\iint\limits_D xy\,\mathrm{d}x\mathrm{d}y=\int_0^2\int_0^{x^2}xy\,\mathrm{d}y\mathrm{d}x=\int_0^2\left(\frac{1}{2}xy^2\right)\Big|_0^{x^2}\mathrm{d}x=\frac{16}{3}.$

例 7.12　求位于两圆 $r=2\sin\theta$ 和 $r=4\sin\theta$ 之间的均匀质量薄板的质心(见图 7－23)。

解　因为区域 D 关于 y 轴对称，所以质心位于 y 轴上，于是 $\bar{x}=0$. 又因 D 的面积 $A=3\pi$，而积分

$$\begin{aligned}\iint\limits_D y\,\mathrm{d}A&=\iint\limits_D r\sin\theta\cdot r\mathrm{d}r\mathrm{d}\theta\\&=\int_0^{\pi}\sin\theta\mathrm{d}\theta\int_{2\sin\theta}^{4\sin\theta}r^2\,\mathrm{d}r\\&=\frac{56}{3}\int_0^{\pi}\sin^4\theta\mathrm{d}\theta=7\pi\end{aligned}$$

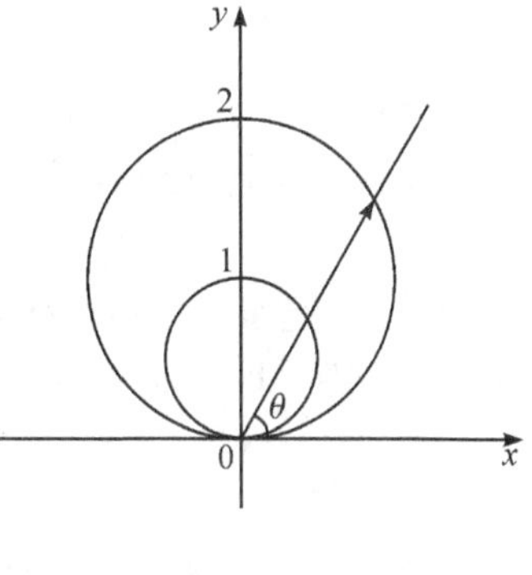

图 7－23

故按形心公式计算，得 $\bar{y}=\frac{7\pi}{3\pi}=\frac{7}{3}$，所求质心坐标$(0,\frac{7}{3})$.

例 7.13　求以原点为圆心的位于 y 轴上部的均质半圆的形心位置.

解　设半圆 D 的直径在 x 轴上，半圆位于 x 轴上方，则 D 可表示为

$$0\leqslant y\leqslant\sqrt{R^2-x^2},\ -R\leqslant x\leqslant R$$

因为 D 关于 y 轴对称，则 $\bar{x}=0$，因此只需计算 $\bar{y}$. 又因为面积 $A=\frac{1}{2}\pi R^2$，

$$\iint\limits_D y\,\mathrm{d}A=\int_{-R}^{R}\mathrm{d}x\int_0^{\sqrt{R^2-x^2}}y\mathrm{d}y=\int_{-R}^{R}\frac{R^2-x^2}{2}\mathrm{d}x=\frac{2}{3}R^3$$

所以
$$\bar{y}=\frac{1}{A}\iint\limits_D y\,\mathrm{d}A=\frac{2R^3/3}{\pi R^2/2}=\frac{4R}{3\pi}$$

故所求形心坐标为$\left(0,\ \frac{4R}{3\pi}\right)$.

7.5　三重积分

7.5.1　“先单后重”法

现在我们考虑一个如图 7－24(a)所示的空间立体结构的质量问题. 假设该物体占据了空

间中的区域

$$V=\{(x,y,z)\,|\,z_1(x,y)\leqslant z\leqslant z_2(x,y),(x,y)\in D\subset \mathbf{R}^2\},$$

其质量是非均匀分布的，体密度函数为 $\mu=\mu(x,y,z)$，要计算这个物体的质量，可采用与二重积分类似的思想方法.

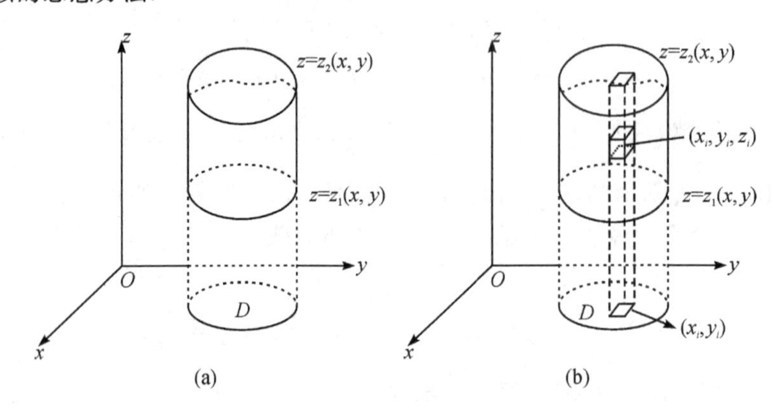

图 7-24

先将物体划分成若干个小长方体，其质量为

$$\Delta m_i\approx\mu(x_i,y_i,z_i)\Delta V_i$$

其中，ΔV_i 是小长方体的体积. 如图 7-24(b)所示，固定这个小长方体，在该小长方体上沿 z 轴方向积分，得到这个小长方体的质量

$$\int_{z_1(x,y)}^{z_2(x,y)}\mu(x,y,z)\mathrm{d}z$$

然后再把所有小长方体的质量累加起来，即在投影域上做二重积分，即可得到整个物体的质量

$$m=\iint_D\left[\int_{z_1(x,y)}^{z_2(x,y)}\mu(x,y,z)\mathrm{d}z\right]\mathrm{d}A$$

一般有如下结论：

如果函数 $f(x,y,z)$ 在 $V=\{(x,y,z)\,|\,z_1(x,y)\leqslant z\leqslant z_2(x,y),(x,y)\in D\subset\mathbf{R}^2\}$ 上连续，则三重积分

$$\iiint_V f(x,y,z)\mathrm{d}V=\iint_D\left[\int_{z_1(x,y)}^{z_2(x,y)}f(x,y,z)\mathrm{d}z\right]\mathrm{d}A$$

也称为“先单后重”的累次积分.

特殊地，当 $f(x,y,z)=1$ 时，空间区域的体积 $=\iiint_V f(x,y,z)\mathrm{d}V$.

例 7.14 计算三重积分 $\iiint_V x\mathrm{d}V$，其中 V 是一个以点(0, 0, 0)，点(1, 0, 0)，点(0, 1, 0)和点(0, 0, 1)为顶点的四面体，如图 7-25(a) 所示.

解 根据积分域的形状，我们把四面体向 xOy 面投影，底面投影域是一个三角形区域，如图 7-25(b)所示，从而把所求的三重积分化为如下形式的先单后重的累次积分.

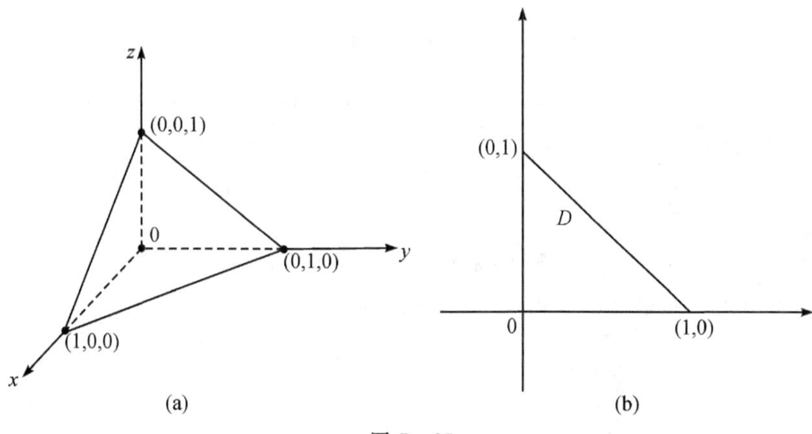

图 7 - 25

$$\iiint\limits_V x\,\mathrm{d}V = \iint\limits_D x\left(\int_0^{1-x-y}\mathrm{d}z\right)\mathrm{d}A = \iint\limits_D xz\Big|_0^{1-x-y}\mathrm{d}x\mathrm{d}y = \iint\limits_D x(1-x-y)\mathrm{d}x\mathrm{d}y$$

$$= \int_0^1\mathrm{d}x\int_0^{1-x}x(1-x-y)\mathrm{d}y = \frac{1}{24}$$

例 7.15　计算 $\iiint\limits_V z\,\mathrm{d}V$，其中 $V=\left\{(x, y, z)\,\middle|\,0\leqslant z\leqslant\sqrt{1-x^2-y^2}\right\}$.

解　根据积分域的形状，我们把上半球体向 xOy 面投影，底面投影域为

$$D=\{(x, y)\,|\,x^2+y^2\leqslant 1\}$$

从而把所求的三重积分化为如下形式的先单后重的累次积分.

$$\iiint\limits_V z\,\mathrm{d}V = \iint\limits_D\left(\int_0^{\sqrt{1-x^2-y^2}}z\mathrm{d}z\right)\mathrm{d}A = \frac{1}{2}\iint\limits_D(1-x^2-y^2)\mathrm{d}x\mathrm{d}y = \frac{1}{2}\int_0^{2\pi}\mathrm{d}\theta\int_0^1(1-r^2)r\mathrm{d}r = \frac{\pi}{4}$$

7.5.2 “先重后单”法

如果把空间立体结构看成是一个一个薄片摞起来的，如图 7 - 26 所示.

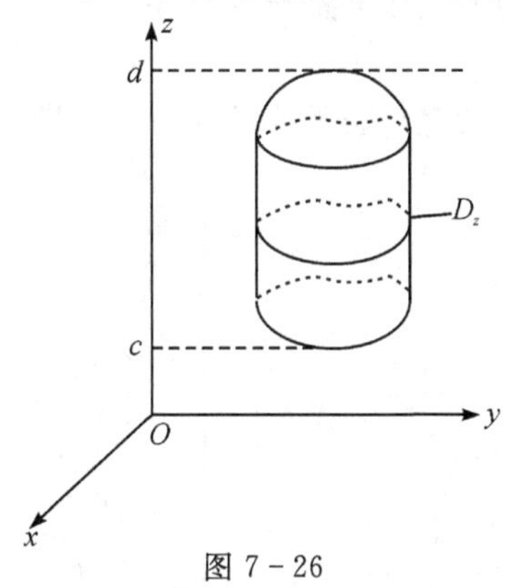

图 7 - 26

可以先固定 z，计算对应薄片 D_z 的质量 $\iint\limits_{D_z}\mu(x, y, z)\mathrm{d}A$，然后再沿 z 轴方向累加起来，就得到了整个物体的质量

$$m=\int_c^d\left[\iint_{D_z}\mu(x,y,z)\,\mathrm{d}x\mathrm{d}y\right]\mathrm{d}z$$

一般地，如果函数 $f(x, y, z)$在 V 上连续，则三重积分

$$\iiint_V f(x,y,z)\,\mathrm{d}V=\int_c^d\left[\iint_{D_z}f(x,y,z)\,\mathrm{d}A\right]\mathrm{d}z$$

这种积分也称为“先重后单”的累次积分，这种方法又叫“切片法”.

特殊地，如果 $f(x, y, z)=f(z)$，则

$$\iiint_V f(x,y,z)\,\mathrm{d}V=\int_c^d f(z)A_z\,\mathrm{d}z$$

其中，A_z 为 D_z 的面积.

例 7.16 利用切片法计算例 7.15.

解 $\iiint_V z\,\mathrm{d}V=\int_0^1 zA_z\,\mathrm{d}z=\int_0^1 z\pi(1-z^2)\,\mathrm{d}z=\frac{\pi}{4}$

7.5.3 楔形法

如果把空间立体结构看成是由一个一个楔形累加起来的(比如买西瓜时，卖瓜人可以切出一块楔形让顾客品尝)，可以先计算出其中一个楔形块的质量，采用这个方法，需要了解球面坐标系. 图 7-27 中空间一点 P 可以用三个曲面的交点来表示：半径为 r 的球面，顶角为 φ 的圆锥面，通过 x 轴且对 xOy 坐标面的转角为 θ 的半平面的交点，数组 (r, φ, θ) 称为点 $P(x, y, z)$的球面坐标，易见，直角坐标和球坐标的变换公式为 $\begin{cases}x=r\sin\varphi\cos\theta\\y=r\sin\varphi\sin\theta\\z=r\cos\varphi\end{cases}$. 我们先用球面坐标系下的三组坐标面划分楔形块，图 7-28(a)表示的是其中一个楔形块，根据图 7-28(b)，楔形物质块体积为 $r^2\sin\varphi\mathrm{d}r\mathrm{d}\varphi\mathrm{d}\theta$.

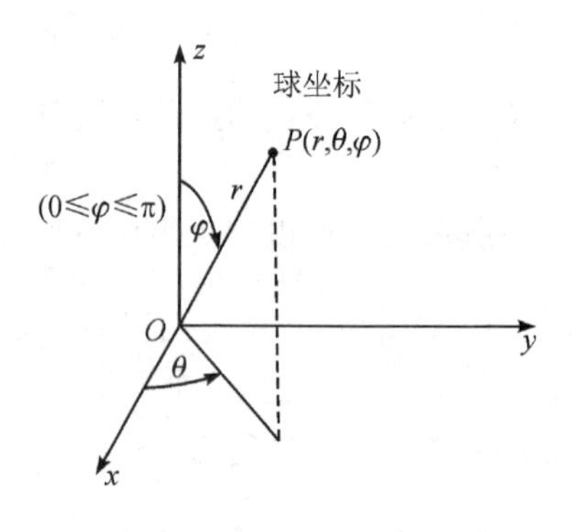

图 7-27

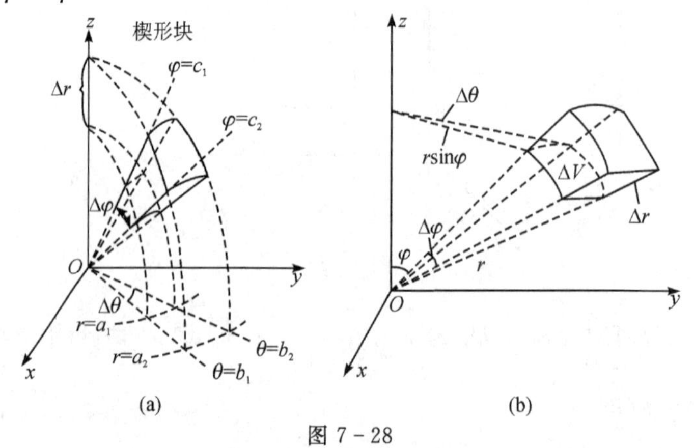

图 7-28

从而该楔形块的质量为

$$f(r\sin\varphi\cos\theta, r\sin\varphi\sin\theta, r\cos\varphi)r^2\sin\varphi \mathrm{d}r\mathrm{d}\varphi\mathrm{d}\theta$$

通过三重积分可以得到整个空间立体结构的质量

一般地，如果函数 $f(x, y, z)$ 在 V 上连续，则三重积分

$$\iiint_V f(x, y, z)\mathrm{d}V = \iiint_V f(r\sin\varphi\cos\theta, r\sin\varphi\sin\theta, r\cos\varphi)r^2\sin\varphi \mathrm{d}r\mathrm{d}\varphi\mathrm{d}\theta$$

注：空间立体结构的质心的求法与平面薄板的质心求法类似，这里不再赘述.

例 7.17　求半径为 a 的球体的质量，其密度与球体上一点到原点的距离成正比.

解　$m = \iiint_V \mu \mathrm{d}V = k\iiint_V \sqrt{x^2+y^2+z^2}\mathrm{d}V = k\int_0^{2\pi}\mathrm{d}\theta\int_0^{\pi}\mathrm{d}\varphi\int_0^a r\cdot r^2\sin\varphi \mathrm{d}r$

$= 2\pi\dfrac{ka^4}{4}\int_0^{\pi}\sin\varphi \mathrm{d}\varphi = k\pi a^4$

例 7.18　求由圆锥面 $z=\sqrt{x^2+y^2}$ 和以原点为中心、半径为 1 的球体所围的立体的体积和形心.

解　$V = \iiint_V \mathrm{d}V = \int_0^{2\pi}\mathrm{d}\theta\int_0^{\frac{\pi}{4}}\mathrm{d}\varphi\int_0^1 r^2\sin\varphi \mathrm{d}r = \dfrac{2}{3}\pi\int_0^{\frac{\pi}{4}}\sin\varphi \mathrm{d}\varphi = \dfrac{2}{3}\pi\left(1-\dfrac{\sqrt{2}}{2}\right)$

由于球体关于 z 轴对称，所以 $\bar{x}=\bar{y}=0$. 设球体对 xOy 平面的静力矩为 M_{xy}，则

$$\bar{z}=\frac{M_{xy}}{V}=\frac{\iiint_V z\mathrm{d}V}{V}=\frac{\int_0^{2\pi}\mathrm{d}\theta\int_0^{\frac{\pi}{4}}\mathrm{d}\varphi\int_0^1 r\cos\varphi\cdot r^2\sin\varphi \mathrm{d}r}{\frac{2}{3}\pi\left(1-\frac{\sqrt{2}}{2}\right)}$$

$$=\frac{\frac{\pi}{8}}{\frac{2}{3}\pi\left(1-\frac{\sqrt{2}}{2}\right)}=\frac{2}{8}\left(1+\frac{\sqrt{2}}{2}\right)$$

习题 7

1. 计算下列所给区域上的二重积分：(a) 先对 x 积分；(b) 先对 y 积分.

(1) $\iint_D (12x-8y)\mathrm{d}A$，$D=[-1,4]\times[2,3]$；

(2) $\iint_D \cos x\sin y\mathrm{d}A$，$D=\left[\dfrac{\pi}{6},\dfrac{\pi}{4}\right]\times\left[\dfrac{\pi}{4},\dfrac{\pi}{3}\right]$.

2. 计算下列矩形域上的二重积分.

(1) $\iint_D (6y\sqrt{x}-2y^3)\mathrm{d}A$，$D=[1,4]\times[0,3]$；

(2) $\iint_D \left(\sin 2x-\dfrac{1}{1+6y}\right)\mathrm{d}A$，$D=\left[\dfrac{\pi}{4},\dfrac{\pi}{2}\right]\times[0,1]$；

(3) $\iint_D (y\mathrm{e}^{y^2-4x})\mathrm{d}A$，$D=[0, 2]\times[0, 2\sqrt{2}]$；

(4) $\iint\limits_D xy\cos(x^2y)\,\mathrm{d}A$, $D=[-2,3]\times[-1,1]$.

3. 计算位于 xOy 平面的矩形域 $[-1,1]\times[0,2]$ 之上和曲面 $f(x,y)=9x^2+4xy+4$ 下方的立体的体积.

4. 试用二重积分的几何意义说明：

(1) $\iint\limits_D k\,\mathrm{d}A=kA$，$k\in\mathbf{R}$ 为常数，A 表示区域 D 的面积；

(2) $\iint\limits_D \sqrt{R^2-x^2-y^2}\,\mathrm{d}A=\frac{2}{3}\pi R^3$，$D$ 是以原点为中心,半径为 R 的圆；

(3) 若积分域关于 y 轴对称,则

① 当 $f(x,y)$ 是 x 的奇函数时，即 $f(x,y)=-f(-x,y)$，$\iint\limits_D f(x,y)\,\mathrm{d}A=0$；

② 当 $f(x,y)$ 是 y 的偶函数时，即 $f(x,y)=f(-x,y)$，$\iint\limits_D f(x,y)\,\mathrm{d}A=2\iint\limits_{D_1} f(x,y)\,\mathrm{d}A$，其中 D_1 为 D 在右半平面 $x\geqslant 0$ 的部分区域；

(4) 若积分域关于轴对称,被积函数分别具有怎样的奇偶性时有

$$\iint\limits_D f(x,y)\,\mathrm{d}A=0,\ \iint\limits_D f(x,y)\,\mathrm{d}A=2\iint\limits_{D_1} f(x,y)\,\mathrm{d}A$$

其中 D_1 为 D 在上半平面 $y\geqslant 0$ 的部分区域.

5. 计算下列二重积分：

(1) $\iint\limits_D (7x^2+14y)\,\mathrm{d}A$，其中 D 是由 $x=2y^2$ 和 $x=8$ 围成；

(2) $\iint\limits_D x(y-1)\,\mathrm{d}A$，其中 D 是由 $y=1-x^2$ 和 $y=x^2-3$ 围成；

(3) $\iint\limits_D 5x^3\cos(y^3)\,\mathrm{d}A$，其中 D 是由 $y=2$，$y=\frac{1}{4}x^2$ 和 y 轴围成；

(4) $\iint\limits_D \frac{1}{\sqrt[3]{y}(x^3+1)}\,\mathrm{d}A$，其中 D 是由 $x=-\sqrt[3]{y}$，$x=3$ 和 x 轴围成.

6. 交换积分次序：

(1) $\int_0^2 \mathrm{d}x\int_0^{\sqrt{x}} f(x,y)\,\mathrm{d}y$；

(2) $\int_0^4 \mathrm{d}y\int_{2y}^8 f(x,y)\,\mathrm{d}x$；

(3) $\int_0^2 \mathrm{d}y\int_1^{\mathrm{e}^y} f(x,y)\,\mathrm{d}x$；

(4) $\int_1^{\mathrm{e}} \mathrm{d}x\int_0^{\ln x} f(x,y)\,\mathrm{d}y$；

(5) $\int_0^1 \mathrm{d}y\int_{\arcsin y}^{\frac{\pi}{2}} f(x,y)\,\mathrm{d}x$；

(6) $\int_0^1 \mathrm{d}y\int_{y^2}^{\sqrt{y}} f(x,y)\,\mathrm{d}x$.

7. 交换积分次序，并计算二重积分的值.

(1) $\int_0^1 \mathrm{d}x\int_{4x}^4 \mathrm{e}^{-y^2}\,\mathrm{d}y$；

(2) $\int_0^2 \mathrm{d}y\int_{\frac{y}{2}}^1 \cos x^2\,\mathrm{d}x$；

(3) $\int_0^4 \mathrm{d}y\int_{\sqrt{y}}^2 \mathrm{e}^{x^3}\,\mathrm{d}x$；

(4) $\int_1^3 \mathrm{d}x\int_0^{\ln x} x\,\mathrm{d}y$.

8. 计算下列累次积分：

(1) $\int_0^{\frac{\pi}{2}} d\theta \int_0^{\sin\theta} r\cos\theta dr$；　(2) $\int_0^{\pi} \left[\int_0^{1+\cos\theta} r dr\right] d\theta$；

(3) $\int_0^{\frac{\pi}{2}} d\theta \int_0^{a\sin\theta} r dr$；　(4) $\int_0^{\frac{\pi}{6}} \left[\int_0^{\cos 3\theta} r dr\right] d\theta$.

9. 利用极坐标计算下列二重积分：

(1) $\iint\limits_D (y^2+3x) dA$，其中 D 是第三象限介于曲线 $x^2+y^2=1$ 和 $x^2+y^2=9$ 之间的部分；

(2) $\iint\limits_D \sqrt{1+4x^2+4y^2} dA$，其中 D 是圆 $x^2+y^2=16$ 的下半部分；

(3) $\iint\limits_D (4xy-7) dA$，其中 D 是圆 $x^2+y^2=2$ 位于第一象限的部分；

(4) $\int_0^3 \int_{-\sqrt{9-x^2}}^0 e^{x^2+y^2} dy dx$.

10. 利用二重积分求解下列问题.

(1) 求由曲线 $y=1-x^2$ 和 $y=x^2-3$ 所围的平面区域的面积；

(2) 求由心脏线 $r=1-\cos\theta$ 所围图形的面积；

(3) 求由 $r=1$ 和 $r=\sin 2\theta \left(\frac{\pi}{4} \leqslant \theta \leqslant \frac{\pi}{2}\right)$ 所围的位于第一象限的图形的面积；

(4) 求由曲面 $z=6-5x^2$，平面 $y=2x$，$y=2$，$x=0$ 以及 xOy 面所围的立体的体积；

(5) 求由曲面 $z=2x^2+2y^2$，柱面 $x^2+y^2=16$ 和 xOy 面所围立体的体积；

(6) 求由曲面 $z=8-x^2-y^2$ 和 $z=3x^2+3y^2-4$ 所围的立体的体积.

11. 利用三重积分求解下列问题.

(1) 求由抛物面 $z=x^2+y^2$ 和平面 $z=4$ 所围立体的体积；

(2) 求由锥面 $z=\sqrt{x^2+y^2}$ 与球面 $x^2+y^2+z^2=16$ 所围的立体的体积；

(3) 求由曲面 $z=4-x^2-y^2$、柱面 $x^2+y^2=2x$ 和 xOy 坐标面所围的立体的体积；

(4) 设一物质块由锥面 $z=\sqrt{x^2+y^2}$ 和平面 $z=0$ 所围，其密度为 $\mu=3-z$，求它的质量；

(5) 求由曲面 $z=12-2x^2-2y^2$ 和曲面 $z=x^2+y^2$ 所围立体的形心；

(6) 求半径为 a 的半球体的重心，其体密度与球体中心到球体上一点的距离成正比；

(7) 求由曲面 $z=1-x^2-y^2$ 和 xOy 平面所围且密度为 $\mu=x^2+y^2+z^2$ 的物质块的质心坐标.

第 8 章　曲线积分与曲面积分

8.1　曲线积分

我们知道，第 7 章中的二重积分和三重积分是把定积分 $\int_a^b f(x)\mathrm{d}x$ 的积分范围 从一维区间推广到二维平面和三维空间，本节我们把积分区间推广到 xOy 平面上的一条曲线 C，相应的积分 $\int_C f(x, y)\mathrm{d}s$ 称为**曲线积分**.

8.1.1　曲线积分的定义

设曲线 C 为一条平面光滑曲线，其参数方程为

$$x = x(t),\ y = y(t)\quad a \leqslant t \leqslant b$$

其中，$x(t)$和 $y(t)$在(a, b)内均连续且不同时为零. 考虑在参数取值区间$[a, b]$内任意插入 $n-1$ 个分点，

$$a = t_0 < t_1 < t_2 < \cdots < t_n = b$$

从而相应地将曲线 C 分成 n 个子弧段$\widehat{P_{i-1}P_i}$，点 P_i 对应于参数 t_i. 用 Δs_i 表示该子弧段的长度，令 $d = \max\limits_{a\leqslant t\leqslant b}\{\Delta s_i\}$表示子弧段长度的最大者. 最后，在子弧段$\widehat{P_{i-1}P_i}$上选取一个样本点 $Q_i(\bar{x}_i, \bar{y}_i)$，如图 8－1 所示.

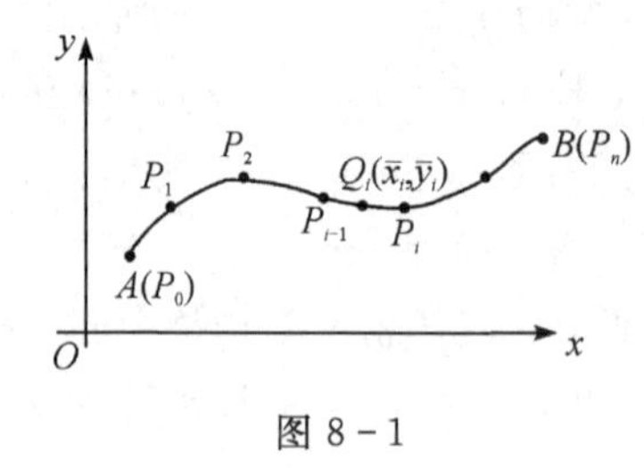

图 8－1

现在考虑黎曼和

$$\sum_{i=1}^{n} f(\bar{x}_i, \bar{y}_i)\Delta s_i$$

如果 $f(x, y)$是非负的，这个和近似等于张在曲线上的垂直放置的弯曲的帘幕的面积，如图8－2所示.

如果 $f(x, y)$在包含曲线 C 的平面区域 D 上连续，则当 $d\to 0$ 时黎曼和有极限，该极限值称为函数 $f(x, y)$沿曲线 C 从 A 到 B 的**曲线积分**. 即

$$\int_C f(x, y)\mathrm{d}s = \lim_{d\to 0}\sum_{i=1}^{n} f(\bar{x}_i, \bar{y}_i)\Delta s_i$$

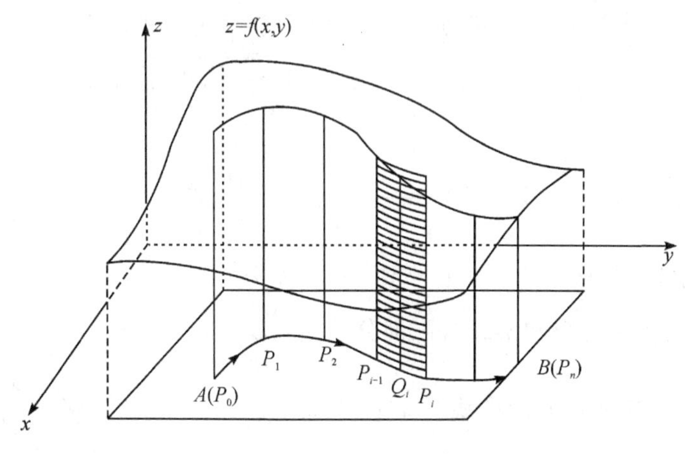

图 8-2

8.1.2　曲线积分的计算

曲线积分的定义没有给出曲线积分的计算方法，当曲线 C 以参数方程形式给出时，我们可以利用 2.3 节的弧长微分

$$\widehat{P_{i-1}P_i} \approx \mathrm{d}s = \sqrt{(\mathrm{d}x)^2 + (\mathrm{d}y)^2}$$

把曲线积分的计算转化为通常的定积分.

$$\int_C f(x, y)\mathrm{d}s = \int_a^b f[x(t), y(t)]\sqrt{[x'(t)]^2 + [y'(t)]^2}\mathrm{d}t$$

注：

(1) 如果 $f(x, y)=1$，则曲线 C 的长度为 $\int_C \mathrm{d}s = \int_a^b \sqrt{[x'(t)]^2 + [y'(t)]^2}\mathrm{d}t$.

(2) 因为弧长总是正的，所以公式中的积分下限必须小于积分上限.

以上结论很容易推广到空间曲线. 特别地，如果空间曲线 C 的参数方程为

$$x = x(t),\ y = y(t),\ z = z(t)\quad (a \leqslant t \leqslant b)$$

则

$$\int_C f(x, y, z)\mathrm{d}s = \int_a^b f[x(t), y(t), z(t)]\sqrt{[x'(t)]^2 + [y'(t)]^2 + [z'(t)]^2}\mathrm{d}t$$

例 8.1　计算 $\int_C x^2 y\mathrm{d}s$，其中 C 由参数方程 $x = \cos t$，$y = \sin t$，$0 \leqslant t \leqslant \frac{\pi}{2}$ 所确定，并证明当曲线以 $x = \sqrt{1-y^2}$，$y = y$，$0 \leqslant y \leqslant 1$ 的形式给出时，计算结果相同.

解　利用第一种参数形式

$$\int_C x^2 y\mathrm{d}s = \int_0^{\frac{\pi}{2}} \cos^2 t\sin t\sqrt{(-\sin t)^2 + \cos^2 t}\mathrm{d}t$$

$$= \int_0^{\frac{\pi}{2}} \cos^2 t\sin t\mathrm{d}t = \left(-\frac{1}{3}\cos^3 t\right)\Big|_0^{\pi/2} = \frac{1}{3}$$

对于第二种形式的参数方程，根据 2.3 节的弧长微元表达式，可得

$$ds = \sqrt{(dx)^2 + (dy)^2} = \sqrt{1 + \left(\frac{dx}{dy}\right)^2}\,dy$$

因此

$$\int_C x^2 y ds = \int_0^1 (1-y^2) y \frac{1}{\sqrt{1-y^2}} dy = \int_0^1 \sqrt{1-y^2}\, y dy = \frac{1}{3}$$

例 8.2　求椭圆柱面$\frac{x^2}{5}+\frac{y^2}{9}=1$被平面 $z=y$ 和平面 $z=0$ 所截得的在第一卦限和第二卦限部分的侧面积(见图 8-3).

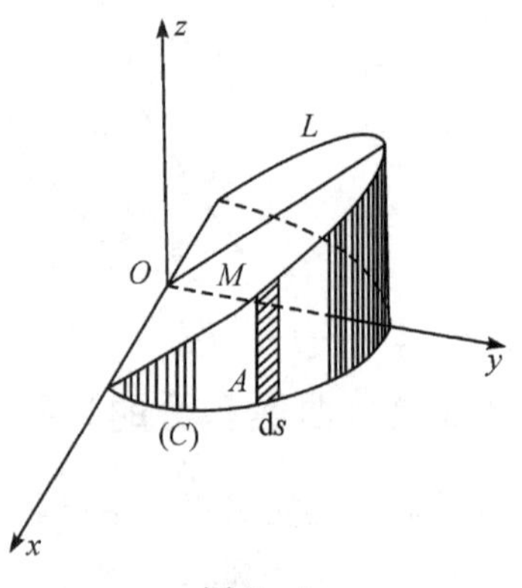

图 8-3

解　很容易看到椭圆柱面的准线是 xOy 平面上的半圆，故所求侧面积为

$$A = \int_C y ds$$

把曲线 C 写成参数方程形式，

$$x = \sqrt{5}\cos t,\ y = 3\sin t,\ 0 \leqslant t \leqslant \pi$$

于是

$$A = \int_C y ds = \int_0^\pi 3\sin t \sqrt{5\sin^2 t + 9\cos^2 t}\,dt$$

$$= \int_0^\pi 3\sin t \sqrt{5 + 4\cos^2 t}\,dt = -3\int_0^\pi \sqrt{4 + 5\cos^2 t}\,d\cos t = 9 + \frac{15}{4}\ln 5$$

例 8.3　设 $y=\frac{2}{3}x^{\frac{3}{2}}$，求介于点(0, 0)和点$\left(4,\frac{16}{3}\right)$之间的曲线的长度.

解　把曲线写成如下的参数形式

$$x = x,\ y = \frac{2}{3}x^{\frac{3}{2}},\ 0 \leqslant x \leqslant 4$$

于是，有

$$L = \int_L ds = \int_0^4 \sqrt{1 + \left[\left(\frac{2}{3}x^{\frac{3}{2}}\right)'\right]^2}\,dx = \int_0^4 \sqrt{1+x}\,dx = \frac{2}{3}(5\sqrt{5}-1)$$

例 8.4　一根细电线弯成半圆形

$$x = a\cos t,\ y = a\sin t,\ 0 \leqslant t \leqslant \pi$$

如果电线在一点的密度与该点到 x 轴的距离成正比，求它的质量和质心.

解　此题求解方法与求薄板和物质块的质量和质心的方法类似.

令 $\mu(x, y)=ky$(k 为常数)为点(x, y)的密度，则整个电线的质量为

$$m = \int_C ky ds = \int_0^\pi ka\sin t \sqrt{a^2\sin^2 t + a^2\cos^2 t}\,dt = ka^2\int_0^\pi \sin t dt = 2ka^2$$

电线到 x 轴的静力矩为

$$M_x = \int_C y \cdot ky ds = \int_0^\pi ka^3 \sin^2 t dt = \frac{ka^3}{2}\int_0^\pi (1-\cos 2t)dt = \frac{ka^3}{2}\left(t - \frac{1}{2}\sin 2t\right)\Big|_0^\pi = \frac{ka^3\pi}{2}$$

于是有

$$\bar{y} = \frac{M_x}{m} = \frac{\frac{1}{2}ka^3\pi}{2ka^2} = \frac{1}{4}\pi a$$

由于电线关于 x 轴对称，所以 $\bar{x}=0$，从而电线的质心为 $\left(0,\dfrac{\pi a}{4}\right)$.

例 8.5　求密度为 $\mu=kz$ 的螺旋形电线 C 的质量，其参数方程为

$$x=3\cos t,\ y=3\sin t,\ z=4t,\ 0\leqslant t\leqslant \pi$$

解　$m=\int_C \mu\,\mathrm{d}s=\int_0^\pi k\cdot 4t\sqrt{9\sin^2 t+9\cos^2 t+16}\,\mathrm{d}t=20k\int_0^\pi t\,\mathrm{d}t=10k\pi^2$

8.2　曲面积分

8.2.1　曲面面积

我们已知一些特殊曲面的面积，例如半径为 r 的球面面积为 $4\pi r^2$. 本节我们将探讨定义在一个区域上的曲面 $z=f(x, y)$的面积的计算公式.

设 Σ 为一曲面，其投影为 xOy 平面上的有界区域 D，假定 $f_x(x, y)$和 $f_y(x, y)$ 在 D 上连续. 我们先划分区域 D，然后选取一个代表区域向上做柱体截取曲面，相应地，柱体和曲面相交得到了一个小的曲面. 在点 $P_k(x_k, y_k, z_k)$处做切平面，可以用切平面的面积近似计算小曲面的面积. 为了清楚，我们把图形做了局部放大，见图 8－4.

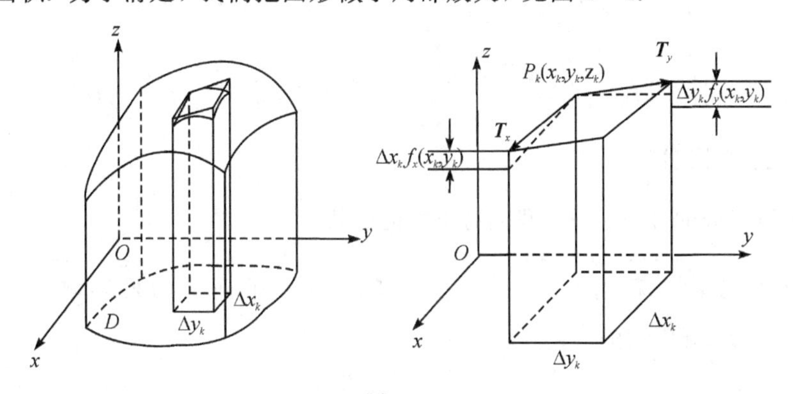

图 8－4

接下来，求以 $|\boldsymbol{T}_x|$ 和 $|\boldsymbol{T}_y|$ 为邻边的平行四边形的面积. 由于

$$\boldsymbol{T}_x=\Delta x_k\boldsymbol{i}+f_x(x_k, y_k)\Delta x_k\boldsymbol{k}$$

$$\boldsymbol{T}_y=\Delta y_k\boldsymbol{j}+f_y(x_k, y_k)\Delta y_k\boldsymbol{k}$$

利用《基础线性代数》5.2 节的结论，切平面的面积为 $|\boldsymbol{T}_x\times\boldsymbol{T}_y|$，其中

$$\boldsymbol{T}_x\times\boldsymbol{T}_y=\begin{vmatrix}\boldsymbol{i} & \boldsymbol{j} & \boldsymbol{k}\\ \Delta x_k & 0 & f_x(x_k, y_k)\Delta x_k\\ 0 & \Delta y_k & f_y(x_k, y_k)\Delta y_k\end{vmatrix}=\Delta x_k\Delta y_k[-f_x(x_k, y_k)\boldsymbol{i}-f_y(x_k, y_k)\boldsymbol{j}+\boldsymbol{k}]$$

所以平行四边形面积为

$$|\boldsymbol{T}_x\times\boldsymbol{T}_y|=\Delta x_k\Delta y_k\sqrt{1+[f_x(x_k, y_k)]^2+[f_y(x_k, y_k)]^2}$$

然后把所有切平面面积相加，取极限，就得到了曲面 Σ 的面积

$$\text{曲面 }\Sigma\text{ 的面积}=\lim_{n\to\infty}\sum_{k=1}^{n}\sqrt{1+[f_x(x_k, y_k)]^2+[f_y(x_k, y_k)]^2}\,\Delta x_k\Delta y_k$$

从而得到曲面面积公式.

$$\text{曲面}\ \Sigma\ \text{的面积} = \iint_D \sqrt{1+[f_x(x,y)]^2+[f_y(x,y)]^2}\,\mathrm{d}x\mathrm{d}y$$

例 8.6 设圆柱面 Σ: $z=\sqrt{4-x^2}$ 在 xOy 平面上的投影域 D 为以直线 $x=0$, $x=1$, $y=0$ 和 $y=2$ 所围的矩形域，求该圆柱面的面积.

解 令 $f(x,y)=\sqrt{4-x^2}$，则 $f_x=\dfrac{-x}{\sqrt{4-x^2}}$, $f_y=0$，并且

$$\Sigma\ \text{的面积}=\iint_D \sqrt{1+f_x^2+f_y^2}\,\mathrm{d}x\mathrm{d}y=\iint_D \sqrt{\frac{x^2}{4-x^2}+1}\,\mathrm{d}x\mathrm{d}y=2\iint_D \frac{1}{\sqrt{4-x^2}}\mathrm{d}x\mathrm{d}y$$

$$=\int_0^1 \frac{1}{\sqrt{4-x^2}}\mathrm{d}x\int_0^2 2\mathrm{d}y=4\int_0^1 \frac{1}{\sqrt{4-x^2}}\mathrm{d}x=4\arcsin\frac{x}{2}\Big|_0^1=\frac{2}{3}\pi$$

例 8.7 求位于平面 $z=\sqrt{20}$ 下方的曲面 $z=x^2+y^2$ 的面积.

解 令 $f(x,y)=x^2+y^2$，则 $f_x=2x$, $f_y=2y$，并且

$$\Sigma\ \text{的面积}=\iint_D \sqrt{1+f_x^2+f_y^2}\,\mathrm{d}x\mathrm{d}y=\iint_D \sqrt{1+4x^2+4y^2}\,\mathrm{d}x\mathrm{d}y$$

$$=\int_0^{2\pi}\mathrm{d}\theta\int_0^{\sqrt{20}} \sqrt{1+4r^2}\,r\mathrm{d}r=2\pi\times\frac{1}{8}\times\frac{2}{3}\,(1+4r^2)^{\frac{3}{2}}\Big|_0^{\sqrt{20}}=\frac{243}{2}\pi$$

8.2.2 曲面积分

类似于曲线积分的计算公式，我们有下面的公式

设平面 Σ 的方程为 $z=z(x,y)$，其在 xOy 平面上的投影域为 D. 如果 $f(x,y,z)$, $\dfrac{\partial z}{\partial x}$ 和 $\dfrac{\partial z}{\partial y}$ 均连续，则

$$\iint_\Sigma f(x,y,z)\mathrm{d}S=\iint_D f[x,y,z(x,y)]\sqrt{1+z_x^2+z_y^2}\,\mathrm{d}x\mathrm{d}y$$

例 8.8 计算 $\iint_\Sigma xz\mathrm{d}S$，其中 Σ 是锥面 $z=\sqrt{x^2+y^2}$ 下方被柱面 $x^2+y^2=2ax\,(a>0)$ 所截的部分.

解 曲面 Σ 在 xOy 平面上的投影为 D: $x^2+y^2\leqslant 2ax$，则

$$\iint_\Sigma xz\mathrm{d}S=\iint_D x\sqrt{x^2+y^2}\,\sqrt{1+z_x^2+z_y^2}\,\mathrm{d}x\mathrm{d}y$$

$$=\sqrt{2}\iint_D x\sqrt{x^2+y^2}\,\mathrm{d}x\mathrm{d}y$$

$$=\sqrt{2}\int_{-\frac{\pi}{2}}^{\frac{\pi}{2}}\mathrm{d}\theta\int_0^{2a\cos\theta} r^3\cos\theta\mathrm{d}r$$

$$=4\sqrt{2}a^4\int_{-\frac{\pi}{2}}^{\frac{\pi}{2}}\cos^5\theta\mathrm{d}\theta=\frac{64\sqrt{2}}{15}a^4$$

例 8.9　设曲面 Σ 的方程为

$$z = \sqrt{9-x^2-y^2}$$

在 x，y 满足 $x^2+y^2\leqslant 4$ 的范围内镀上一层薄金属，其密度为 $\mu(x, y, z)=z$. 求薄金属的质量.

解　镀薄金属部分曲面在 xOy 平面上的投影为 D：$x^2+y^2\leqslant 4$，则

$$\begin{aligned} m &= \iint_{\Sigma}\mu(x, y, z)\mathrm{d}S = \iint_{D} z\sqrt{1+z_x^2+z_y^2}\mathrm{d}x\mathrm{d}y \\ &= \iint_{D} z\sqrt{1+\frac{x^2}{9-x^2-y^2}+\frac{y^2}{9-x^2-y^2}}\mathrm{d}x\mathrm{d}y \\ &= 3\iint_{D}\mathrm{d}x\mathrm{d}y = 12\pi \end{aligned}$$

例 8.10　计算半径为 a 的球体的表面积.

解　设球面方程为 $x^2+y^2+z^2=a^2$，则由对称性，只需计算上半球面的面积，其方程为 $z=\sqrt{a^2-x^2-y^2}$，它在 xOy 面上的投影区域 D_{xy} 是圆：$x^2+y^2\leqslant a^2$.

又
$$z_x=\frac{-x}{\sqrt{a^2-x^2-y^2}},\ z_y=\frac{-y}{\sqrt{a^2-x^2-y^2}}$$

故
$$\mathrm{d}S=\sqrt{1+z_x^2+z_y^2}\mathrm{d}x\mathrm{d}y=\frac{a}{\sqrt{a^2-x^2-y^2}}\mathrm{d}x\mathrm{d}y$$

由曲面面积公式，得球面的面积

$$\begin{aligned} S &= 2\iint_{D}\frac{a}{\sqrt{a^2-x^2-y^2}}\mathrm{d}x\mathrm{d}y = 2\iint_{D}\frac{a}{\sqrt{a^2-r^2}}r\mathrm{d}r\mathrm{d}\theta \\ &= 2a\int_0^{2\pi}\mathrm{d}\theta\int_0^a\frac{r\mathrm{d}r}{\sqrt{a^2-r^2}} \\ &= 4\pi a\left(-\sqrt{a^2-r^2}\right)\Big|_0^a = 4\pi a^2 \end{aligned}$$

例 8.11 求球面 $x^2+y^2+z^2=4a^2$ 被圆柱面 $x^2+y^2=2ax(a>0)$ 所截得的(含圆柱面内的部分)曲面的面积，在第一卦限中如图 8－5 所示.

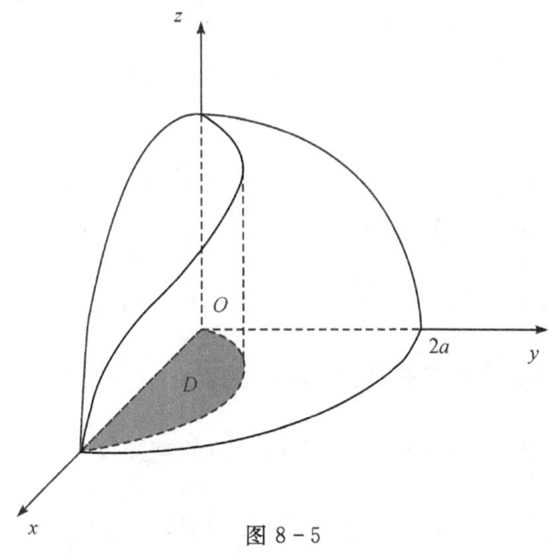

图 8－5

解 由对称性，所求曲面面积等于第一卦限面积的 4 倍. 由于曲面方程为 $z=\sqrt{4a^2-x^2-y^2}$，它在 xOy 面上的投影区域 D_{xy} 是圆：$x^2+y^2\leqslant 2ax$ $(y\geqslant 0)$，故曲面的面积

$$S=4\iint_D \sqrt{1+z_x^2+z_y^2}\,\mathrm{d}\sigma=4\iint_D \frac{2a}{\sqrt{4a^2-x^2-y^2}}\mathrm{d}x\mathrm{d}y$$

$$=8a\int_0^{\frac{\pi}{2}}\mathrm{d}\theta\int_0^{2a\cos\theta}\frac{r}{\sqrt{4a^2-r^2}}\mathrm{d}r$$

$$=-8a\int_0^{\frac{\pi}{2}}\sqrt{4a^2-r^2}\Big|_0^{2a\cos\theta}\mathrm{d}\theta$$

$$=-16a^2\int_0^{\frac{\pi}{2}}(\sin\theta-1)\mathrm{d}\theta=4(\pi-4)a^2$$

例 8.12 求以$(a,0,0)$，$(0,a,0)$ 和 $(0,0,a)$为顶点的等边三角形的质心坐标.

解 设三角形质心坐标为$(\bar{x},\bar{y},\bar{z})$，三角形所在平面方程为

$$x+y+z=a$$

方法一
$$\bar{x}=\frac{\iint_\Sigma x\,\mathrm{d}S}{\iint_\Sigma \mathrm{d}S}=\frac{\iint_D x\sqrt{1+z_x^2+z_y^2}\,\mathrm{d}x\mathrm{d}y}{\frac{1}{2}\sqrt{2}a\cdot\sqrt{2}a\cdot\frac{\sqrt{3}}{2}}$$

$$=\frac{\sqrt{3}\int_0^a\mathrm{d}x\int_0^{a-x}x\,\mathrm{d}y}{\frac{\sqrt{3}}{2}a^2}=\frac{\frac{\sqrt{3}}{6}a^2}{\frac{\sqrt{3}}{2}a}=\frac{1}{3}a$$

方法二 由于$\iint_\Sigma x\,\mathrm{d}S=\iint_\Sigma y\,\mathrm{d}S=\iint_\Sigma z\,\mathrm{d}S$，所以$\iint_\Sigma x\,\mathrm{d}S=\frac{1}{3}\iint_\Sigma(x+y+z)\mathrm{d}S=\frac{a}{3}\iint_\Sigma\mathrm{d}S$，故

$$\bar{x}=\frac{\iint_\Sigma x\,\mathrm{d}S}{\iint_\Sigma\mathrm{d}S}=\frac{\frac{1}{3}a\iint_\Sigma\mathrm{d}S}{\iint_\Sigma\mathrm{d}S}=\frac{1}{3}a$$

根据图形的对称有，有

$$\bar{x}=\bar{y}=\bar{z}=\frac{1}{3}a$$

习题 8

1. 计算下列曲线积分：

(1) $\int_C(x+y)\sqrt{x^2+y^2}\,\mathrm{d}s$，其中 C：$x^2+y^2=a^2(y\geqslant 0)$；

(2) $\int_C(x^3+y)\mathrm{d}s$，其中 C：$x=3t$，$y=t^3$，$0\leqslant t\leqslant 1$；

(3) $\int_C(\sin x+\cos y)\mathrm{d}s$，其中 C 是从点$(0,0)$到点$(\pi,2\pi)$的直线段；

(4) $\int_C x\mathrm{e}^y\mathrm{d}s$，其中 C 是从点$(-1,2)$到点$(1,1)$的直线段；

(5) $\int_C(2x+9z)\mathrm{d}s$，其中 C：$x=t$，$y=t^2$，$z=t^3$，$0\leqslant t\leqslant 1$；

(6) $\int_C (x^2+y^2+z^2)\mathrm{d}s$，其中 C：$x=4\cos t$，$y=4\sin t$，$z=3t$，$0\leqslant t\leqslant 2\pi$.

2. 计算 $I=\oint_C (x+y)\mathrm{d}s$，其中 C 是由直线 $y=2x$，$y=2$ 及 $x=0$ 所围平面区域的边界曲线.

3. 计算半径为 R、中心角为 2α 的均匀圆弧形构件 L 的质量和质心位置(设线密度为 μ).

4. 设一根螺旋线型电线可用方程 $x=a\cos t$，$y=a\sin t$，$z=bt$，$0\leqslant t\leqslant 3\pi$ 表示，其密度为常数，求它的质量和质心.

5. 当 $g(x, y, z)$分别为以下情况时，计算$\iint_\Sigma g(x, y, z)\mathrm{d}S$.

(1) $g(x, y, z)=x^2+y^2+z$，Σ：$z=x+y+1$，$0\leqslant x\leqslant 1$，$0\leqslant y\leqslant 1$；

(2) $g(x, y, z)=x$，Σ：$x+y+2z=4$，$0\leqslant x\leqslant 1$，$0\leqslant y\leqslant 1$；

(3) $g(x, y, z)=x+y$，Σ：$z=\sqrt{4-x^2}$，$0\leqslant x\leqslant\sqrt{3}$，$0\leqslant y\leqslant 1$；

(4) $g(x, y, z)=2y^2+z$，Σ：$z=x^2-y^2$，$0\leqslant x^2+y^2\leqslant 1$；

(5) $g(x, y, z)=y$，Σ：$z=4-y^2$，$0\leqslant x\leqslant 3$，$0\leqslant y\leqslant 2$；

(6) $g(x, y, z)=x+y$，Σ 是立方体 $0\leqslant x\leqslant 1$，$0\leqslant y\leqslant 1$，$0\leqslant z\leqslant 1$ 的表面.

6. 计算$\iint_\Sigma (xy+z)\mathrm{d}S$，其中 Σ 为平面 $2x-y+z=3$ 位于如图 8－6 所示的三角形 R 之上的一部分.

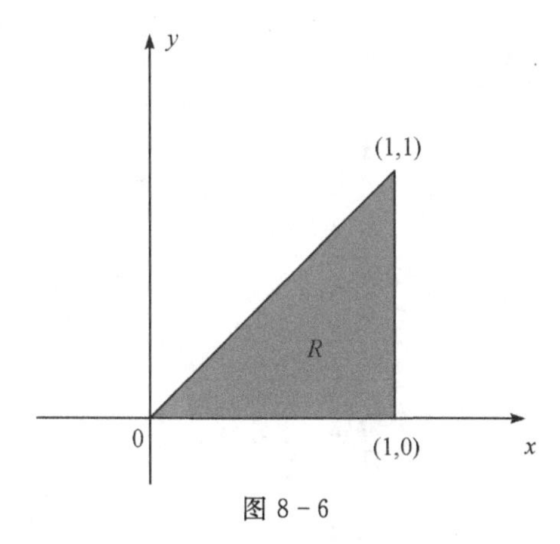

图 8－6

7. 计算$\iint_G xyz\,\mathrm{d}S$，其中 G 为平面 $z^2=x^2+y^2$ 介于 $z=1$ 和 $z=4$ 之间的部分.

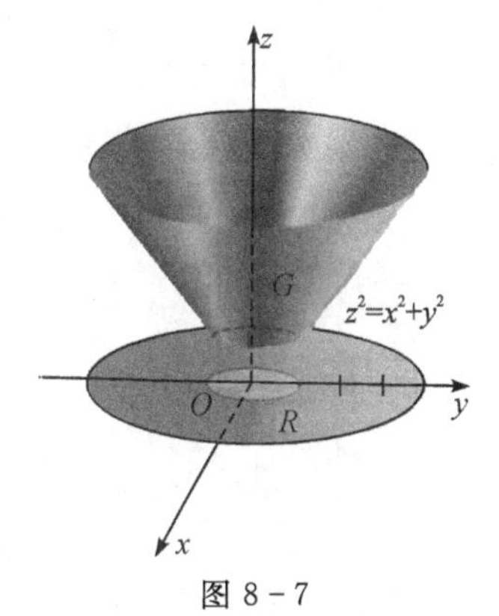

图 8－7

8. 求下列曲面的面积：

(1)求曲面 $z=\sqrt{4-y^2}$ 在第一卦限部分的面积，该部分位于 xOy 平面上的圆 $x^2+y^2=4$ 的上方；

(2)求旋转抛物面 $z=x^2+y^2$ 被平面 $z=4$ 所截部分的面积.

9. 求以 $(a, 0, 0)$，$(0, a, 0)$ 和 $(0, 0, a)$ 为顶点的三角形薄板的质心，该三角形薄板的密度为 $\mu(x, y, z)=kx^2$.

10. 求旋转抛物面 $z=1-(x^2+y^2)/2$ 位于平面区域 $0\leqslant x\leqslant 1$，$0\leqslant y\leqslant 1$ 之上部分的质量，其密度为 $\mu(x, y, z)=kxy$.

第 9 章　常微分方程

在科学研究和解决实际问题时，经常需要寻找描述这些问题的函数关系. 但我们需要寻找的函数往往不容易直接找到，却比较容易建立所找函数与其导数或微分之间的关系，我们把这样的关系式称为**微分方程**.

9.1　基本概念

9.1.1　引例

为了介绍微分方程的一些术语，首先考虑下面的例子.

例 9.1　设 xOy 平面上的曲线过点(1，2)且在任一点(x, y)处切线的斜率为 $2x$，求该曲线方程.

解　根据导数的几何意义，所求曲线应满足

$$\frac{\mathrm{d}y}{\mathrm{d}x} = 2x \quad 或 \quad \mathrm{d}y = 2x\mathrm{d}x \tag{9-1}$$

式(9－1)是一个包含未知函数导数的方程，称为微分方程。为了求得未知函数，对式(9－1)两边积分，得

$$y = \int 2x\mathrm{d}x = x^2 + C \tag{9-2}$$

其中，C 是任意常数. 由于曲线通过点(1，2)，故 $y=y(x)$满足

$$y\Big|_{x=1} = 2 \tag{9-3}$$

把式(9－3)代入式(9－2)，则 $C=1$. 因此，所求曲线的方程为

$$y = x^2 + 1 \tag{9-4}$$

9.1.2　几个术语

定义 9.1　**微分方程**　我们把含有未知函数导数的方程叫作**微分方程**.

定义 9.2　**阶**　微分方程中未知函数导数的最高阶数，称为微分方程的**阶**.

例 9.1 中，微分方程的阶数是 1.

定义 9.3　**解**　如果把函数 $y=y(x)$代入微分方程可使之成为恒等式，则 $y=y(x)$称为微分方程的**解**.

定义 9.4　**通解**　如果一个解中含有任意常数，且方程的全部解都可以通过改变任意常数导出，这样的解称为微分方程的**通解**.

例如，式(9－2)就是例 9.1 中微分方程的通解.

定义 9.5 特解 确定了任意常数的解称为微分方程的**特解**.

例如，式(9-4)是例 9.1 中微分方程的特解.

定义 9.6 初值问题 当一个应用问题转化为求解微分方程的问题时，通常需要一些条件来确定任意常数的值. 通常，一个 n 阶微分方程需要 n 个条件来确定 n 个任意常数的值. 对于一个一阶微分方程，就需要一个条件，即 $y(x_0)=y_0$，来确定任意常数的取值. 这个条件称为**初始条件**. 带有初始条件的一阶微分方程的求解问题称为一阶**初值问题**. 几何上，利用初始条件可以从积分曲线族中分离出过点(x_0, y_0)的那一条曲线.

在例 9.1 中，$\frac{dy}{dx}=2x$，$y\Big|_{x=1}=2$ 是一个初值问题，$y\Big|_{x=1}=2$ 是初始条件. 通解 $y=x^2+C$ 表示积分曲线族，特解 $y=x^2+1$ 是曲线族中过(1, 2)这一点的曲线，见图 9-1.

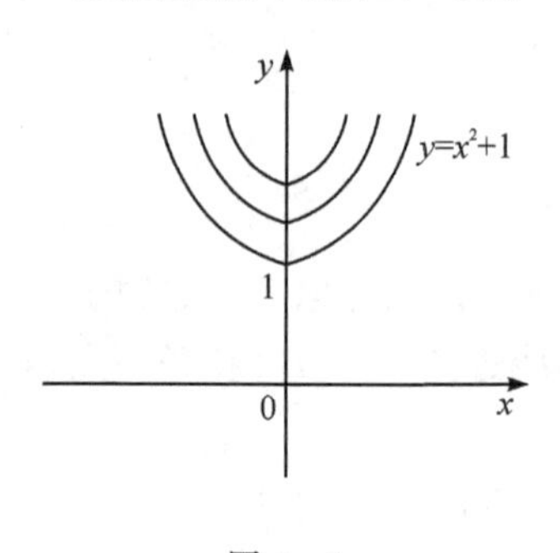

图 9-1

9.2 一阶微分方程

9.2.1 一阶可分离变量的微分方程

如果一阶微分方程可写成如下形式

$$\frac{dy}{dx}=h(x)g(y) \tag{9-5}$$

则称为**可分离变量的微分方程**. 之所以称为可分离变量的微分方程，是因为当 $g(y)\neq 0$ 时，该方程可以写成微分形式

$$\frac{1}{g(y)}dy=h(x)dx \tag{9-6}$$

上述形式中含 x 的表达式出现在等式的右端，含 y 的表达式出现在等式的左端，这个过程称为**变量分离**. 当 $g(y)=0$ 时，式(9-5)有一个解 $y=C$(C 是任意常数).

可分离变量微分方程的求解方法就是对式(9-6)两端进行积分，下面通过举例来说明这类方程的求解方法.

例 9.2 求微分方程$\frac{dy}{dx}=2xy$ 的通解.

解 当 $y\neq 0$ 时，变量分离得

$$\frac{dy}{y}=2x\,dx \tag{9-7}$$

对式(9－7)两端积分

$$\int \frac{\mathrm{d}y}{y} = \int 2x\mathrm{d}x$$

得

$$\ln|y| = x^2 + C_1$$

则

$$y = \pm \mathrm{e}^{x^2 + C_1} = \pm \mathrm{e}^{C_1}\mathrm{e}^{x^2} = C\mathrm{e}^{x^2}\ (C = \pm \mathrm{e}^{C_1} \neq 0)$$

显然，$y=0$ 也是所给微分方程的解. 因此方程的通解为

$$y = C\mathrm{e}^{x^2}\ (C\text{ 是任意常数})$$

例 9.3　求解初值问题

$$\frac{\mathrm{d}y}{\mathrm{d}x} = -\frac{xy^2}{1+x^2},\ y\Big|_{x=0} = 1$$

解　$y\neq 0$ 时，可以把方程写为

$$\frac{1}{y^2}\frac{\mathrm{d}y}{\mathrm{d}x} = -\frac{x}{1+x^2} \tag{9-8}$$

对式(9－8)两端积分

$$-\int \frac{1}{y^2}\mathrm{d}y = \int \frac{x}{1+x^2}\mathrm{d}x$$

得

$$\frac{1}{y} = \frac{1}{2}\ln(1+x^2) + C$$

根据初始条件，把 $x=0$，$y=1$ 代入上式，得 $C=1$. 因此，初值问题的解为

$$\frac{1}{y} = \frac{1}{2}\ln(1+x^2) + 1$$

例 9.4　求平面上过点(0，3)且在点$(x，y)$处切线斜率为 $2x/y^2$ 的曲线方程.

解　因为曲线在点$(x，y)$切线的斜率为$\frac{\mathrm{d}y}{\mathrm{d}x}$，即

$$\frac{\mathrm{d}y}{\mathrm{d}x} = \frac{2x}{y^2} \tag{9-9}$$

并且曲线过点(0，3)，即初始条件为

$$y(0) = 3 \tag{9-10}$$

方程(9－9)是可分离变量的，而且可写成

$$y^2\mathrm{d}y = 2x\mathrm{d}x$$

故

$$\int y^2\mathrm{d}y = \int 2x\mathrm{d}x \quad \text{或} \quad \frac{1}{3}y^3 = x^2 + C$$

根据初始条件式(9－10)有，当 $x=0$ 时，$y=3$，代入上式得 $C=9$，从而所求曲线方程为

$$\frac{1}{3}y^3 = x^2 + 9 \quad \text{或} \quad y = (3x^2+27)^{\frac{1}{3}}$$

9.2.2　一阶线性微分方程

若一阶微分方程

$$\frac{\mathrm{d}y}{\mathrm{d}x} + P(x)y = Q(x) \tag{9-11}$$

其中 $Q(x)\equiv 0$，即

$$\frac{\mathrm{d}y}{\mathrm{d}x}+P(x)y=0 \tag{9-12}$$

则该方程称为**一阶线性齐次微分方程**；如果 $Q(x)\neq 0$，则式(9－11)称为**一阶线性非齐次微分方程**.

例如，$\frac{\mathrm{d}y}{\mathrm{d}x}+x^2y=\mathrm{e}^x$，$\frac{\mathrm{d}y}{\mathrm{d}x}+5y=2$ 和 $\frac{\mathrm{d}y}{\mathrm{d}x}+(\sin x)y+x^3=0$ 都是一阶线性非齐次微分方程.

9.2.3 利用常数变异法求解一阶线性非齐次微分方程

容易看出式(9－12)是可以进行变量分离的

$$\frac{\mathrm{d}y}{y}=-P(x)\mathrm{d}x \tag{9-13}$$

对式(9－13) 两端积分，得

$$\ln y=-\int P(x)\mathrm{d}x+\ln C$$

则有

$$y=C\mathrm{e}^{-\int P(x)\mathrm{d}x}(C\text{ 为任意常数})$$

这是方程(9－13)的通解，而且有理由推测式(9－11)有形如下式的解

$$y=C(x)\mathrm{e}^{-\int P(x)\mathrm{d}x} \tag{9-14}$$

为了进一步求解，需要确定 $C(x)$. 把式(9－14)代入式(9－11)，我们有

$$C'(x)\mathrm{e}^{-\int P(x)\mathrm{d}x}-C(x)P(x)\mathrm{e}^{-\int P(x)\mathrm{d}x}+P(x)C(x)\mathrm{e}^{-\int P(x)\mathrm{d}x}=Q(x)$$

或

$$C'(x)=Q(x)\mathrm{e}^{\int P(x)\mathrm{d}x}$$

因此

$$C(x)=\int Q(x)\mathrm{e}^{\int P(x)\mathrm{d}x}\mathrm{d}x+C$$

把上面的 $C(x)$表达式代入式(9－14)，可以得到

$$y=\mathrm{e}^{-\int P(x)\mathrm{d}x}\left(\int Q(x)\mathrm{e}^{\int P(x)\mathrm{d}x}\mathrm{d}x+C\right) \tag{9-15}$$

或

$$y=C\mathrm{e}^{-\int P(x)\mathrm{d}x}+\mathrm{e}^{-\int P(x)\mathrm{d}x}\int Q(x)\mathrm{e}^{\int P(x)\mathrm{d}x}\mathrm{d}x \tag{9-16}$$

由于解(9－15)含有任意常数，所以它是方程(9－11)的通解. 利用式(9－16)可以得到一个结论，**非齐次线性微分方程的通解等于其相应的齐次微分方程的通解加上非齐次微分方程的一个特解**.

例 9.5 求解微分方程

$$\frac{\mathrm{d}y}{\mathrm{d}x}+y\cos x=\mathrm{e}^{-\sin x}$$

解 把 $P(x)=\cos x$，$Q(x)=\mathrm{e}^{-\sin x}$代入式(9－15)，从而得到所求微分方程的通解

$$y=e^{-\int\cos x\mathrm{d}x}\left(\int e^{-\sin x}e^{\int\cos x\mathrm{d}x}\mathrm{d}x+C\right)$$
$$=e^{-\sin x}\left(\int e^{-\sin x}e^{\sin x}\mathrm{d}x+C\right)$$
$$=e^{-\sin x}(x+C)=xe^{-\sin x}+Ce^{-\sin x}$$

例 9.6　求解初值问题

$$x\mathrm{d}y+y\mathrm{d}x=xe^{x},\ y\Big|_{x=1}=1$$

解　把方程变形为

$$y'+\frac{1}{x}y=e^{x}$$

利用公式(9-15)得方程的通解为

$$y=e^{-\int\frac{1}{x}\mathrm{d}x}\left(\int e^{x}e^{\int\frac{1}{x}\mathrm{d}x}\mathrm{d}x+C\right)$$
$$=\frac{1}{x}\left(\int xe^{x}\mathrm{d}x+C\right)$$
$$=\frac{1}{x}(xe^{x}-e^{x}+C)$$

由初始条件知，$x=1$ 时，$y=1$，代入最后一个方程，得 $C=1$，所求方程的解为

$$y=\frac{1}{x}(xe^{x}-e^{x}+1)$$

例 9.7　求微分方程的通解

$$\frac{\mathrm{d}y}{\mathrm{d}x}=\frac{1}{x+y} \tag{9-17}$$

解　所给方程不是一个关于 x 的线性微分方程. 如果把方程(9-17)化成如下形式

$$\frac{\mathrm{d}x}{\mathrm{d}y}-x=y \tag{9-18}$$

则式(9-18)是关于 y 的线性非齐次微分方程.

$$x=e^{\int\mathrm{d}y}\left(\int ye^{-\int\mathrm{d}y}\mathrm{d}y+C\right)$$
$$=e^{y}\left(\int ye^{-y}\mathrm{d}y+C\right)$$
$$=e^{y}(-ye^{-y}-e^{-y}+C)=Ce^{y}-y-1$$

9.3　二阶可求解的微分方程

本节介绍三种类型的二阶微分方程.

9.3.1　$y''=f(x)$

这种类型的微分方程比较简单，只要对 $f(x)$ 连续积分两次即可求解.

例 9.8　求下列方程的通解

$$y'''=e^{-x}-\sin x$$

解 对方程两端连续积分三次

$$y''=-e^{-x}+\cos x+C_1$$
$$y'=e^{-x}+\sin x+C_1x+C_2$$

于是得方程的通解为

$$y=-e^{-x}-\cos x+\frac{C_1}{2}x^2+C_2x+C_3$$

9.3.2 $y''=f(x, y')$

这类方程的特点是 f 不含有未知函数 y.

令 $y'=p(x)=p$，则 $y''=p'=\frac{dp}{dx}$. 将其代入 $y''=f(x, y')$，得

$$p'=f(x, p) \tag{9-19}$$

如果求得方程(9-19)的通解，且记为

$$p=\frac{dy}{dx}=\varphi(x, C_1) \tag{9-20}$$

对式(9-20)再进行积分，可得所求方程的通解为

$$y=\int\varphi(x, C_1)dx+C_2$$

例 9.9 求下列方程的通解

$$y''=y'+x$$

解 令 $y'=p$，则 $y''=\frac{dp}{dx}$，代入所给方程，有

$$p'-p=x$$

根据式(9-15)，得

$$p=e^{\int dx}\left[\int xe^{-\int dx}dx+C_1\right]=e^x\left(\int xe^{-x}dx+C_1\right)$$
$$=e^x(-xe^{-x}-e^{-x}+C_1)=C_1e^x-x-1$$

即

$$y'=C_1e^x-x-1$$

再次积分，得所求方程的通解为

$$y=C_1e^x-\frac{1}{2}x^2-x+C_2$$

例 9.10 求解初值问题

$$y''=\frac{2x}{1+x^2}y', \ y\Big|_{x=0}=1, \ y'\Big|_{x=0}=3$$

解 令 $y'=p$，则 $y''=\frac{dp}{dx}$. 代入所给方程，有

$$\frac{dp}{dx}=\frac{2x}{1+x^2}p$$

这是一个可分离变量的微分方程，变量分离并积分，得

$$\int\frac{dp}{p}=\int\frac{2x}{1+x^2}dx$$
$$\ln p=\ln(1+x^2)+\ln C_1$$

即

$$y' = p = C_1(1 + x^2)$$

根据初始条件，当 $x=0$ 时，$y'=3$，代入上式，可求得 $C_1=3$，于是有

$$\frac{dy}{dx} = 3(1 + x^2)$$

再次积分，得

$$y = x^3 + 3x + C_2$$

利用另一个初始条件 $y\Big|_{x=0}=1$，可得 $C_2=1$，则所求方程的特解为

$$y = x^3 + 3x + 1$$

9.3.3　$y''=f(y, y')$

这个方程的特点是 f 不包含自变量 x.

令 $y'=p(x)=p$，利用链式求解法则，有

$$y'' = \frac{dy'}{dx} = \frac{dp}{dx} = \frac{dp}{dy} \cdot \frac{dy}{dx} = p\frac{dp}{dy} \tag{9-21}$$

将式(9－21)代入 $y''=f(y, y')$，方程可转化为一个一阶微分方程

$$p\frac{dp}{dy} = f(y, p) \tag{9-22}$$

如果方程(9－22)的通解为

$$p = \frac{dy}{dx} = \varphi(y, C_1)$$

变量分离后，两端积分，可得方程 $y''=f(y, y')$ 的通解

$$\int \frac{dy}{\varphi(y, C_1)} = x + C_2$$

例 9.11　求下列方程的通解

$$yy'' - y'^2 = 0$$

解　令 $y'=p$，则 $y''=p\dfrac{dp}{dy}$. 将其代入所给方程，得

$$yp\frac{dp}{dy} - p^2 = 0$$

当 $y\neq0$ 且 $p\neq0$ 时，有

$$\frac{dp}{p} = \frac{dy}{y} \tag{9-23}$$

对式(9－23)两端积分，得

$$\ln p = \ln y + \ln C_1$$

即

$$\frac{dy}{dx} = p = C_1 y$$

变量分离，再积分，得所求方程的通解为

$$\ln|y| = C_1 x + C$$

或

$$y = C_2 e^{C_1 x}\ (C_2 = \pm e^C)$$

9.4 一阶微分方程的应用

例 9.4 中，我们已经看到了微分方程在几何上的应用，下面的例子给出了微分方程在其他问题中的应用.

9.4.1 混合问题

例 9.11 在 $t=0$ 时刻容器内有盐溶液 10 L，其含盐量为 1 kg. 假设现以 3 L/min 的速率注入清水，快速搅拌后，溶液以同样的速率流出，求 1 h 后容器内的含盐量.

解 令 $y(t)$表示 t 分钟后容器内的含盐量，由题设条件知 $y(0)=1$，需要求的是 $y(60)$. 首先，需要建立关于 $y(t)$的导数的微分方程，即 t 时刻容器内含盐量的变化率为

$$\frac{\mathrm{d}y}{\mathrm{d}t}=\text{盐流入的速率}-\text{盐流出的速率} \tag{9-24}$$

因为注入的是清水，所以盐流入的速率为 0. 又因为溶液流入和流出的速率相同，所以体积不变，始终为 10 L.

因此，t 分钟后容器内含盐量为 $y(t)$时，盐流出的速率为

$$\text{盐流出的速率}=\frac{y(t)}{10}\times 3\ (\mathrm{L/min})$$

从而式(9-24)可以写成

$$\frac{\mathrm{d}y}{\mathrm{d}t}=-\frac{y(t)}{10}\times 3$$

这是一个一阶可分离变量的微分方程，根据初始条件 $y(0)=1$，可以通过求解初值问题得到 $y(t)$. 即由

$$\frac{\mathrm{d}y}{\mathrm{d}t}=-\frac{y(t)}{10}\times 3,\ y(0)=1$$

变量分离，两端积分，得

$$y(t)=\mathrm{e}^{-0.3t}(\mathrm{kg})$$

当 $t=60$ 时，容器内的含盐量为

$$y(60)=\mathrm{e}^{-18}(\mathrm{kg})$$

9.4.2 带有空气阻力的自由落体运动模型

例 9.12 假设质量为 m 的降落伞从跳伞塔下落后，所受空气阻力与速度成正比，并设降落伞离开跳伞塔时速度为零. 求降落伞下落速度与时间 t 的函数关系.

解 在自由落体的情况下，受空气阻力，作用在降落伞上的合力为

$$F=F_{\mathrm{G}}-F_{\mathrm{R}}$$

其中：F_{G} 表示重力；F_{R} 表示空气阻力.

根据**牛顿第二定律**，知

$$ma=F_{\mathrm{G}}-F_{\mathrm{R}} \quad \text{或} \quad m\frac{\mathrm{d}v}{\mathrm{d}t}=mg-kv \tag{9-25}$$

由于当降落伞离开跳伞塔时速度为 0，所以初始条件为 $v(0)=0$.

式(9-25) 是一个可分离变量的一阶方程，重新改写为

$$\frac{\mathrm{d}v}{\mathrm{d}t}+\frac{k}{m}v=g$$

利用通解公式(9－15)，得方程的通解

$$v(t)=\mathrm{e}^{-\int\frac{k}{m}\mathrm{d}t}\left(\int g\mathrm{e}^{\int\frac{k}{m}\mathrm{d}t}\mathrm{d}t+C\right)=\mathrm{e}^{-\frac{k}{m}t}\left(\int g\mathrm{e}^{\frac{k}{m}t}\mathrm{d}t+C\right)$$

$$=\mathrm{e}^{-\frac{k}{m}t}\left(\frac{mg}{k}\mathrm{e}^{\frac{k}{m}t}+C\right)=\frac{mg}{k}+C\mathrm{e}^{-\frac{k}{m}t}$$

根据初始条件，当 $t=0$ 时，$v=0$，代入上式，得 $C=-\frac{mg}{k}$，于是所求特解为

$$v(t)=\frac{mg}{k}(1-\mathrm{e}^{-\frac{k}{m}t}) \tag{9-26}$$

式(9－26)表明当时间充分大后，速度趋向于$\frac{mg}{k}$.

换句话说，跳伞以加速运动开始，但逐渐接近匀速运动. 所以理论上来说，跳伞者从高空跳到地面是安全的.

9.5　二阶常系数齐次线性微分方程

9.5.1　二阶线性微分方程的一般形式

具有下列形式的方程，称为**二阶线性微分方程**

$$\frac{\mathrm{d}^2y}{\mathrm{d}x^2}+p(x)\frac{\mathrm{d}y}{\mathrm{d}x}+q(x)y=Q(x)$$

或者可以写成这种形式

$$y''+p(x)y'+q(x)y=Q(x) \tag{9-27}$$

如果 $Q(x)=0$，则式(9－27)简化为

$$y''+p(x)y'+q(x)y=0 \tag{9-28}$$

这个方程称为**二阶齐次线性微分方程**.

9.5.2　函数的线性相关和线性无关

为了讨论方程(9－27) 的解，首先介绍两个术语. 如果定义在区间 I 上的两个函数 f 和 g 成比例，则称它们在区间 I 上是**线性相关**的，否则，称为**线性无关**的. 由此知，

$$f(x)=\mathrm{e}^x \text{和} g(x)=2\mathrm{e}^x$$

是线性相关的，但

$$f(x)=x \text{ 和 } g(x)=x^2$$

是线性无关的. 下面的定理是研究二阶齐次线性微分方程的解的重要定理.

定理 9.1　如果 $y_1(x)$和 $y_2(x)$是方程(9－28)在区间 I 上的两个线性无关的解，则该方程的通解为

$$y(x)=C_1y_1(x)+C_2y_2(x) \tag{9-29}$$

也就是说，方程(9－28)的每一个解都可以通过在式(9－29)中选取适当的常数 C_1和 C_2

得到；反过来，当 C_1 和 C_2 任意取值时，式(9－29)给出了方程的全部的解.

9.5.3 二阶常系数齐次线性微分方程的求解方法

我们主要研究形如下面形式的二阶常系数齐次线性微分方程的求解方法

$$y''+py'+qy=0 \tag{9-30}$$

其中 p 和 q 为常数. 根据定理 9.1，为了确定方程(9－30)的通解，我们仅仅需要找到两个线性无关的解 $y_1(x)$ 和 $y_2(x)$. 于是方程的通解可以表示为

$$y(x)=C_1y_1(x)+C_2y_2(x)\ (C_1 \text{ 和 } C_2 \text{ 为任意常数})$$

由于指数函数 e^{rx} 的一阶和二阶导数仍为指数函数，所以我们推断方程(9－30)有形如 $y=e^{rx}$ 的解，将其代入方程(9－30)，可通过选择合适的 r，找到方程的解. 于是我们将

$$y=e^{rx},\ y'=re^{rx},\ y''=r^2e^{rx}$$

代入方程(9－30)，得

$$(r^2+pr+q)e^{rx}=0$$

因为 $e^{rx}\neq0$，故有

$$r^2+pr+q=0 \tag{9-31}$$

式(9－31)称为微分方程(9－30)的**特征方程**. 特征方程(9－31)的根 r_1 和 r_2 可通过因式分解或求根公式得到，即

$$r_{1,2}=\frac{-p\pm\sqrt{p^2-4q}}{2} \tag{9-32}$$

当 p^2-4q 取值为正、零、负，特征方程的根可分别取两个不同实根、一对相等实根和一对共轭复根. 下面分别对这三种情形来讨论.

1. 两个不同实根

如果 r_1 和 r_2 是两个不同实根，则方程(9－30)有两个解

$$y_1=e^{r_1x} \text{ 和 } y_2=e^{r_2x}$$

并且这两个解不成比例，即线性无关，于是方程(9－30)的通解为

$$y(x)=C_1e^{r_1x}+C_2e^{r_2x}$$

2. 两个相等实根

如果 r_1 和 r_2 为两个相等实根，即 $r_1=r_2=r$，则只能得到方程(9－30)的一个解，

$$y_1(x)=e^{rx}$$

我们可以验证

$$y_2(x)=xe^{rx}$$

是方程(9－30)的另一个解，且与 $y_1(x)$线性无关，所以方程(9－30)的通解为

$$y=C_1e^{rx}+C_2xe^{rx}=(C_1+C_2x)e^{rx}$$

3. 一对共轭复根

如果特征方程的根为 $r_1=\alpha+i\beta$ 和 $r_2=\alpha-i\beta$(α，β 为实数，且 $\beta\neq0$)，则我们可以验证 $y_1=e^{\alpha x}\cos\beta x$和 $y_2=e^{\alpha x}\sin\beta x$ 是方程(9－30)的线性无关的解. 因此方程(9－30)的通解为

$$y=e^{\alpha x}(C_1\cos\beta x+C_2\sin\beta x)$$

综上所述，方程(9－30)的求解可分为以下三步。

第一步　写出(9－30)的特征方程 $r^2+pr+q=0$；

第二步　求特征方程的根 r_1，r_2；

第三步　通过表 9－1 确定方程(9－30)的通解.

表 9－1

特征方程 $r^2+pr+q=0$ 的根	微分方程 $y''+py'+qy=0$ 的通解
两个不相等实根 $r_1\neq r_2$	$y(x)=C_1e^{r_1x}+C_2e^{r_2x}$
两个相等实根 $r_1=r_2=r$	$y=(C_1+C_2x)e^{rx}$
一对共轭复根 $r_{1,2}=\alpha\pm i\beta$	$y=e^{\alpha x}(C_1\cos\beta x+C_2\sin\beta x)$

例 9.13　求方程 $y''-2y'-3y=0$ 的通解.

解　特征方程为

$$r^2-2r-3=0$$

所以特征根为 $r_1=-1$，$r_2=3$，根据表 9－1，方程的通解为

$$y=C_1e^{-x}+C_2e^{3x}$$

例 9.14　求微分方程 $y''-8y'+16y=0$ 的通解.

解　特征方程为

$$r^2-8r+16=0$$

所以特征根为 $r_1=r_2=4$，根据表 9－1，微分方程的通解为

$$y=C_1e^{4x}+C_2xe^{4x}$$

例 9.15　求解初值问题

$$\frac{d^2s}{dt^2}+2\frac{ds}{dt}+s=0,\ s|_{t=0}=4,\ s'|_{t=0}=-2$$

解　特征方程为

$$r^2+2r+1=0$$

所以特征根为 $r_1=r_2=-1$，根据表 9－1，微分方程的通解为

$$s=(C_1+C_2t)e^{-t} \tag{9-33}$$

解的一阶导数为

$$s'=(C_2-C_1-C_2t)e^{-t} \tag{9-34}$$

把 $t=0$ 代入式(9－33)和式(9－34)，并利用初始条件$s|_{t=0}=4$ 和$s'|_{t=0}=-2$，得

$$C_1=4\ 和\ C_2=2$$

从而，所求方程的特解为

$$s=(4+2t)e^{-t}$$

例 9.16　求方程 $y''+6y'+13y=0$ 的通解.

解　特征方程为

$$r^2+6r+13=0$$

故特征根为 $r_{1,2}=-3\pm 2i$，这是一对共轭复根，根据表 9－1，方程的通解为

$$y=e^{-3x}(C_1\cos 2x+C_2\sin 2x)$$

习题 9

1. 求下列微分方程的通解.

(1) $xy'-y\ln y=0$；　(2) $3x^2+5x-5y'=0$；

(3) $\sqrt{1-x^2}y'=\sqrt{1-y^2}$；　(4) $y\mathrm{d}x+(x^2-4x)\mathrm{d}y=0$.

2. 求下列微分方程满足所给初始条件的特解.

(1) $y'=\mathrm{e}^{2x-y}$，$y|_{x=0}=0$；

(2) $xy'=y\ln y$，$y|_{x=\frac{\pi}{2}}=\mathrm{e}$；

(3) $\cos y\mathrm{d}x+(1+\mathrm{e}^{-x})\sin y\mathrm{d}y=0$，$y|_{x=0}=\dfrac{\pi}{4}$.

3. 求下列线性微分方程的通解.

(1) $\dfrac{\mathrm{d}y}{\mathrm{d}x}+y=\mathrm{e}^{-x}$；　(2) $xy'+y=x^2+3x+2$；

(3) $y'+y\tan x=\sin 2x$；　(4) $(y^2-6x)\dfrac{\mathrm{d}y}{\mathrm{d}x}+2y=0$.

4. 求一曲线的方程. 该曲线通过原点，并且在点(x, y)处的切线斜率等于$2x+y$.

5. 质量为 1 g 的质点受外力作用做直线运动，该外力和时间成正比、和质点运动的速度成反比. 在$t=10$ s 时，速度等于 50 cm/s，外力为 4 g · cm/s^2，问从运动开始经过了一分钟后的速度是多少？

6. 设有连接点$O(0, 0)$和$A(1, 1)$的一段向上凸的弧线$\overset{\frown}{OA}$，对于$\overset{\frown}{OA}$上任一点$P(x, y)$，弧线$\overset{\frown}{OP}$与直线段$\overline{OP}$所围图形的面积为x^2，求弧线$\overset{\frown}{OA}$的方程.

7. 求下列各微分方程的通解.

(1) $\dfrac{\mathrm{d}^2y}{\mathrm{d}x^2}-\dfrac{9}{4}x=0$；　(2) $y'''=x\mathrm{e}^x$；

(3) $(1+x^2)y''=2xy'$；　(4) $y''-\dfrac{2}{1-y}y'^2=0$.

8. 求下列各微分方程满足所给初始条件的特解.

(1) $y'''=\mathrm{e}^x$，$y|_{x=1}=y'|_{x=1}=y''|_{x=1}=0$；

(2) $y''=3\sqrt{y}$，$y|_{x=0}=1$，$y'|_{x=0}=2$；

(3) $y''-\mathrm{e}^{2y}=0$，$y|_{x=0}=y'|_{x=0}=0$；

(4) $y^3y''+1=0$，$y|_{x=1}=1$，$y'|_{x=1}=0$.

9. 下列函数组在其定义区间内哪些是线性无关的？

(1) $\cos x$，x^2；　(2) x^2，$5x^2$；

(3) $2x$，x^3；　(4) e^{2x}，$3\mathrm{e}^{2x}$.

10. 验证$y_1=\mathrm{e}^{-2x}$及$y_2=\mathrm{e}^{-6x}$都是方程$y''+8y'+12y=0$的解，并写出该方程的通解.

11. 验证$y_1=\sin x$及$y_2=\cos x$都是方程$y''+y=0$的解，并写出该方程的通解.

12. 求下列微分方程的通解.

(1) $y''-3y'-10y=0$；　(2) $y''-4y'=0$；

(3) $y''+2y=0$；　(4) $y''+8y'+16y=0$；

(5) $\dfrac{\mathrm{d}^2x}{\mathrm{d}t^2}-6\dfrac{\mathrm{d}x}{\mathrm{d}t}+9x=0$；　(6) $y''+2y'+2y=0$.

13. 求下列微分方程满足所给初始条件的特解.

(1) $y''-6y'+8y=0$，$y|_{x=0}=1$，$y'|_{x=0}=6$；

(2) $4y''+4y'+y=0$，$y|_{x=0}=2$，$y'|_{x=0}=0$.

14. 设一车间的容积为 10800 m^3，开始时空气中的 CO_2 的体积分数为 0.12%，为降低 CO_2 的含量，用一台风量为 1500 m^3/min 的鼓风机通入 CO_2 体积分数为 0.04%的新鲜空气. 假定通入的新鲜空气与车间原有空气能很快混合均匀，且以相同风速排出，问鼓风机开动 10 min后，车间中 CO_2 的体积分数为多少？

参考答案

第1章

1. (1)0；(2) $\frac{1}{2}$；(3) $\frac{1}{2}$；(4) $\frac{2}{3}a^{-\frac{1}{3}}$；(5) $-\frac{1}{3}$；(6) $\frac{1}{4\sqrt{3a}}$；(7) $\sqrt{2}$；(8) $\frac{1}{2}$；(9) 2.

2. (1) $\frac{3}{4}$；(2) $\frac{5}{3}$；(3) 0；(4) $\frac{a}{2}$.

3. (1) $\frac{m}{n}$；(2) $\frac{a}{b}$；(3) $\frac{a}{c}$；(4) 2；(5) $a-b$；(6) $\frac{1}{2}$；(7) 2；(8) $\cos y$；(9) $-\sqrt{2}$；

   (10) $\frac{\sec c}{2\sqrt{c}}$.

4. (1) 6；(2) 1；(3) $a-b$；(4) $\ln ab$；(5) 1；(6) -1.

5. (1)连续；(2) 间断；(3) 间断；(4) 间断.

6. (1)连续；(2) 间断.

7. (2) 不连续，$f(x)=\begin{cases}2x-3, & x<2\\ 1, & x=2.\\ 3x-5, & x>2\end{cases}$

8. (1) 1；(2) 12.

第2章

1. (1) $2x$；(2) $-\frac{1}{x^2}$；(3) $\frac{1}{2\sqrt{x}}$；(4) $2x^{\frac{3}{4}}+3x^{\frac{1}{2}}+x^{\frac{1}{4}}+x^{-\frac{1}{4}}$；(5) $4x^3(27x^5+25x-9)$；

   (6) $\frac{x^2+2x-2}{(x+1)^2}$.

2. (1) $4x(8x^2-3)$；(2) $\frac{4}{3}(4x+5)^{-\frac{2}{3}}$；(3) $8(3x^2+2x-1)^3(3x+1)$；

   (4) $\frac{-5}{2\sqrt{8-5x}}$；(5) $\frac{2ax+b}{2(ax^2+bx+c)}$；(6) $\sqrt{\frac{x^2+a^2}{x^2-a^2}}$.

3. (1) 14；(2) $36x^2-2$；(3) $\frac{18}{x^4}$；(4) $\frac{8}{(2x+1)^3}$.

4. (1) $-\frac{x}{y}$；(2) $\frac{a^2}{b^2}\frac{x}{y}$；(3) $\frac{2x(1+2y)}{3y^2-2x^2}$；(4) $\frac{y}{x}$.

5. (1) $-2\sin 4x$；(2) $\frac{\cos 2x}{\sqrt{\sin 2x}}$；(3) $10x\sec^2(5x^2+6)$；(4) $-\frac{2}{x^2}\sec^2\frac{1}{x}\tan\frac{1}{x}$；

(5) $-5\sec^2(\cos 5x)\sin 5x$；(6) $12\csc^3(\cot 4x)\cot(\cot 4x)\csc^2 4x$；

(7) $\frac{1}{\sqrt{x}}\sec^2(\tan\sqrt{x})\tan(\tan\sqrt{x})\sec^2\sqrt{x}$；(8) $-6\sin(2\cos 6x)\cdot\sin 6x$.

6. (1) $(2x+3)\sin 5x+5(x^2+3x)\cos 5x$；

(2) $(1+2\cos 2x)\sec 3x^2+6x(x+\sin 2x)\cdot\sec 3x^2\cdot\tan 3x^2$；

(3) $\frac{1}{2x}\left(\cos\sqrt{x}-\frac{1}{\sqrt{x}}\sin\sqrt{x}\right)$；(4) $\frac{n(ax-b)\sec nx\tan x-a\sec nx}{ax-b}$；

(5) $(m+n)\sin 2(m+n)x-(m-n)\sin^2(m-n)x$；(6) $4\cos 8x-\cos 2x$.

7. (1) $\sec x\tan x$；(2) $2n\sec^2 nx$；(3) $\tan\frac{1}{2}x\sec^2\frac{1}{2}x$；(4) $\sec x\tan x-\sec^2 x$；

(5) $\sec^2\left(\frac{\pi}{4}+x\right)$；(6) $\cot\frac{1}{2}x\cos^2\frac{1}{2}x$；(7) $2\sec x\ (\sec x+\tan x)^2$；

(8) $\sec^2\left(\frac{\pi}{4}+x\right)$.

8. (1) $\frac{3}{\sqrt{1-(3x-4)^2}}$；(2) $\frac{-2\sqrt{3}}{\sqrt{4-3x^2}}$；(3) $\frac{2}{1+x^2}$；(4) $\frac{2x}{1+(1-x^2)^2}$；(5) 1；(6) $\frac{1}{\sqrt{1-x^2}}$.

9. (1) $\frac{1+\sin(x-y)}{\sin(x-y)-1}$；(2) $\frac{2x-y\cos xy}{x\cos xy-2y}$；(3) $\frac{2xy^2-a\sec^2(ax+by)}{b\sec^2(ax+by)-2x^2y}$；

(4) $\frac{y-2x\sec^2(x^2+y^2)}{2y\sec^2(x^2+y^2)-x}$.

10. (1) $-\frac{b}{a}$；(2) $-\frac{b}{a}\tan 2t$；(3) $\tan\theta$；(4) $\tan t$；(5) $-\frac{1}{t}$.

11. (1) $\cot x$；(2) $\frac{1+\sec^2 x}{x+\tan x}$；(3) $\frac{5e^{5x}}{1+e^{5x}}$；(4) $\frac{1}{x\ln x}$；(5) $\tan x$；(6) $\frac{\sin 2x}{1+\sin^2 x}$.

12. (1) $-\frac{e^{\sqrt{\cos x}}\sin x}{2\sqrt{\cos x}}$；(2) $\frac{e^{1+\ln x}}{x}$；(3) $\frac{1}{x}e^{\sin(\ln x)}\cos(\ln x)$；(4) $\frac{1}{x}\sec^2(\ln x)$；

(5) $-\tan x\cdot\sin(\ln\sec x)$；(6) $\frac{\sec(\ln\tan x)\tan(\ln\tan x)\sec^2 x}{\tan x}$.

13. (1) $\frac{y^2}{x(1-y\ln x)}$；(2) $-\frac{e^{\sin x}\cos x}{e^{\sin y}\cos y}$；(3) $-\frac{y(y+x\ln y)}{x(y\ln x+x)}$；(4) $\frac{y(x\cos x\ln x-\sin y)}{x(y\cos y\ln y-\sin x)}$.

第3章

1. (1) $(1,+\infty)$单调增加，$(-\infty,1)$单调减少；

(2) $\left(\frac{1}{4},+\infty\right)$单调增加，$\left(-\infty,\frac{1}{4}\right)$单调减少；

(3) $(-\infty,-1)\cup(2,+\infty)$单调减少，$(-1,2)$单调增加；

(4) $[-3,-2)\cup(2,5]$单调增加，$(-2,2)$单调减少.

2. (1) 最大值$=20$，最小值$=0$；

(2) 最大值$=42$，最小值$=33$.

3. (1) 在$x=1$处取得极小值$=0$. 没有拐点；

(2) 在$x=-1$处取得极大值$=16$，在$x=4$处取得极小值$=-109$；

(3) 在 $x=2$ 处取得极小值，极小值为 3；在 $x=\frac{1}{2}$ 处取得极大值，极大值为 $\frac{39}{4}$，拐点为 $(\frac{5}{4},\frac{3}{4})$；

(4) 在 $x=-10$ 处取得极小值，极小值为 -25；在 $x=10$ 处取得极大值，极大值为 15，没有拐点.

5. (1) $x>1$ 和 $x<0$ 时曲线下凹，$0<x<1$ 时曲线上凸；

(2) $x>0$ 时曲线凹，$x<0$ 时曲线凸.

6. $1296\ \mathrm{m}^2$.

8. 当半径为 $\left(\frac{26}{\pi}\right)^{1/3}$，高为 $2\left(\frac{26}{\pi}\right)^{1/3}$ 时表面积最小.

9. 5，5.

10. 29 m/s，$4\ \mathrm{m/s^2}$.

11. (1) $S'(t)=\frac{9000t}{(t^2+50)^2}$；(2) $S(10)=60$，$S'(10)=4$；(3) 大约 64000 张光盘.

12. (1) $C'(t)=\frac{0.14-0.14t^2}{(t^2+1)^2}$；(2) $C'(0.5)=0.0672$，$C'(3)=-0.0112$.

13. $C'(x)=0.02x+10$，$L'(x)=-0.02x+20$，1000 件产品.

14. $R'(x)=\frac{1}{5}x+100$，当销量小于 500 时，再增加销售可使总收入增加，但销量超过 500 时，收益会减少.

15. (1) $693\ \mathrm{cm^3/min}$；(2) $\frac{9}{32\pi}$ cm/min；(3) $\frac{8}{27\pi}$ cm/min.

16. (1) $\frac{9}{128\pi}$ cm/s.

17. 10 cm/s.

18. (1)10.08 km/h；(2) 9 cm/s.

第 4 章

1. (1) $2x^3+\frac{7}{2}x^2+2x+C$；

(2) $\frac{1}{3}x^3+\frac{1}{x}+C$；

(3) $\frac{2}{3}x^{3/2}-2x^{1/2}+C$；

(4) $\frac{3}{2}x^2-5x+2\ln x+C$；

(5) $\frac{3}{8}x^{8/3}+\frac{9}{5}x^{5/3}+\frac{15}{2}x^{2/3}+C$；

(6) $-\frac{(a-bx)^6}{6b}+C$；

(7) $\sqrt{2x+7}+C$；

(8) $3x+5\ln(x-2)+C$；

(9) $\frac{1}{3a}[(x+a)^{3/2}+(x-a)^{3/2}]+C$；

(10) $\frac{2}{25}[(5x+3)^{3/2}+(5x+3)^{1/2}]+C$；

(11) $\frac{1}{2}x^2-\frac{1}{x+3}+C$；

(12) $\frac{1}{2}x^2+2x+\ln|x+1|+C$；

(13) $\frac{e^{px}}{p}-\frac{e^{-qx}}{q}+C$；

(14) $\frac{1}{3}e^{3x}+e^x+C$.

2. (1) $\frac{\sin(a^2x+b)}{a^2}+C$;
(2) $\frac{1}{2}\tan(2x+3)+C$;
(3) $\frac{1}{2}\left(x-\frac{\sin 2ax}{2a}\right)+C$;
(4) $\frac{1}{a}\tan ax-x+C$;
(5) $\frac{1}{32}(12x-8\sin 2x+\sin 4x)+C$;
(6) $\tan x-\cot x+C$;
(7) $-\cot x-\frac{3}{2}x-\frac{1}{4}\sin 2x+C$;
(8) $\frac{1}{a}(\sin ax-\cos ax)+C$;
(9) $\frac{1}{a}(\tan ax+\sec ax)+C$;
(10) $\frac{\sin 2x}{4}-\frac{\sin 12x}{24}+C$.

3. (1) $\frac{1}{4}(x^3+1)^4+C$;
(2) $-\frac{1}{4(3x^2+4x+1)^2}+C$;
(3) $\frac{2}{3}\sqrt{x^3+3x+4}+C$;
(4) $\frac{1}{2}(\ln x)^2+C$;
(5) $\frac{1}{8}\cos^8 x-\frac{1}{6}\cos^6 x+C$;
(6) $\frac{1}{4a}(a\sin x-b)^4+C$;
(7) $\frac{1}{4}[\ln(\sin x)]^4+C$;
(8) $\frac{1}{3}\tan^3\theta+\frac{1}{5}\tan^5\theta+C$;
(9) $\frac{1}{4}\tan^4 x+\frac{1}{6}\tan^6 x+C$;
(10) $\frac{1}{2}\tan^2 x+\ln|\cos x|+C$;
(11) $e^{\sin x\cos x}+C$;
(12) $e^{x+1/x}+C$;
(13) $-2\cos\sqrt{x}+C$;
(14) $e^x-\ln(1+e^x)+C$;
(15) $2\ln(e^{x/2}+e^{-x/2})+C$.

4. (1) $\frac{2}{5}(x+1)^{\frac{5}{2}}-\frac{2}{3}(x+1)^{\frac{3}{2}}+C$;
(2) $\frac{1}{10}(2x+3)^{\frac{5}{2}}+\frac{1}{6}(2x+3)^{\frac{3}{2}}+C$;
(3) $\frac{x}{a^2\sqrt{a^2-x^2}}+C$;
(4) $\frac{a^2}{2}\sin^{-1}\frac{x}{a}-\frac{1}{2}x\sqrt{(a^2-x^2)}+C$;
(5) $\ln|x+\sqrt{x^2-4}|+C$;
(6) $-\frac{\sqrt{x^2+1}}{x}+C$;
(7) $a\arcsin\frac{x}{a}-\sqrt{a^2-x^2}+C$;
(8) $a\arcsin\sqrt{\frac{x}{a}}-\sqrt{x(a-x)}+C$.

5. (1) $\frac{1}{4}x^2(2\ln x-1)+C$;
(2) $\frac{1}{25}(5x-1)e^{5x}+C$;
(3) $x\tan x+\ln\cos x+C$;
(4) $\sin x-x\cos x+C$;
(5) $\frac{1}{2}\sec x\tan x+\frac{1}{2}\ln|\sec x+\tan x|+C$;
(6) $x\arcsin x+\sqrt{1-x^2}+C$;
(7) $x\sec x-\ln|\sec x+\tan x|+C$;
(8) $\frac{1}{4}x^2-\frac{1}{4}x\sin 2x-\frac{1}{8}\cos 2x+C$.

第 5 章

1. (1)$b-\frac{b^2}{2}$; (2) $\frac{4c^3}{3}$; (3) e^a-1.

2. (1) $\frac{b^3-a^3}{12b}$; (2) 240; (3) 57; (4) $2(9-2\sqrt{6})$; (5) $\frac{1}{a}(e^{ca}-e^{ba})$; (6) $2\ln 2-1$.

3. (1) $\frac{32\sqrt{2}}{3}a^2$; (2) $\frac{4}{3}\sqrt{a}(h-a)^{3/2}$; (3) $\frac{b^2}{12}$.

4. (1) $5\frac{1}{3}$; (2) $\frac{1}{4}$; (3) $25\frac{3}{5}$; (4) $\frac{16}{3}a^2$; (5) $\frac{4}{3}$.

5. (1) $13\frac{1}{6}$; (2) $-\ln 2$; (3) $\frac{1}{12}(3\sqrt{3}-1)$; (4) $\frac{1}{4}e(e^6-1)$; (5) $\frac{\sqrt{2}-1}{\sqrt{2}a}$; (6) $\ln\frac{4}{3}$;

(7) $1-\frac{\pi}{4}$; (8) $\frac{1}{2}$; (9) $\frac{2}{3}$; (10) $\sqrt{2}$; (11) 2; (12) $\frac{3}{5}$; (13) $\frac{7}{120}\sqrt{2}$;

(14) $\frac{1}{2}(1-\ln 2)$; (15) $\frac{8}{15}$; (16) $-\frac{\pi}{6}$; (17) $\frac{\pi}{2}-1$; (18) $\frac{1}{4}(e^2+1)$.

6. $\frac{\pi}{4}\ln 5$. 7. $\frac{8}{5}\pi$. 10. (1) $\frac{1}{4}\pi$; (2) $\frac{8}{35}\pi$. 11. 2π.

第6章

1. (1) $2x+4y\geqslant 1$; (2) $x>y$; (3) $-|y|\leqslant x\leqslant |y|$, $x\neq 0$, $y\neq 0$; (4) $x+y\geqslant 0$, $x\geqslant 3$.

3. (1) 0; (2) $-\frac{4}{289}$; (3)不存在; (4) 不存在.

4. (1) $\frac{\partial f}{\partial x}=y+\frac{1}{y}, \frac{\partial f}{\partial y}=x-\frac{x}{y^2}$;

(2) $\frac{\partial f}{\partial x}=\frac{y}{1+x^2y^2}, \frac{\partial f}{\partial y}=\frac{x}{1+x^2y^2}$;

(3) $\frac{\partial f}{\partial x}=(1+y^2)^x\ln(1+y^2)$, $\frac{\partial f}{\partial y}=2xy(1+y^2)^{x-1}$;

(4) $\frac{\partial f}{\partial x}=\frac{x}{x^2+y^2+z^2}, \frac{\partial f}{\partial y}=\frac{y}{x^2+y^2+z^2}, \frac{\partial f}{\partial z}=\frac{z}{x^2+y^2+z^2}$.

5. (1) $f_x(1,1)=2$, $f_x(-1,-1)=-2$, $f_y(1,2)=\frac{3}{4}$, $f_y(2,1)=0$;

(2) $f_x(1,1)=\frac{1}{2}$, $f_x(-1,-1)=-\frac{1}{2}$, $f_y(1,2)=\frac{1}{5}$, $f_y(2,1)=\frac{2}{5}$;

(3) $f_x(1,1)=4\ln 2$, $f_x(-1,-1)=-\ln 2$, $f_y(1,2)=1$, $f_y(2,1)=4$.

6. 1.

7. 3.

8. (1) $\frac{\partial^2 f}{\partial x^2}=0, \frac{\partial^2 f}{\partial y^2}=0, \frac{\partial^2 f}{\partial x\partial y}=0, \frac{\partial^2 f}{\partial y\partial x}=0$;

(2) $\frac{\partial^2 f}{\partial x^2}=2, \frac{\partial^2 f}{\partial y^2}=2, \frac{\partial^2 f}{\partial x\partial y}=0, \frac{\partial^2 f}{\partial y\partial x}=0$;

(3) $\frac{\partial^2 f}{\partial x^2}=6x, \frac{\partial^2 f}{\partial y^2}=6y, \frac{\partial^2 f}{\partial x\partial y}=1, \frac{\partial^2 f}{\partial y\partial x}=1$;

(4) $\frac{\partial^2 f}{\partial x^2}=8+12y, \frac{\partial^2 f}{\partial y^2}=6y+12x^2, \frac{\partial^2 f}{\partial x\partial y}=24xy, \frac{\partial^2 f}{\partial y\partial x}=24xy$;

(5) $\frac{\partial^2 f}{\partial x^2}=0,\frac{\partial^2 f}{\partial y^2}=0,\frac{\partial^2 f}{\partial x\partial y}=1,\frac{\partial^2 f}{\partial y\partial x}=1.$

9. (1)0, 0, 0；(2) 2, 2, 0；(3) 2, -12, 1；(4)-28, 36, -24.

11. (1) $\frac{\partial f}{\partial x}=\sin(xt)+2xt+xt\cos(xt),\frac{\partial^2 f}{\partial x\partial t}=2x\cos(xt)-x^2t\sin(xt)+2x$；

(2) $\frac{\partial f}{\partial x}=zt-te^{xt},\frac{\partial^2 f}{\partial x\partial t}=z-e^{xt}-xte^{xt}$；

(3) $\frac{\partial f}{\partial x}=-6x\sin(t+x^2),\frac{\partial^2 f}{\partial x\partial t}=-6x\cos(t+x^2)$.

12. (1) (0, 0) 为极小值点，(0, 2)为鞍点；

(2) (0, 0)为鞍点，(-10, 0)为极大值点；

(3) (0, 0)为极大值点；

(4) (-1, 3)为鞍点；

(5) (0, 0)为鞍点，(1, 0)为极小值点，(-1, 0)为极小值点；

(6) $f(x, y)=(x^2+y^2+1)^2$，(0, 0)为极小值点.

13. 劳动力成本 200 万美元，设备成本 300 万美元，最小成本为 $C(2, 3)=1500$ 万美元.

14. $6\times4\times2$.

第 7 章

1. (1)-135；(2)$\frac{3-2\sqrt{2}}{4}$.

2. (1) $\frac{9}{2}$；(2)$\frac{1}{2}-\frac{\pi}{24}\ln7$；(3)$\frac{1}{8}(e^8+e^{-8}-2)$；(4) 0.

3. 28.

5. (1) 4096；(2) 0；(3)$\frac{20}{3}\sin8$；(4) $-\frac{1}{2}\ln28$.

6. (1) $\int_0^{\sqrt{2}}dy\int_{y^2}^{x}f(x, y)dx$；(2) $\int_0^8 dx\int_0^{\frac{x}{2}}f(x, y)dy$；(3) $\int_1^{e^2}dx\int_{\ln x}^{e^2}f(x, y)dy$；

(4) $\int_0^1 dy\int_{e^y}^{e}f(x, y)dx$；(5) $\int_0^{\frac{\pi}{2}}dx\int_0^{\sin x}f(x, y)dy$；(6) $\int_0^1 dx\int_{\sqrt{x}}^{x^2}f(x, y)dy$.

7. (1) $\frac{1}{8}(1-e^{-16})$；(2) $\sin1$；(3) $\frac{1}{3}(e^8-1)$；(4)$\frac{1}{2}\left(9\ln3-\frac{9}{2}\right)$.

8. (1) 1/6；(2)$\frac{3\pi}{4}$；(3)$\frac{1}{8}\pi a^2$；(4)$\frac{\pi}{24}$.

9. (1) $5\pi-26$；(2) $\frac{\pi}{12}(65^{\frac{3}{2}}-1)$；(3) 256π；(4)$\frac{\pi}{4}(e^9-1)$.

10. (1) $\frac{16}{3}\sqrt{2}$；(2)$\frac{3\pi}{2}$；(3)$\frac{\pi}{16}$；(4)$\frac{31}{6}$；(5) 256π；(6) 18π.

11. (1) 8π；(2) $\frac{64\pi(2-\sqrt{2})}{3}$；(3)$\frac{5\pi}{2}$；(4) $\frac{27\pi}{4}$；(5) $(0, 0,\frac{16}{3})$；(6) $(0, 0,\frac{3a}{8})$；

(7) $(0, 0,\frac{11}{30})$.

第 8 章

1. (1) $2a^3$; (2) $14(2\sqrt{2}-1)$; (3) $2\sqrt{5}$; (4) $4(\mathrm{e}-\mathrm{e}^2)$; (5) $\frac{1}{6}(14\sqrt{14}-1)$;

(6) $160\pi+120\pi^2$.

2. $I=I_1+I_2+I_3=\frac{3}{2}\sqrt{5}+\frac{5}{2}+2=\frac{3}{2}(3+\sqrt{5})$.

3. 质量：$2\mu\alpha R$；质心坐标：$\left(0,\frac{R\sin\alpha}{\alpha}\right)$.

4. 质量：$3\pi k\sqrt{a^2+b^2}$；质心坐标：$(0,\frac{2a}{3\pi},\frac{3\pi b}{2})$

5. (1) $\frac{8\sqrt{3}}{3}$; (2) $\frac{\sqrt{6}}{4}$; (3) $\frac{\pi}{3}+2$; (4) $\frac{\pi(25\sqrt{5}+1)}{60}$; (5) $\frac{1}{4}(17\sqrt{17}-1)$; (6) 6.

6. $\frac{9\sqrt{6}}{8}$.

7. 0.

8. (1) 4; (2) $\frac{\pi}{6}(17\sqrt{17}-1)$.

9. $\frac{\sqrt{3}ka^4}{12}$.

10. $\frac{k(9\sqrt{3}-8\sqrt{2}+1)}{15}$.

第 9 章

1. (1) $y=\mathrm{e}^{Cx}$; (2) $y=\frac{1}{5}x^3+\frac{1}{2}x^2+C$;

   (3) $\arcsin y=\arcsin x+C$; (4) $(x-4)y^4=Cx$.

2. (1) $\mathrm{e}^y=\frac{1}{2}(\mathrm{e}^{2x}+1)$; (2) $y=\mathrm{e}^{\frac{2}{\pi}x}$; (3) $(1+\mathrm{e}^x)\sec y=2\sqrt{2}$.

3. (1) $y=\mathrm{e}^{-x}(x+C)$; (2) $y=\frac{1}{3}x^2+\frac{3}{2}x+2+\frac{C}{x}$;

   (3) $y=C\cos x-2\cos^2 x$; (4) $x=Cy^3+\frac{1}{2}y^2$.

4. $y=2(\mathrm{e}^x-x-1)$.

5. $v=\sqrt{72500}\approx 269.3$ cm/s.

6. $y=x(1-4\ln x)$.

7. (1) $y=\frac{3}{8}x^3+C_1x+C_2$; (2) $y=x\mathrm{e}^x-3\mathrm{e}^x+C_1x^2+C_2x+C_3$;

   (3) $y=C_1(x+\frac{1}{3}x^3)+C_2$; (4) $(y-1)^3=C_1x+C_2$.

8. (1) $y=\mathrm{e}^x-\frac{\mathrm{e}}{2}x^2-\frac{\mathrm{e}}{2}$; (2) $y=\left(\frac{1}{2}x+1\right)^4$;

(3) $y=\ln\sec x$; (4) $y=\sqrt{2x-x^2}$.

9. (1) 线性无关; (2) 线性相关; (3) 线性无关; (4) 线性相关.

10. $y=C_1\mathrm{e}^{-2x}+C_2\mathrm{e}^{-6x}$.

11. $y=C_1\sin x+C_2\cos x$.

12. (1) $y=C_1\mathrm{e}^{-2x}+C_2\mathrm{e}^{5x}$; (2) $y=C_1+C_2\mathrm{e}^{4x}$;

(3) $y=C_1\cos\sqrt{2}x+C_2\sin\sqrt{2}x$; (4) $y=(C_1+C_2x)\mathrm{e}^{-4x}$;

(5) $x=(C_1+C_2t)\mathrm{e}^{3t}$; (6) $y=\mathrm{e}^{-x}(C_1\cos x+C_2\sin x)$.

13. (1) $y=-\mathrm{e}^{2x}+2\mathrm{e}^{4x}$; (2) $y=(2+x)\mathrm{e}^{-\frac{x}{2}}$.

14. $\begin{cases}\dfrac{\mathrm{d}x}{\mathrm{d}t}=1500\times 0.04\%-1500\times\dfrac{x}{10800}\\ x(0)=10800\times 0.12\%=12.96\end{cases}$, $\dfrac{x(10)}{10800}=0.06\%$.

English Section

Chapter 1 Limit and Continuity

1.1 Function

Limit is a tool for studying function, which can be used to study continuity, derivability, differentiability and integrability of function. In the elementary mathematics section, we have discussed functions and their graphs. Before defining of the limit of a function $f(x)$, it's important to review the concept of a function.

1.1.1 Definition of Function

Let A and B be two non-empty real number sets. Then a function f from A to B is a rule which assigns a unique element of B to each element of A. The unique element of B which assigns corresponding to an element $x \in A$ is denoted by $f(x)$. So, we also write $y = f(x)$. The symbol $f: A \to B$ usually means f is a function from A to B. The element $f(x)$ of B is called the image of x under the function f.

1.1.2 Value of the Function

If f is a function from A to B and $x = a$ is an element in the domain of f, then the image $f(a)$ corresponding to $x = a$ is said to be the value of the function at $x = a$.

If the value of the function $f(x)$ at $x = a$ denoted by $f(a)$ is a finite number, then we say that $f(x)$ is defined at $x = a$. Otherwise, $f(x)$ is not defined at $x = a$. For example:

(1) $y = f(x) = 3x + 5$ is defined at $x = 2$, as $f(2) = 3 \times 2 + 5 = 11$ is a finite number.

(2) $y = f(x) = \dfrac{1}{x-1}$ is not defined at $x = 1$, as $f(1) = \dfrac{1}{0}$ is not a finite number.

(3) Consider the function $y = f(x) = \dfrac{x^2 - 1}{x - 1}$ and $x = 1$.

When $x = 1$, $y = f(1) = \dfrac{0}{0}$. This does not give any number. It simply indicates that each number in the numerator and denominator is zero. So, there are some functions that take the form $\dfrac{0}{0}$ for some value of x. Such form is said to be an **indeterminate form.** Other indeterminate forms which a function may take for some values of x are $\dfrac{\infty}{\infty}$, $\infty - \infty$, 1^∞ and 0^∞.

1.2 Limit

1.2.1 Meaning of $x \to x_0$

We consider an example to illustrate the meaning of $x \to 2$. Let the variable x to take the values 1.9, 1.99, 1.999, 1.9999, …. As the number of 9 increases, the value of x will be getting closer to 2, but will never be 2. In such a situation also, the numerical difference between x and 2 will be sufficiently small. Thus, if x takes the value greater than 2 or less than 2, but the numerical difference between x and 2 is sufficiently small, then we say that x approaches 2 or x tends to 2, and can be written as $x \to 2$.

Generally, let x be a variable and x_0 a constant number. The meaning of $x \to x_0$ is that the numerical difference between x and x_0 is sufficiently small.

1.2.2 Intuitive Idea of Limit

In this section we try to give a clear understanding of limit by means of some examples. Consider a regular polygon inscribed in a circle as shown in figure 1 - 1. Keeping the circle fixed, as increaing the number of edges of the polygon, the area (or the perimeter) of the polygon also increases. But, no matter however large may be the number of edges, the area (or the perimeter) of the polygon can never be greater than that of the circle. But by taking the number of the edges of the polygon sufficiently large, we can make the difference between the area (or the perimeter) of the polygon and the area (or the perimeter) of the circle sufficiently small (or as small as we please). In such a case we say that the area (or the perimeter) of the circle is the limit of the area (or perimeter) of the polygon.

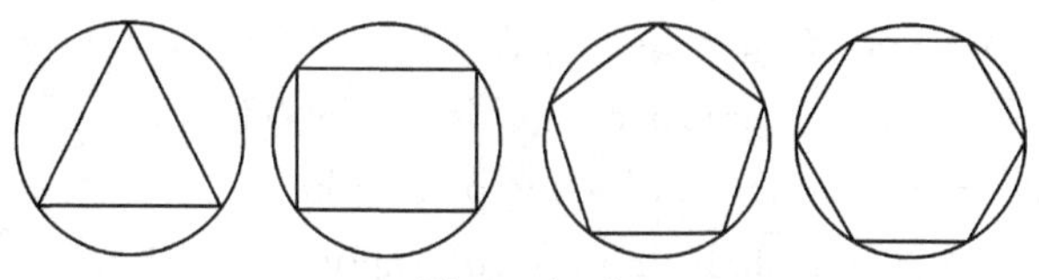

Figure 1 - 1

Now, let us put the above example in a functional notation. For this, we shall denote the area of a regular polygon with n edges inscribed in a fixed circle by A_n ($n > 2$). This functional relation can also be written as $f(n) = A_n$. When n tends to infinity, $f(n)$ or A_n gets close to (or approaches) the area of the circle, that is, the limiting value of $f(n)$ or A_n is the area of the circle and is denoted by $\lim\limits_{n \to \infty} f(n)$ or $\lim\limits_{n \to \infty} A_n$.

The next example we take is the sequence of numbers 0.9, 0.99, 0.999, 0.9999, …. In this sequence the terms are gradually increasing but remain always less than 1. But by making a proper choice of the term, we can make the term sufficiently close to 1 or make the difference between 1 and the term sufficiently small (or as small as we please). In such a case we say that the sequence tends to the limiting value 1.

The above sequence can also be put in a functional notation by defining the function f as

$f(1)=0.9$, the 1st term

$f(2)=0.99$, the 2nd term

$f(3)=0.999$, the 3rd term

⋮ ⋮

$f(n)=0.99\cdots9$, the nth term

When n tends to infinity, $f(n)$ almost equals to 1. So the limiting value of $f(n)$ is 1 and it is denoted by

$$\lim_{n\to\infty} f(n)=1$$

Now we conclude this section with one more example. This example is the sequence a_n,

$$1,\ 1-\frac{1}{2},\ 1-\frac{1}{2^2},\ \cdots,\ 1-\frac{1}{2^{n-1}},\ \cdots$$

It is obvious that when n is sufficiently large, $\frac{1}{2^{n-1}}$ is sufficiently small. So when n tends to infinity, i. e. , the limiting value of this sequence is 1 and it is denoted by

$$\lim_{n\to\infty} a_n=1$$

1.2.3 Limit of a Function

We use the concept of the limit of a sequence to help understanding the meaning of the limit of a function.

First, we consider the function $y=f(x)=2x+3$.

Considering the sequence of values of x to be 0.5, 0.75, 0.9, 0.99, 0.999, 0.9999, ⋯ whose limit is 1, the corresponding values of $f(x)$ are 4, 4.5, 4.8, 4.98, 4.998, 4.9998, ⋯, which go closer to 5, when x approaches 1 from the left.

Again if we consider the sequence of value of x to be 2, 1.5, 1.25, 1.1, 1.01, 1.001, 1.0001,⋯, whose limit is 1, we shall find the corresponding values of $f(x)$ to be 7, 6, 5.5, 5.2, 5.02, 5.002,5.0002,⋯ which get closer to 5, when x is sufficiently close to 1 from the right.

That is, when $x\to 1$, $f(x)\to 5$. In symbol, we write

$$\lim_{x\to 1} f(x)=\lim_{x\to 1}(2x+3)=5$$

Hence, we have the following definition of limit of a function.

Definition 1.1 A function $f(x)$ is said to tend to a limit "a" when $x\to x_0$ if the numerical difference between $f(x)$ and a can be made as small as we wish by making x sufficiently close to x_0, and we write

$$\lim_{x\to x_0} f(x)=a$$

It is easy to obtain $\lim\limits_{x\to 1}(x+1)=2$, $\lim\limits_{x\to 0}\sin x=0$, $\lim\limits_{x\to 0}\cos x=1$.

In addition, we can deduce the following conclusion by the definition of limit.

$$\lim_{x \to x_0} f(x) = a \Leftrightarrow \lim_{x \to x_0^-} f(x) = \lim_{x \to x_0^+} f(x) = a$$

For example, we observe the following limit of the function at $t = 0$.

$$u(t) = \begin{cases} 1, & t \geqslant 0 \\ 0, & t < 0 \end{cases}$$

We draw the graph of $u(t)$ as shown in figure 1 - 2.

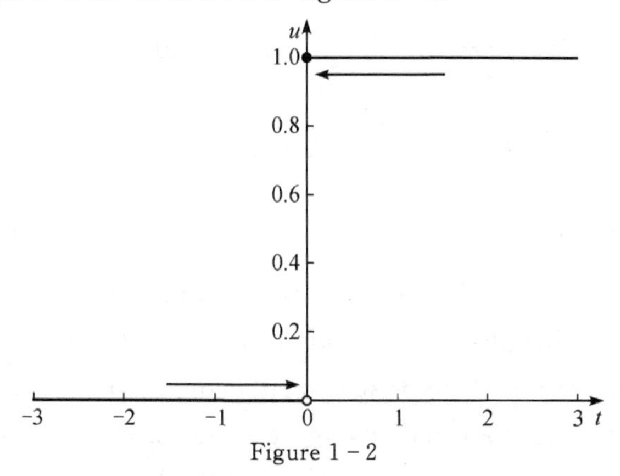

Figure 1 - 2

As shown in the graph, the corresponding value of the function tends to 1 as t tends to 0 from the right, and the corresponding value of the function tends to 0 as t tends to 0 from the left. That is

$$\lim_{t \to 0^+} u(t) = 1, \quad \lim_{t \to 0^-} u(t) = 0$$

Hence $\lim\limits_{t \to 0} u(t)$ does not exist.

Another example, observe $\lim\limits_{x \to 0} \sin \dfrac{\pi}{x}$. Using the graphing tool, we can draw the graph of this function, see figure 1 - 3. However, it is difficult to see whether the limit exist or not.

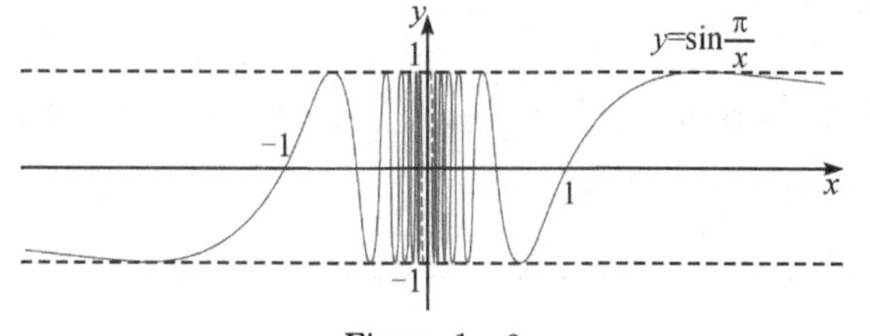

Figure 1 - 3

But we may investigate the varying tendence of the function by calculating the value of the function at $x = 0$.

$$f(1) = \sin\pi = 0, \ f\left(\frac{1}{2}\right) = \sin 2\pi = 0, \ f\left(\frac{1}{3}\right) = \sin 3\pi = 0, \ f\left(\frac{1}{4}\right) = \sin 4\pi = 0$$

$$f(0.1) = \sin 10\pi = 0, \ f(0.01) = \sin 100\pi = 0, \ \cdots, \ f(0.001) = f(0.0001) = 0$$

On the basis of this information we might be tempted to guess that $\lim\limits_{x \to 0} \sin \dfrac{\pi}{x} = 0$, but this time we are wrong. Note that although $f\left(\frac{1}{n}\right)=\sin n\pi=0$ for any integer n, it is also true that $f\left(\frac{1}{n+\frac{1}{2}}\right)=\sin(n+\frac{1}{2})\pi=\pm 1$.

The dashed lines indicate that the values of $\sin\frac{\pi}{x}$ oscillate between 1 and -1 infinitely often as x approaches 0. Therefore, $\lim\limits_{x\to 0}\sin\frac{\pi}{x}$ does not exist.

1.2.4 Meaning of Infinity

Let us consider the function $y=f(x)=\frac{1}{x}$.

If we consider the sequence of values of x to be 1, 0.5, 0.1, 0.01, 0.001, 0.0001, ⋯, whose limit is 0, we see that the corresponding values of $f(x)$ are 1, 2, 10, 100, 1000, 10000, ⋯, which are increasing. If we take x small enough, the corresponding value of $f(x)$ will be large enough. Taking the value of x to be sufficiently close to 0 from the right, the value of $f(x)$ will be greater than any positive number, however large. In such a case, we say that as x tends to 0, $f(x)$ tends to positive infinity and is indicated by the symbol $f(x)\to+\infty$ as $x\to 0^+$, or $\lim\limits_{x\to 0^+}\frac{1}{x}=+\infty$.

Definition 1.2 Let $f(x)$ be a function of x. Making x sufficiently close to x_0, if the value of $f(x)$ obtained is greater than any pre-assigned number, however large, we say that the limit of $f(x)$ is positive infinity as x tends to x_0. Symbolically, we write

$$\lim_{x\to x_0} f(x)=+\infty$$

1.2.5 Meaning of $x\to\infty$

Let us consider the function $f(x)=\frac{1}{x^2}$ and we see the value of x and the corresponding value of $f(x)$.

x	1	10	100	1000	⋯
$f(x)$	1	0.01	0.0001	0.000001	⋯

From the table above, we see that when the value of x increases, the corresponding value of $f(x)$ decreases. When the value of x is large enough, the corresponding value of $f(x)$ becomes small enough. That is, taking the value of x to be sufficiently large, i. e., the value is greater than any positive number, however large, the value of $f(x)$ can be made sufficiently close to 0. In such situation, we say that as x tends to infinity, $f(x)$ tends to 0 and is indicated by symbol $f(x) \to 0$ when $x \to \infty$, or $\lim_{x\to\infty} f(x) = \lim_{x\to\infty} \frac{1}{x^2} = 0$. See figure 1 - 4.

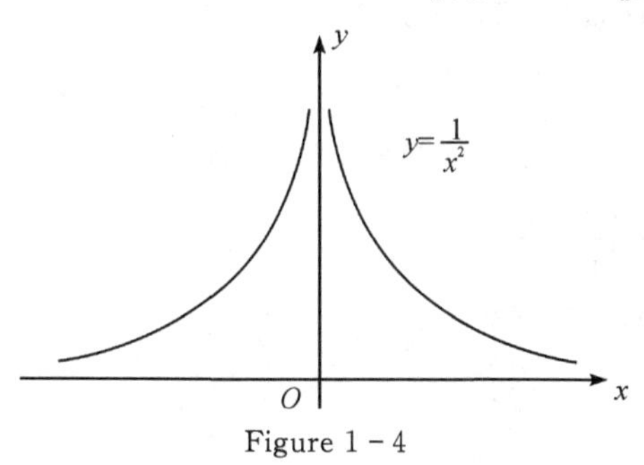

Figure 1 - 4

Definition 1.3 A function $f(x)$ is said to tend to a when $x \to +\infty$, if $f(x)$ can be made close to a when x is greater than any pre-assigned number, however large. Symbolically, we write

$$\lim_{x\to+\infty} f(x) = a$$

Similarly, we can define

$$\lim_{x\to-\infty} f(x) = a$$

Obviously,

$$\lim_{x\to-\infty} f(x) = \lim_{x\to+\infty} f(x) = a \Leftrightarrow \lim_{x\to\infty} f(x) = a$$

Limit Calculation Rule

Let $\lim_{x\to x_0} f(x) = a$, $\lim_{x\to x_0} g(x) = b$, then

(1) $\lim_{x\to x_0}[f(x) \pm g(x)] = \lim_{x\to x_0} f(x) \pm \lim_{x\to x_0} g(x) = a \pm b$.

(2) $\lim_{x\to x_0}[f(x) \cdot g(x)] = \lim_{x\to x_0} f(x) \cdot \lim_{x\to x_0} g(x) = a \cdot b$.

(3) $\lim_{x\to x_0} \frac{f(x)}{g(x)} = \frac{\lim_{x\to x_0} f(x)}{\lim_{x\to x_0} g(x)} = \frac{a}{b}$, where b can not be zero.

(4) $\lim_{x\to x_0}[kf(x)] = k\lim_{x\to x_0} f(x) = ka$.

(5) $\lim_{x\to x_0}[f(x)]^n = [\lim_{x\to x_0} f(x)]^n = a^n$, $n \in \mathbf{Z}^+$.

(6) $\lim_{x\to x_0} \sqrt[n]{f(x)} = \sqrt[n]{\lim_{x\to x_0} f(x)} = \sqrt[n]{a}$, $a > 0, n \in \mathbf{Z}^+$.

Example 1.1 Find $\lim_{x\to 3}(3x - 2)$.

Solution When x approaches 3, $3x$ approaches $3\times 3 = 9$, then $3x - 2$ approaches $9-2=7$. Therefore,

$$\lim_{x\to 3} f(x) = \lim_{x\to 3}(3x-2) = 7$$

Example 1.2 Evaluate $\lim_{x\to 2}(3x^2-5x+6)$.

Solution $\lim_{x\to 2}(3x^2-5x+6) = \lim_{x\to 2} 3x^2 - \lim_{x\to 2} 5x + \lim_{x\to 2} 6 = 3\lim_{x\to 2} x^2 - 5\lim_{x\to 2} x + 6$

$$= 3\times 4 - 5\times 2 + 6 = 8$$

Example 1.3 Evaluate $\lim_{x\to 3}\frac{4x-5}{2x+3}$.

Solution $\lim_{x\to 3}\frac{4x-5}{2x+3} = \frac{\lim_{x\to 3}(4x-5)}{\lim_{x\to 3}(2x+3)} = \frac{12-5}{6+3} = \frac{7}{9}$.

Example 1.4 Compute $\lim_{x\to 0}\frac{5x^2+3x}{x}$.

Solution $\lim_{x\to 0}\frac{5x^2+3x}{x} = \lim_{x\to 0}\frac{x(5x+3)}{x} = \lim_{x\to 0}(5x+3) = 3$.

Example 1.5 Find $\lim_{x\to a}\frac{x^5-a^5}{x^4-a^4}$.

Solution $\lim_{x\to a}\frac{x^5-a^5}{x^4-a^4} = \lim_{x\to a}\frac{(x-a)(x^4+x^3a+x^2a^2+xa^3+a^4)}{(x-a)(x^3+x^2a+xa^2+a^3)}$

$$= \lim_{x\to a}\frac{x^4+x^3a+x^2a^2+xa^3+a^4}{x^3+x^2a+xa^2+a^3}$$

$$= \frac{5a^4}{4a^3} = \frac{5}{4}a$$

Example 1.6 Evaluate $\lim_{x\to a}\frac{x^{\frac{1}{3}}-a^{\frac{1}{3}}}{x^{\frac{1}{2}}-a^{\frac{1}{2}}}$.

Solution $\lim_{x\to a}\frac{x^{\frac{1}{3}}-a^{\frac{1}{3}}}{x^{\frac{1}{2}}-a^{\frac{1}{2}}} = \lim_{x\to a}\frac{(x^{1/6})^2-(a^{1/6})^2}{(x^{1/6})^3-(a^{1/6})^3}$

$$= \lim_{x\to a}\frac{(x^{1/6}-a^{1/6})(x^{1/6}+a^{1/6})}{(x^{1/6}-a^{1/6})(x^{2/6}+x^{1/6}a^{1/6}+a^{2/6})}$$

$$= \lim_{x\to a}\frac{x^{1/6}+a^{1/6}}{x^{2/6}+x^{1/6}a^{1/6}+a^{2/6}} = \frac{2a^{1/6}}{3a^{1/3}} = \frac{2}{3a^{1/6}}$$

Example 1.7 Evaluate $\lim_{x\to a}\frac{\sqrt{x+a}-\sqrt{3x-a}}{x-a}$.

Solution $\lim_{x\to a}\frac{\sqrt{x+a}-\sqrt{3x-a}}{x-a} = \lim_{x\to a}\frac{(\sqrt{x+a}-\sqrt{3x-a})(\sqrt{x+a}+\sqrt{3x-a})}{(x-a)(\sqrt{x+a}+\sqrt{3x-a})}$

$$= \lim_{x\to a}\frac{x+a-3x+a}{(x-a)(\sqrt{x+a}+\sqrt{3x-a})}$$

$$= \lim_{x\to a}\frac{-2(x-a)}{(x-a)(\sqrt{x+a}+\sqrt{3x-a})}$$

$$= \lim_{x\to a}\frac{-2}{\sqrt{x+a}+\sqrt{3x-a}} = \frac{-2}{\sqrt{2a}+\sqrt{2a}} = \frac{-1}{\sqrt{2a}}$$

Example 1.8 Evaluate $\lim_{x\to\infty}(\sqrt{x+a}-\sqrt{x})$

Solution $\lim_{x\to+\infty}\sqrt{x+a}-\sqrt{x}=\lim_{x\to+\infty}\dfrac{(\sqrt{x+a}-\sqrt{x})(\sqrt{x+a}+\sqrt{x})}{(\sqrt{x+a}+\sqrt{x})}$

$$=\lim_{x\to+\infty}\frac{x+a-x}{\sqrt{x+a}+\sqrt{x}}$$

$$=\lim_{x\to+\infty}\frac{a}{\sqrt{x+a}+\sqrt{x}}=\frac{a}{\infty}=0$$

1.2.6 Two Important Results and Two Limits

We give the other two important results without proof.

Theorem 1.2

If $f(x)\leqslant g(x)$ when x is near x_0 (except possibly at x_0) and the limit of $f(x)$ and $g(x)$ both exists as approaches x_0, then $\lim_{x\to x_0}f(x)\leqslant\lim_{x\to x_0}g(x)$.

Theorem 1.3 (Squeeze Theorem)

If $f(x)\leqslant g(x)\leqslant h(x)$ when x is near x_0 (except possibly at x_0) and

$$\lim_{x\to x_0}f(x)=\lim_{x\to x_0}h(x)=L$$

then

$$\lim_{x\to x_0}g(x)=L$$

Important Limit 1 $\lim_{x\to 0}\dfrac{\sin x}{x}=1$.

Proof Figure 1 - 5 shows that in this unit circle, area of $\triangle OAB\leqslant$ area of sector $OAB\leqslant$ area of $\triangle OPA$, i. e. ,

$$\frac{1}{2}OA\cdot BC\leqslant\frac{1}{2}OA^2\cdot\theta\leqslant\frac{1}{2}OA\cdot AP$$

Again $OA=OB=1$, $BC=\sin\theta$, $AP=\tan\theta$, thus we have

$$\sin\theta\leqslant\theta\leqslant\tan\theta$$

$$1\leqslant\frac{\theta}{\sin\theta}\leqslant\frac{1}{\cos\theta}$$

i. e. ,

$$\cos\theta\leqslant\frac{\sin\theta}{\theta}\leqslant 1$$

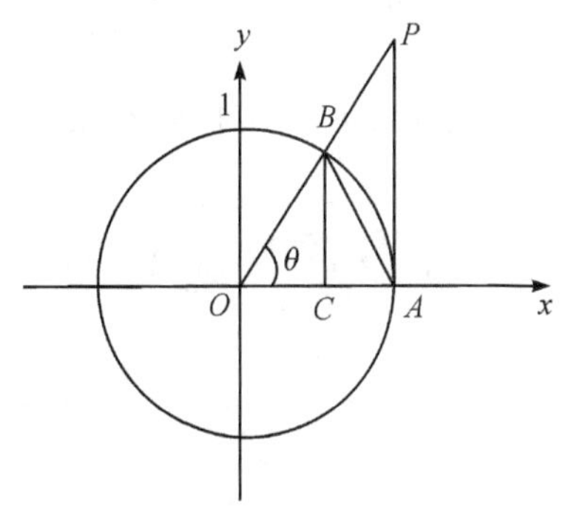

Figure 1 - 5

Applying the Squeeze Theorem,

$$\lim_{x\to 0^+}\frac{\sin x}{x}=1$$

Because $\dfrac{\sin x}{x}$ is an even function,

$$\lim_{x\to 0^-}\frac{\sin x}{x}=1$$

Therefore,

$$\lim_{x\to 0}\frac{\sin x}{x}=1$$

Important Limit 2 $\lim\limits_{x\to\infty}\left(1+\dfrac{1}{x}\right)^x=\mathrm{e}$.

Since the proof of this limit is beyond the requirements of this textbook, we verify the above conclusion by an approximate method.

x	1	10	100	1000	10000	100000	1000000
$\left(1+\frac{1}{x}\right)^x$	≈ 2.000000	≈ 2.563742	≈ 2.704814	≈ 2.716924	≈ 2.718146	≈ 2.718268	≈ 2.718280

By substitution, it is easy to have

$$\lim_{x\to 0}(1+x)^{\frac{1}{x}}=\mathrm{e}$$

Example 1.9 Find the following limits.

(1) $\lim\limits_{x\to 0}\dfrac{\ln(1+x)}{x}$; (2) $\lim\limits_{x\to 0}\dfrac{\mathrm{e}^x-1}{x}$; (3) $\lim\limits_{x\to 0}\dfrac{a^x-1}{x}$;

(4) $\lim\limits_{x\to 0}\dfrac{\mathrm{e}^{3x}-1}{x\cdot 5^x}$; (5) $\lim\limits_{x\to 0}\dfrac{a^x-b^x}{x}$; (6) $\lim\limits_{x\to 0}\dfrac{\tan x}{x}$;

(7) $\lim\limits_{x\to 0}\dfrac{1-\cos 3x}{3x^2}$; (8) $\lim\limits_{x\to a}\dfrac{\sqrt{x}-\sqrt{a}}{\tan(x-a)}$.

Solution

(1) $\lim\limits_{x\to 0}\dfrac{\ln(1+x)}{x}=\lim\limits_{x\to 0}\dfrac{1}{x}\ln(1+x)=\lim\limits_{x\to 0}\ln(1+x)^{\frac{1}{x}}=\ln\lim\limits_{x\to 0}(1+x)^{\frac{1}{x}}=\ln\mathrm{e}=1$

(2) Let $\mathrm{e}^x-1=t\Rightarrow x=\ln(1+t)$, and when $x\to 0$, $t\to 0$. Hence, $\lim\limits_{x\to 0}\dfrac{\mathrm{e}^x-1}{x}=$ $\lim\limits_{t\to 0}\dfrac{t}{\ln(1+t)}=\lim\limits_{t\to 0}\dfrac{1}{\dfrac{\ln(1+t)}{t}}=\dfrac{1}{1}=1$.

(3) Let $a^x-1=t\Rightarrow x\ln a=\ln(1+t)\Rightarrow x=\dfrac{\ln(1+t)}{\ln a}$, and when $x\to 0$, $t\to 0$. So, $\lim\limits_{x\to 0}\dfrac{a^x-1}{x}=\lim\limits_{t\to 0}\dfrac{t}{\dfrac{\ln(1+t)}{\ln a}}=\ln a$.

(4) $\lim\limits_{x\to 0}\dfrac{\mathrm{e}^{3x}-1}{x\cdot 5^x}=\lim\limits_{x\to 0}\dfrac{\mathrm{e}^{3x}-1}{3x}\times 3\times\dfrac{1}{5^x}=3\lim\limits_{x\to 0}\dfrac{\mathrm{e}^{3x}-1}{3x}\cdot\lim\limits_{x\to 0}\dfrac{1}{5^x}=3$.

(5) $\lim\limits_{x\to 0}\dfrac{a^x-b^x}{x}=\lim\limits_{x\to 0}\dfrac{(a^x-1)-(b^x-1)}{x}=\lim\limits_{x\to 0}\dfrac{a^x-1}{x}\cdot\lim\limits_{x\to 0}\dfrac{b^x-1}{x}=\ln\dfrac{a}{b}$.

(6) $\lim\limits_{x\to 0}\dfrac{\tan x}{x}=\lim\limits_{x\to 0}\dfrac{\sin x}{\cos x}\dfrac{1}{x}=\lim\limits_{x\to 0}\dfrac{\sin x}{x}\cdot\lim\limits_{x\to 0}\dfrac{1}{\cos x}=1.$

(7) $\lim\limits_{x\to 0}\dfrac{1-\cos 3x}{3x^2}=\lim\limits_{x\to 0}\dfrac{2\left(\sin\dfrac{3x}{2}\right)^2}{3\times\dfrac{4}{9}\left(\dfrac{3x}{2}\right)^2}=\dfrac{3}{2}\lim\limits_{x\to 0}\left(\dfrac{\sin\dfrac{3x}{2}}{\dfrac{3x}{2}}\right)^2=\dfrac{3}{2}.$

(8) $\lim\limits_{x\to a}\dfrac{\sqrt{x}-\sqrt{a}}{\tan(x-a)}=\lim\limits_{x\to a}\left[\dfrac{\sqrt{x}-\sqrt{a}}{\tan(x-a)}\cdot\dfrac{\sqrt{x}+\sqrt{a}}{\sqrt{x}+\sqrt{a}}\right]$, using(6), $\lim\limits_{x\to 0}\dfrac{\tan x}{x}=1\Rightarrow\lim\limits_{x\to 0}\dfrac{x}{\tan x}=1$

$\Rightarrow\lim\limits_{x\to a}\dfrac{x-a}{\tan(x-a)}=1$, so, $\lim\limits_{x\to a}\dfrac{\sqrt{x}-\sqrt{a}}{\tan(x-a)}=\lim\limits_{x\to a}\dfrac{x-a}{\tan(x-a)}\cdot\lim\limits_{x\to a}\dfrac{1}{\sqrt{x}+\sqrt{a}}=\dfrac{1}{2\sqrt{a}}.$

1.3 Continuity

1.3.1 Continuity at a Point

The intuitive idea of a continuous function f in the interval $[a,b]$ gives the impression that the graph of the function f in this interval is a smooth curve without any break in it. Actually this curve is such that it can be drawn by the continuous motion of pencil without lifting the pencil from the paper. Similarly, a discontinuous function gives the picture consisting of disconnected curves. Let us look at their graphs shown in figure 1 - 6 and discuss their nature.

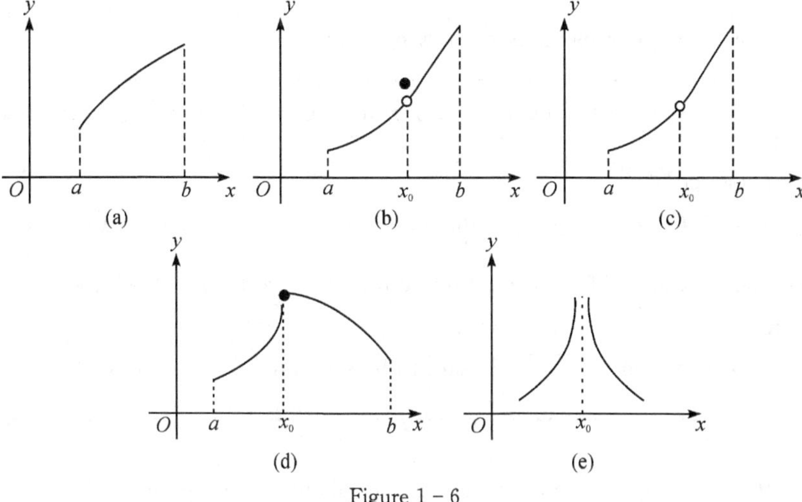

Figure 1 - 6

Now it is obvious that leaving aside the graph of the function in figure 1 - 6(a), all the other graphs in the remaining four figures are discontinuous. In figure 1 - 6(b), $\lim\limits_{x\to x_0}f(x)$ exists and $f(x_0)$ is also defined, but $\lim\limits_{x\to x_0}f(x)\neq f(x_0)$. In figure 1 - 6(c), $\lim\limits_{x\to x_0}f(x)$ exists but $f(x)$ is not defined at $x=x_0$. In figure 1 - 6(d), $\lim\limits_{x\to x_0}f(x)$ does not exist, while $f(x)$ is defined at $x=x_0$. But in figure 1 - 6(e) neither $\lim\limits_{x\to x_0}f(x)$ exists nor $f(x)$ is defined at $x=x_0$.

So the relation $\lim\limits_{x \to x_0} f(x) = f(x_0)$ can be considered the necessary and sufficient condition for a curve or the function to be continuous at $x = x_0$.

Definition 1.4 The function $f(x)$ is said to be continuous at the point $x = x_0$, if and only if

$$\lim_{x \to x_0} f(x) = f(x_0)$$

This definition implies that

> The function $f(x)$ is said to be continuous at the point $x = x_0$ means
> (1) $\lim\limits_{x \to x_0} f(x)$ exists, i. e., $\lim\limits_{x \to x_0^-} f(x)$ and $\lim\limits_{x \to x_0^+} f(x)$ exist and equal.
> (2) $f(x)$ is defined at $x = x_0$.
> (3) $\lim\limits_{x \to x_0} f(x) = f(x_0)$.

Hence, $f(x)$ will be continuous at $x = x_0$ if

$$\lim_{x \to x_0^-} f(x) = \lim_{x \to x_0^+} f(x) = f(x_0)$$

If any of the above conditions is not satisfied then the function is said to be discontinuous at that point.

1.3.2 Types of Discontinuous Point

A discontinuous point may be of the following types:

> (1) If $\lim\limits_{x \to x_0} f(x)$ does not exist and $\lim\limits_{x \to x_0^-} f(x) \neq \lim\limits_{x \to x_0^+} f(x)$, then $x = x_0$ is said to be a jump discontinuous point.
> (2) If $\lim\limits_{x \to x_0} f(x) \neq f(x_0)$ or is not defined at $x = x_0$, then $x = x_0$ is said to be a removable discontinuous point. This type of discontinuity can be removed by redefining the function.
> (3) If $\lim\limits_{x \to x_0} f(x) = \infty$, then $x = x_0$ is said to be a infinite discontinuous point.

Let us see the following examples to get the idea about the continuity and the different type of discontinuities.

(1) The graph of $f(x)$ is shown in figure 1 - 7. It is continuous at every point.

(2) A function is defined as $f(x) = \begin{cases} 1, x > 0 \\ 2, x < 0 \end{cases}$. Its graph is shown in figure 1 - 8. Therefore it is discontinuous at $x = 0$, and $x = 0$ is a jumpt discontinuous point.

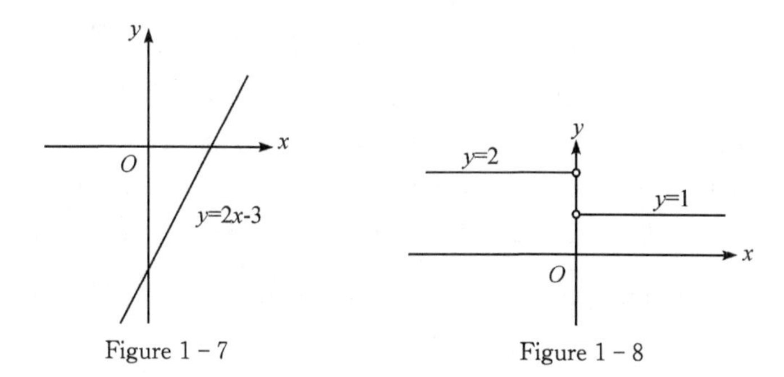

Figure 1 - 7 Figure 1 - 8

(3) The graph of $f(x) = \dfrac{x^2 - 4}{x - 2}$ is shown in figure 1 - 9. But $f(2) = \dfrac{0}{0}$ which is an indeterminate form. So, $f(2)$ does not exist, i. e. , $f(x)$ is not defined at $x = 2$. Hence $x = 2$ is removable discontinuous point.

(4) The graph of $f(x)$ is shown in figure 1 - 10. Because $\lim\limits_{x\to 2^-} f(x) = -\infty$ and $\lim\limits_{x\to 2^+} f(x) = +\infty$, $x = 2$ is a infinite discontinuous point.

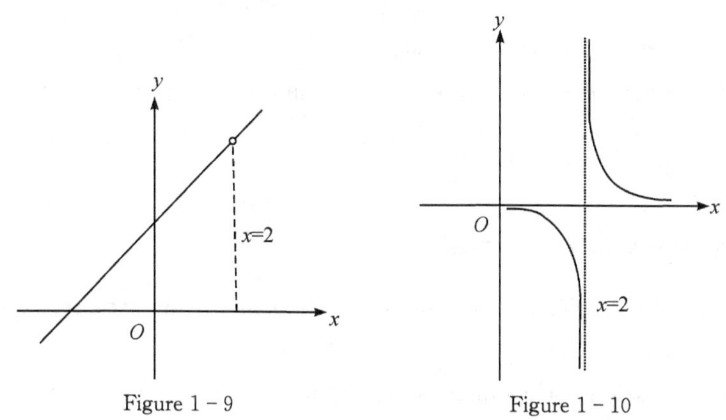

Figure 1 - 9 Figure 1 - 10

Example 1.10 Test the continuity or discontinuity of the following function by calculating the left limit, the right limit and the value of the function at the points mentioned.

(1) $f(x) = 2x^2 - 3x + 10$ at $x = 1$;

(2) $f(x) = \dfrac{1}{x - 2}$ at $x = 2$.

Solution (1) Left limit at $x = 1$ is $\lim\limits_{x\to 1^-} f(x) = \lim\limits_{x\to 1^-} (2x^2 - 3x + 10) = 9$, right limit at $x = 1$ is $\lim\limits_{x\to 1^+} f(x) = \lim\limits_{x\to 1^+} (2x^2 - 3x + 10) = 9$. Also, $f(1) = 9$, $\lim\limits_{x\to 1} f(x) = f(1)$, hence $f(x)$ is continuous at $x = 1$.

(2) Left limit at $x = 2$ is $\lim\limits_{x\to 2^-} f(x) = \lim\limits_{x\to 2^-} \dfrac{1}{x - 2} = -\infty$ which does not exist. Hence $f(x)$ is discontinuous at $x = 2$.

Example 1.11 A function $f(x)$ is difined as follows

$$f(x) = \begin{cases} 2x+3, & x<1 \\ 4, & x=1 \\ 6x-1, & x>1 \end{cases}$$

Is the function continuous at $x=1$? If not, how can you make it continuous?

Solution $\lim\limits_{x\to 1^-} f(x) = \lim\limits_{x\to 1^-}(2x+3) = 5$ and $\lim\limits_{x\to 1^+} f(x) = \lim\limits_{x\to 1^+}(6x-1) = 5$, So, $\lim\limits_{x\to 1} f(x) = 5$. But $f(1) = 4$, so $\lim\limits_{x\to 1} f(x) \neq f(1)$. Hence $f(x)$ is not continuous at $x=1$. This is a case of removable discontinuity.

The given function will be continuous if $f(1)$ is also equal to 5. Thus the given function can be made continuous by defining as:

$$f(x) = \begin{cases} 2x+3, x<1 \\ 5,\ x=1 \\ 6x-1, x>1 \end{cases}$$

Example 1.12 A function $f(x)$ is defined as follows

$$f(x) = \begin{cases} x^2-1, & x<3 \\ 2kx, & x\geqslant 3 \end{cases}$$

Find the values of k so that $f(x)$ is continuous at $x=3$.

Solution $\lim\limits_{x\to 3^-} f(x) = \lim\limits_{x\to 3^-}(x^2-1) = 8$, $\lim\limits_{x\to 3^+} f(x) = \lim\limits_{x\to 3^+} 2kx = 6k$ and $f(3) = 6k$. Since $f(x)$ is continuous at $x=3$, so $6k = 8 \Rightarrow k = \frac{4}{3}$.

1.3.3 Continuity in an Interval

Definition 1.5 A function $f(x)$ is said to be continuous on an **open interval** (a,b), if it is continuous at every point in (a,b).

A function $f(x)$ is said to be continuous in an **closed interval** $[a,b]$, if it is continuous at every point in (a,b) and continuous at the point $x=a$ from the right and continuous at the point $x=b$ from the left. i. e.,

$$\lim_{x\to a^+} f(x) = f(a) \text{ and } \lim_{x\to b^-} f(x) = f(b)$$

Exercise 1

1. Find the following limits.

(1) $\lim\limits_{x\to 2}(2x^2+3x-14)$;

(2) $\lim\limits_{x\to 1}\dfrac{3x^2+2x-4}{x^2+5x-4}$;

(3) $\lim\limits_{x\to 0}\dfrac{4x^3-x^2+2x}{3x^2+4x}$;

(4) $\lim\limits_{x\to a}\dfrac{x^{\frac{2}{3}}-a^{\frac{2}{3}}}{x-a}$;

(5) $\lim\limits_{x\to 2}\dfrac{x^2-5x+6}{x^2-x-2}$;

(6) $\lim\limits_{x\to a}\dfrac{\sqrt{3x}-\sqrt{2x+a}}{2(x-a)}$;

(7) $\lim\limits_{x\to 1}\dfrac{\sqrt{2x}-\sqrt{3-x^2}}{x-1}$;　　(8) $\lim\limits_{x\to 0}\dfrac{\sqrt[6]{x}-2}{\sqrt[3]{x}-4}$;

(9) $\lim\limits_{x\to 1}\dfrac{x-\sqrt{2-x^2}}{2x-\sqrt{2+2x^2}}$.

2. Compute the following limits.

(1) $\lim\limits_{x\to\infty}\dfrac{3x^2-4}{4x^2}$;　　(2) $\lim\limits_{x\to\infty}\dfrac{5x^2+2x-7}{3x^2+5x+2}$;

(3) $\lim\limits_{x\to+\infty}(\sqrt{x-a}-\sqrt{x-b})$;　　(4) $\lim\limits_{x\to+\infty}\sqrt{x}(\sqrt{x}-\sqrt{x-a})$.

3. Evaluate the following limits.

(1) $\lim\limits_{x\to 0}\dfrac{\sin mx}{\sin nx}$;　　(2) $\lim\limits_{x\to 0}\dfrac{\tan ax}{\tan bx}$;

(3) $\lim\limits_{x\to 0}\dfrac{\sin ax\cos bx}{\sin cx}$;　　(4) $\lim\limits_{x\to 0}\dfrac{1-\cos 2x}{x^2}$;

(5) $\lim\limits_{x\to 0}\dfrac{\sin ax-\sin bx}{x}$;　　(6) $\lim\limits_{x\to 0}\dfrac{\tan x-\sin x}{x^3}$;

(7) $\lim\limits_{x\to\frac{\pi}{4}}\dfrac{\sec^2 x-2}{\tan x-1}$;　　(8) $\lim\limits_{x\to y}\dfrac{\sin x-\sin y}{x-y}$;

(9) $\lim\limits_{\theta\to\frac{\pi}{4}}\dfrac{\cos\theta-\sin\theta}{\theta-\dfrac{\pi}{4}}$;　　(10) $\lim\limits_{x\to c}\dfrac{\sqrt{x}-\sqrt{c}}{\sin x-\sin c}$.

4. Find the limits of the followings.

(1) $\lim\limits_{x\to 0}\dfrac{e^{6x}-1}{x}$;　　(2) $\lim\limits_{x\to 0}\dfrac{e^{2x}-1}{x\cdot 2^{x+1}}$;

(3) $\lim\limits_{x\to 0}\dfrac{e^{ax}-e^{bx}}{x}$;　　(4) $\lim\limits_{x\to 0}\dfrac{a^x+b^x-2}{x}$;

(5) $\lim\limits_{x\to 2}\dfrac{x-2}{\ln(x-1)}$;　　(6) $\lim\limits_{x\to\frac{\pi}{2}}\dfrac{\cos x}{\ln(x-\dfrac{\pi}{2}+1)}$.

5. Test the continuity or discontinuity of the following functions by calculating the left limits, the right limits and the values of the functions at points specified.

(1) $f(x)=2-3x^2$ at $x=1$;　　(2) $f(x)=\dfrac{1}{2x}$ at $x=0$;

(3) $f(x)=\dfrac{x^2-9}{x-3}$ at $x=3$;　　(4) $f(x)=\dfrac{|x-2|}{x-2}$ at $x=2$.

6. Discuss the continuity of functions at the points specified.

(1) $f(x)=\begin{cases}2x^2+1, & x\leqslant 2\\ 4x+1, & x>2\end{cases}$ at $x=2$;

(2) $f(x)=\begin{cases}2x+1, & x<1\\ 2, & x=1\\ 3x, & x>1\end{cases}$ at $x=1$.

7. (1) A function $f(x)$ is defined as follow

$$f(x)=\begin{cases}x^2+2, & x<5\\ 20, & x=5\\ 3x+12, & x>5\end{cases}$$

Show that $f(x)$ has removable discontinuity at $x=5$.

(2) A function $f(x)$ is defined as follow

$$f(x)=\begin{cases}2x-3, & x<2\\ 2, & x=2\\ 3x-5, & x>2\end{cases}$$

Is the $f(x)$ function continuous at $x=2$? If not, how can the fucntion $f(x)$ be made continuous at $x=2$?

8. (1) A function $f(x)$ is defined as follow

$$f(x)=\begin{cases}kx+3, & x\geqslant 2\\ 3x-1, & x<2\end{cases}$$

Find the value of k so that $f(x)$ is continuous at $x=2$.

(2) A function $f(x)$ is defined as follow

$$f(x)=\begin{cases}\dfrac{2x^2-18}{x-3}, & x\neq 3\\ k, & x=3\end{cases}$$

Find the value of k so that $f(x)$ is continuous at $x=3$.

Chapter 2 The Derivative

Differential calculus is a theory which has its origin in the solution of two old problems—one of drawing a tangent line to a curve and the other of calculating the velocity of non-uniform motion of particle. In both of the problems, the curves involved are continuous curves and the process used is the limiting process. So the objects of study in the differential calculus are continuous functions. These problems were solved in a certain sense by Isaac Newton (English, 1642—1727) and Gottfried Wilhelm Leibniz (German, 1646—1716), and in the process differential calculus is discovered.

2.1 Tangent Line to a Curve

Let AB be a continuous curve given by $y = f(x)$, and P, Q be any two points in it. Let the coordinates of P and Q be (x_0, y_0) and (x, y). When a point moves along the curve from point P to point Q, it moves horizontally through the distance PR and vertically through the distance RQ. See figure 2 - 1.

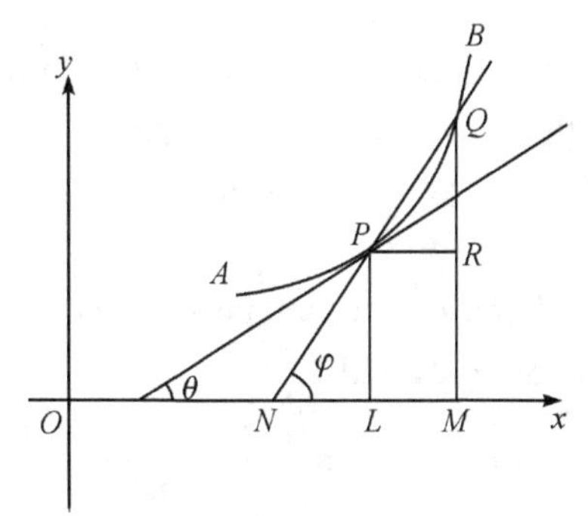

Figure 2 - 1

$$PR = LM = OM - OL = x - x_0$$
$$RQ = QM - RM = y - PL = y - y_0$$

These quantities $x - x_0$ and $y - y_0$ are called the **increment** in x and y respectively, and are denoted by Δx and Δy, i. e.,

$$\Delta x = x - x_0 \text{ and } \Delta y = y - y_0$$

Also

$$\Delta y = y - y_0 = f(x) - f(x_0) = f(x_0 + \Delta x) - f(x_0)$$

If we join the points P and Q, we get secant PQ which makes an angle φ with the x-axis, i. e., $\angle QPR = \angle QNM = \varphi$, and the slope of the secant PQ is

$$\tan\varphi = \frac{RQ}{PR} = \frac{\Delta y}{\Delta x}$$

As Q moves along the curve and approaches P, the secant rotates about P. The limiting position of the secant, when Q ultimately coincides with P, is the tangent at P, making the angle θ with the x-axis. In that situation, Δx tends to zero. So

$$\lim_{\Delta x \to 0} \frac{\Delta y}{\Delta x} = \lim_{\Delta x \to 0} \tan\varphi = \tan\theta \text{ or } \lim_{\Delta x \to 0} \frac{\Delta y}{\Delta x} = \lim_{\Delta x \to 0} \frac{f(x_0 + \Delta x) - f(x_0)}{\Delta x} = \tan\theta$$

Thus, $\lim\limits_{\Delta x \to 0} \frac{\Delta y}{\Delta x}$ or $\lim\limits_{\Delta x \to 0} \frac{f(x_0 + \Delta x) - f(x_0)}{\Delta x}$ gives the slope of a tangent to the curve given by the function f.

2.2 Instantaneous Velocity

Suppose a particle is moving along a straight line AB, then the distance described increase with time. So the distance s can be considered a function f of the time t, that is $s = f(t)$.

A P Q B

At times t and $t + \Delta t$, suppose the particle is at points P and Q respectively such that $AP = s$ and $AQ = s + \Delta s$. Then,

$$PQ = AQ - AP = s + \Delta s - s = \Delta s$$

Also,

$$\Delta s = s + \Delta s - s = f(t + \Delta t) - f(t)$$

So the average velocity $\bar{v}$ during the time interval $(t, t + \Delta t)$ is

$$\bar{v} = \frac{\Delta s}{\Delta t} = \frac{f(t + \Delta t) - f(t)}{\Delta t}$$

Now as $\Delta t \to 0$, Q tends to P. So the instantaneous velocity v of the particle at P (or in time t) is the limit to which $\bar{v}$ tends as $\Delta t \to 0$, and

$$v = \lim_{\Delta t \to 0} \frac{\Delta s}{\Delta t} = \lim_{\Delta t \to 0} \frac{f(t + \Delta t) - f(t)}{\Delta t}$$

2.3 Derivative

Both of the above problems lead us to the calculation of the limit of the ratio of the increment of a function to the corresponding increment of the independent variable as the latter tends to zero. This method of calculating a limit is an operation which is called **differentiation** of the function. The operation is denoted by $\frac{\mathrm{d}}{\mathrm{d}x}$. The result of operation is called the **derivative.**

Definition 2.1 Let the function f be defined in the interval (a, b), $x_0 \in (a, b)$. If the limit $\lim\limits_{\Delta x \to 0} \frac{f(x_0 + \Delta x) - f(x_0)}{\Delta x}$ exists, then f is called **derivable** at $x = x_0$, and the limit is called the derivative of f at $x = x_0$. Denote $f'(x_0)$, $\frac{\mathrm{d}y}{\mathrm{d}x}\big|_{x=x_0}$, $y'\big|_{x=x_0}$, i. e.,

$$f'(x_0)=\lim_{\Delta x\to 0}\frac{f(x_0+\Delta x)-f(x_0)}{\Delta x}$$

Let $x = x_0 + \Delta x$,then $\Delta x = x - x_0$,the expression above can be written as

$$f'(x_0)=\lim_{x\to x_0}\frac{f(x)-f(x_0)}{x-x_0}$$

If function $f(x)$ is derivable at every point in (a,b) , then $f(x)$ is called a derivable funtion in (a,b) , and so

$$f'(x)=\lim_{\Delta x\to 0}\frac{f(x+\Delta x)-f(x)}{\Delta x}$$

The derivable funtion of f can be denoted as

$$f'(x),\ \frac{\mathrm{d}f(x)}{\mathrm{d}x},\ y',\ \frac{\mathrm{d}y}{\mathrm{d}x}$$

In such a situation, $f'(x_0) = f'(x)|_{x=x_0}$.

With this definition, we can state the above problems as follows:

(1) The slope of the tangent at a point (x,y) of a curve given by a function $y = f(x)$ is equal to the derivative of the function with respect to x, i. e. , $f'(x)$ or $\frac{\mathrm{d}f(x)}{\mathrm{d}x}$.

(2) The velocity of a particle describing a path given by $s = f(t)$ at a time t is equal to the derivative of the function with respect to t, i. e. , s' or $\frac{\mathrm{d}f(t)}{\mathrm{d}t}$.

Definition 2. 2 If the function $y = f(x)$ is derivable, then $f'(x)\mathrm{d}x$ is called the differentiation of $f(x)$ and denoted $\mathrm{d}y = f'(x)\mathrm{d}x$.

According to the definition of differentiation, we have $\mathrm{d}x = \Delta x$. Figure 2 - 2 indicate that the differentition of a function $y = f(x)$ at x_0 repesents the increment of the ordinate of the tangent line to the curve $y = f(x)$ at the corresponding point P, i. e. SR.

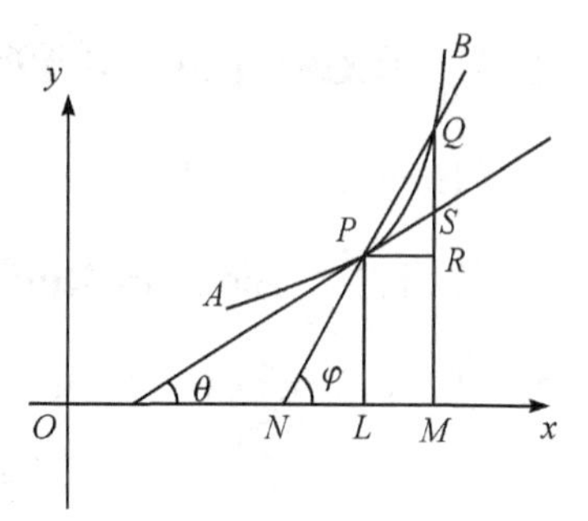

Figure 2 - 2

Triangle PRS is said to be an **arc differential triangle.** It indicates that we can replace the curve $\overset{\frown}{PQ}$ approximately by the tangent line PS in the neighbourhood of the point P, i. e.

$$\overset{\frown}{PQ}\approx \mathrm{d}s=\sqrt{(\mathrm{d}x)^2+(\mathrm{d}y)^2}$$

Now let us calculate the derivatives of the constant functions C and the derivative of $x^n\ (n\in \mathbf{Z}^+)$.

(1) Let $y = f(x) = C$, here C is a constant. Then

$$y+\Delta y=C$$

$$\Delta y=C-y=C-C=0\Rightarrow \frac{\Delta y}{\Delta x}=0$$

So

$$\frac{\mathrm{d}y}{\mathrm{d}x}=\lim_{\Delta x\to 0}\frac{\Delta y}{\Delta x}=0$$

(2) Let $y = f(x) = x^n$. Let Δx be a small increment in x and Δy be the corresponding small increment in y. Then

$$y + \Delta y = (x + \Delta x)^n \Rightarrow \Delta y = (x + \Delta x)^n - y = (x + \Delta x)^n - x^n$$

$$\frac{dy}{dx} = \lim_{\Delta x \to 0} \frac{\Delta y}{\Delta x} = \lim_{\Delta x \to 0} \frac{(x + \Delta x)^n - x^n}{(x + \Delta x) - x} = \lim_{t \to x} \frac{t^n - x^n}{t - x} \text{ (let } t = x + \Delta x,\ \Delta x \to 0,\ t \to x)$$

As

$$\lim_{x \to a} \frac{x^n - a^n}{x - a} = nx^{n-1}$$

Then

$$\frac{dy}{dx} = \lim_{\Delta x \to 0} \frac{\Delta y}{\Delta x} = \lim_{t \to x} \frac{t^n - x^n}{t - x} = nx^{n-1}$$

Note: In the latter example, we will prove that this is true for any real number n.

> **Theorem 2.1**
>
> If $f(x)$ is derivable at $x = x_0$, then $f(x)$ is continuous at $x = x_0$.

Proof Since $f(x)$ is derivable at $f(x)$, $f'(x_0) = \lim\limits_{\Delta x \to 0} \frac{\Delta y}{\Delta x}$ exist. Therefore

$$\lim_{\Delta x \to 0} \Delta y = \lim_{\Delta x \to 0} \frac{\Delta y}{\Delta x} \Delta x = \lim_{\Delta x \to 0} \frac{\Delta y}{\Delta x} \lim_{\Delta x \to 0} \Delta x = 0$$

According to the definition of continuity, $f(x)$ is continuous at $x = x_0$.

2.4 Techniques of Differentiation

Here we shall deduce some fundamental formula of derivative.

2.4.1 The Sum Rule

If $f(x)$ and $g(x)$ be any two derivable functions of x, then

$$\frac{d}{dx}[f(x) \pm g(x)] = \frac{d}{dx} f(x) \pm \frac{d}{dx} g(x)$$

or

$$[f(x) \pm g(x)]' = f'(x) \pm g'(x)$$

Proof Let $h(x) = f(x) + g(x)$, then

$$\begin{aligned}
h'(x) &= \lim_{\Delta x \to 0} \frac{h(x + \Delta x) - h(x)}{\Delta x} \\
&= \lim_{\Delta x \to 0} \frac{[f(x + \Delta x) \pm g(x + \Delta x)] - [f(x) \pm g(x)]}{\Delta x} \\
&= \lim_{\Delta x \to 0} \left[\frac{f(x + \Delta x) - f(x)}{\Delta x} \pm \frac{g(x + \Delta x) - g(x)}{\Delta x}\right] \\
&= \lim_{\Delta x \to 0} \left[\frac{f(x + \Delta x) - f(x)}{\Delta x}\right] \pm \lim_{\Delta x \to 0} \left[\frac{g(x + \Delta x) - g(x)}{\Delta x}\right] \\
&= f'(x) \pm g'(x)
\end{aligned}$$

The sum rule can be extended to the sum of any number of function. For instance, using this rule twice, we get

$$[f(x)+g(x)+p(x)]' = f'(x)+g'(x)+p'(x)$$

Example 2.1 Find the derivative of $5x^3+4x^2-2x+7$.

Solution Let $y=5x^3+4x^2-2x+7$, then

$$\begin{aligned} y' &= (5x^3+4x^2-2x+7)' = (5x^3)'+(4x^2)'+(-2x)'+(7)' \\ &= 5(x^3)'+4(x^2)'-2(x)' = 15x^2+8x-2 \end{aligned}$$

2.4.2 The Product Rule

If $f(x)$ and $g(x)$ be any two derivable functions of x, then

$$\frac{\mathrm{d}}{\mathrm{d}x}[f(x)g(x)] = f(x)\frac{\mathrm{d}}{\mathrm{d}x}[g(x)]+g(x)\frac{\mathrm{d}}{\mathrm{d}x}[f(x)]$$

or

$$[f(x)g(x)]' = f'(x)g(x)+f(x)g'(x)$$

Proof Let $h(x)=f(x)g(x)$, then

$$h'(x)=\lim_{\Delta x\to 0}\frac{h(x+\Delta x)-h(x)}{\Delta x}=\lim_{\Delta x\to 0}\frac{f(x+\Delta x)g(x+\Delta x)-f(x)g(x)}{\Delta x}$$

In order to evaluate this limit, we would like to separate the function and as in the proof of the sum rule. We can achieve this separation by subtracting and adding the term $f(x+\Delta x)g(x)$ in the numerator.

$$\begin{aligned} h'(x) &= \lim_{\Delta x\to 0}\frac{f(x+\Delta x)g(x+\Delta x)-f(x+\Delta x)g(x)+f(x+\Delta x)g(x)-f(x)g(x)}{\Delta x} \\ &= \lim_{\Delta x\to 0}\left[f(x+\Delta x)\frac{g(x+\Delta x)-g(x)}{\Delta x}+g(x)\frac{f(x+\Delta x)-f(x)}{\Delta x}\right] \\ &= \lim_{\Delta x\to 0}f(x+\Delta x)\cdot\lim_{\Delta x\to 0}\frac{g(x+\Delta x)-g(x)}{\Delta x}+\lim_{\Delta x\to 0}g(x)\cdot\lim_{\Delta x\to 0}\frac{f(x+\Delta x)-f(x)}{\Delta x} \\ &= f(x)g'(x)+g(x)f'(x) \end{aligned}$$

Note that because $\lim\limits_{\Delta x\to 0}g(x)=g(x)$ is a constant with respect to the variable x. Also, since f is derivable at x, it is continuous at x and so

$$\lim_{\Delta x\to 0}f(x+\Delta x)=f(x)$$

with the application of the product rule, we can get easily that

$$[Cf(x)]'=Cf'(x), \text{ (where } C \text{ could be any constant)}$$

Example 2.2 Find $f'(x)$ if $f(x)=(3x^2-5x)(2x+3)$.

Solution

$$\begin{aligned} f'(x) &= (3x^2-5x)'(2x+3)+(3x^2-5x)(2x+3)' \\ &= (6x-5)(2x+3)+2(3x^2-5x) = 18x^2-2x-15 \end{aligned}$$

2.4.3 The Quotient Rule

Let $u(x)$ and $v(x)$ be any two derivable functions of x. Let $h(x)=\dfrac{u(x)}{v(x)}$, then

$$h'(x)=\left[\frac{u(x)}{v(x)}\right]'=\frac{u'(x)v(x)-u(x)v'(x)}{v^2(x)}$$

Proof We have, by definition,

$$h'(a)=\lim_{x\to a}\frac{h(x)-h(a)}{x-a}=\lim_{x\to a}\frac{\dfrac{u(x)}{v(x)}-\dfrac{u(a)}{v(a)}}{x-a}$$

$$=\lim_{x\to a}\frac{v(a)u(x)-u(a)v(x)}{(x-a)v(x)v(a)}$$

$$=\lim_{x\to a}\frac{v(a)u(x)-v(a)u(a)+v(a)u(a)-u(a)v(x)}{(x-a)v(x)v(a)}$$

$$=\lim_{x\to a}\frac{1}{v(x)v(a)}\left[v(a)\frac{u(x)-u(a)}{(x-a)}-u(a)\frac{v(x)-v(a)}{(x-a)}\right]$$

$$=\frac{1}{v(a)v(a)}[v(a)u'(a)-u(a)v'(a)]$$

$$=\frac{v(a)u'(a)-u(a)v'(a)}{v^2(a)}$$

Hence, in general, we have

$$h'(x)=\left[\frac{u(x)}{v(x)}\right]'=\frac{u'(x)v(x)-u(x)v'(x)}{v^2(x)}$$

or

$$\frac{d}{dx}\left[\frac{u(x)}{v(x)}\right]=\frac{v(x)\dfrac{d}{dx}u(x)-u(x)\dfrac{d}{dx}v(x)}{v^2(x)}$$

Example 2.3 Find the derivative of $\dfrac{4x^2+3}{3x^2-2}$.

Solution Let $u(x)=4x^2+3$ and $v(x)=3x^2-2$. Then

$$\frac{d}{dx}\left[\frac{u(x)}{v(x)}\right]=\frac{v(x)\dfrac{d}{dx}u(x)-u(x)\dfrac{d}{dx}v(x)}{v^2(x)}$$

$$=\frac{(3x^2-2)\cdot 8x-(4x^2+3)\cdot 6x}{(3x^2-2)^2}$$

$$=\frac{-34x}{(3x^2-2)^2}$$

Example 2.4 Find the derivative of $\dfrac{\sqrt{x}+a}{\sqrt{x}+b}$.

Solution $\left(\dfrac{\sqrt{x}+a}{\sqrt{x}+b}\right)'=\dfrac{(\sqrt{x}+a)'(\sqrt{x}+b)-(\sqrt{x}+a)(\sqrt{x}+b)'}{(\sqrt{x}+b)^2}=\dfrac{b-a}{2\sqrt{x}(\sqrt{x}+b)^2}$

2.4.4 The Chain Rule

Theorem 2.2

If $y=f(u)$ and $u=g(x)$, where f and g are derivable functions, then $\dfrac{dy}{dx}$ exists and

$$\frac{dy}{dx}=\frac{dy}{du}\cdot\frac{du}{dx}=f'(u)u'(x)$$

Proof We have $u = g(x)$. Let Δx be a small increment in x and Δu be the corresponding small increment in u. Then

$$\Delta u = g(x + \Delta x) - g(x)$$

Since $g(x)$ is derivable, it is continuous. So

$$\lim_{\Delta u \to 0} \Delta u = \lim_{\Delta x \to 0} [g(x + \Delta x) - g(x)] = g(x) - g(x) = 0$$

Thus $\Delta u \to 0$, as $\Delta x \to 0$. Then

$$\frac{dy}{dx} = \lim_{\Delta x \to 0} \frac{\Delta y}{\Delta x} = \lim_{\Delta x \to 0} \left[\frac{\Delta y}{\Delta u} \frac{\Delta u}{\Delta x}\right]$$

$$= \lim_{\Delta u \to 0} \frac{\Delta y}{\Delta u} \lim_{\Delta x \to 0} \frac{\Delta u}{\Delta x} = \frac{dy}{du} \cdot \frac{du}{dx}$$

Example 2.5 Find $\frac{dy}{dx}$, if $y = 4u^2 - 3u + 5$ and $u = 2x^2 - 3$.

Solution We have $\frac{dy}{du} = 8u - 3$ and $\frac{du}{dx} = 4x$, so

$$\frac{dy}{dx} = \frac{dy}{du} \cdot \frac{du}{dx} = (8u - 3)4x = [8(2x^2 - 3) - 3]4x = 64x^3 - 108x$$

After we are familiar with the process of using the chain rule, we may skip the step of $\frac{dy}{du}$, differentiating from outside to inside directly.

Example 2.6 Find the derivative of $(2 - 3x)^{1/2}$.

Solution $[(2 - 3x)^{1/2}]' = \frac{1}{2}(2 - 3x)^{\frac{1}{2}-1}(2 - 3x)' = \frac{-3}{2\sqrt{2 - 3x}}$.

Example 2.7 Find the derivative of $\sqrt{\frac{1 - x}{1 + x}}$.

Solution

$$\left(\sqrt{\frac{1-x}{1+x}}\right)' = \frac{1}{2\sqrt{\frac{1-x}{1+x}}} \cdot \left(\frac{1-x}{1+x}\right)'$$

$$= \frac{1}{2\sqrt{\frac{1-x}{1+x}}} \cdot \frac{(1-x)'(1+x) - (1-x)(1+x)'}{(1+x)^2}$$

$$= -\frac{1}{(1+x)\sqrt{1-x^2}}$$

Example 2.8 Find the derivative of $\frac{1}{x + \sqrt{x^2 - a^2}}$.

Solution

$$\left(\frac{1}{x + \sqrt{x^2 - a^2}}\right)' = -\frac{1}{(x + \sqrt{x^2 - a^2})^2}(x + \sqrt{x^2 - a^2})'$$

$$= -\frac{1}{(x + \sqrt{x^2 - a^2})^2}\left(1 + \frac{1}{2\sqrt{x^2 - a^2}}(x^2 - a^2)'\right)$$

$$= -\frac{1}{(x + \sqrt{x^2 - a^2})\sqrt{x^2 - a^2}}$$

The Techniques of Derivative

1. If $f(x)$ and $g(x)$ be any two derivable functions of x, then

(1) $[f(x)\pm g(x)]'=f'(x)\pm g'(x)$,

or $\quad d[f(x)\pm g(x)]=f'(x)dx\pm g'(x)dx$.

(2) $[f(x)g(x)]'=f'(x)g(x)+f(x)g'(x)$,

or $\quad d[f(x)g(x)]=g(x)df(x)+f(x)dg(x)$.

(3) $\left[\dfrac{u(x)}{v(x)}\right]'=\dfrac{u'(x)v(x)-u(x)v'(x)}{v^2(x)}, \quad v(x)\neq 0$

or $\quad d\left[\dfrac{u(x)}{v(x)}\right]=\dfrac{v(x)du(x)-u(x)dv(x)}{v^2(x)}$.

2. If $y=f(u)$ and $u=g(x)$, where f and g are derivable functions, then $\dfrac{dy}{dx}$ exists and

$$\frac{dy}{dx}=\frac{dy}{du}\cdot\frac{du}{dx}=f'(u)u'(x)$$

2.5 Second and Higher Derivatives

Definition 2.3 Suppose $y=f(x)$ be a derivable function, then $\dfrac{dy}{dx}=f'(x)$ is called the first derivative of $f(x)$ with respect to x. If we differentiate $\dfrac{dy}{dx}$ again, we get $\dfrac{d}{dx}\left(\dfrac{dy}{dx}\right)$ written as $\dfrac{d^2y}{dx^2}$ or $f''(x)$, which is called the **second derivative** with respect to x and so on. In this way, you can define higher derivatives.

Example 2.9 Find the second and higher derivatives of $y=5x^4-3x^2+11$.

Solution We have $y=5x^4-3x^2+11$

$\dfrac{dy}{dx}=y'=20x^3-6x$ the first derivative

$\dfrac{d^2y}{dx^2}=y''=60x^2-6$ the second derivative

$\dfrac{d^3y}{dx^3}=y'''=120x$ the third derivative

$\dfrac{d^4y}{dx^4}=y^{(4)}=120$ the fourth derivative

$\dfrac{d^5y}{dx^5}=y^{(5)}=0$ the fifth derivative

All other derivatives are also zero.

If $y = f(x) = a_n x^n + a_{n-1} x^{n-1} + \cdots + a_1 x + a_0$, then

$$\frac{d^n y}{dx^n} = f^{(n)}(x) = n! \cdot a_n$$

$$\frac{d^{n+1} y}{dx^{n+1}} = f^{(n+1)}(x) = 0$$

2.6 Implicit Function and Implicit Differentiation

Definition 2.4 Suppose $F(x, y)$ be an arbitrary function of two variables x and y, if the function $y = f(x)$ is defined by this equation

$$F(x, y) = 0$$

then the equation is called an **implicit function.**

Note: This equation may or may not be solvable for y. But we can differentiate $F(x, y) = 0$ term by term with respect to x and solve for $\frac{dy}{dx}$. This process of finding the value of $\frac{dy}{dx}$ without solving the equation for y is called **implicit differentiation.**

Example 2.10 Use implicit differentiation to find $\frac{dy}{dx}$ of $2x^2 - 3y^2 = 16$.

Solution The given equation is $2x^2 - 3y^2 = 16$, differentiating both sides with respect to x, we get

$$4x - 6y \cdot \frac{dy}{dx} = 0$$

$$\frac{dy}{dx} = \frac{2x}{3y}$$

Example 2.11 Find $\frac{dy}{dx}$ of $x^3 + y^3 - 3axy = 0$.

Solution Differentiating both sides with respect to x, we get

$$3x^2 + 3y^2 \frac{dy}{dx} - 3a \frac{d}{dx}(xy) = 0$$

$$3x^2 + 3y^2 \frac{dy}{dx} - 3a\left(y + x \frac{dy}{dx}\right) = 0$$

$$\frac{dy}{dx} = \frac{ay - x^2}{y^2 - ax}$$

Example 2.12 Find $\frac{dy}{dx}$ of $xy^2 = (x + 2y)^3$.

Solution Differentiating both sides with respect to x, we get

$$y^2 + x \cdot 2y \frac{dy}{dx} = 3(x + 2y)^2 \frac{d}{dx}(x + 2y)$$

$$y^2 + x \cdot 2y \frac{dy}{dx} = 3(x + 2y)^2 \left(1 + 2\frac{dy}{dx}\right)$$

$$\frac{dy}{dx} = \frac{y}{x} \quad (\text{Because } xy^2 = (x + 2y)^3)$$

2.7 Parametric Function and Parametric Differentiation

Definition 2.5 Suppose $x = x(t)$, $y = y(t)$ be two derivable functions with respect to t, and $\frac{dx}{dt} \neq 0$, then the system $\begin{cases} x = x(t) \\ y = y(t) \end{cases}$ is called a **parametric function.**

Note: t may or may not be eliminated . But we can regard t as a intermediate variable and then differentiate it by the Chain Rule.

Example 2.13 Find $\frac{dy}{dx}$ if $x = t + \frac{1}{t}$ and $y = t - \frac{1}{t}$.

Solution $\frac{dy}{dx} = \frac{dy}{dt} \cdot \frac{dt}{dx} = \frac{y'(t)}{x'(t)} = \frac{1 + \frac{1}{t^2}}{1 - \frac{1}{t^2}} = \frac{t^2 + 1}{t^2 - 1}$.

Example 2.14 Differentiate $(3x - 1)^2$ with respect to $2x + 1$.

Solution Let $y = (3x - 1)^2$ and $u = 2x - 1$, then,

$$\frac{dy}{du} = \frac{dy/dx}{du/dx} = \frac{6(3x - 1)}{2} = 3(3x - 1)$$

2.8 Derivatives of the Trigonometrical Functions

2.8.1 Derivatives of sin$x$ and cos$x$

Let

$$y = \sin x \tag{2.1}$$

Let Δx be a small increment in x and Δy be the corresponding increment in y. Then

$$y + \Delta y = \sin(x + \Delta x) \tag{2.2}$$

Now subtracting (2.1) from (2.2) we get

$$\Delta y = \sin(x + \Delta x) - \sin x = 2\sin\frac{\Delta x}{2}\cos\frac{2x + \Delta x}{2}$$

$$\frac{dy}{dx} = \lim_{\Delta x \to 0} \frac{\Delta y}{\Delta x} = \lim_{\Delta x \to 0} \left(\frac{\sin\frac{\Delta x}{2}}{\frac{\Delta x}{2}} \cos\frac{2x + \Delta x}{2} \right) = \cos x$$

With the same process that we followed to derive the derivative of sin$x$, we get the derivative of cos$x$ as

$$\frac{d\cos x}{dx} = -\sin x \text{ or } (\cos x)' = -\sin x$$

2.8.2 Derivatives of tan$x$ and cot$x$

Use the quotient rule, we have

$$\frac{\mathrm{d}\tan x}{\mathrm{d}x}=\left(\frac{\sin x}{\cos x}\right)'=\frac{(\sin x)'\cos x-\sin x(\cos x)'}{\cos^2 x}=\frac{\cos^2 x+\sin^2 x}{\cos^2 x}=\sec^2 x$$

With the same process that we followed to derive the derivative of tan$x$, we can get the derivative of cot$x$, as

$$\frac{\mathrm{d}\cot x}{\mathrm{d}x}=-\csc^2 x$$

2.8.3 Derivatives of sec$x$ and csc$x$

For
$$\sec x=\frac{1}{\cos x}$$

By the quotient rule and the chain rule, we have

$$(\sec x)'=\left(\frac{1}{\cos x}\right)'=-\frac{1}{\cos^2 x}(\cos x)'=\frac{\sin x}{\cos x\cdot\cos x}=\tan x\sec x$$

$$(\sec x)'=\tan x\sec x$$

With the same process that we followed to derive the derivative of sec$x$, we can get the derivative of csc$x$, as

$$(\csc x)'=-\cot x\csc x$$

The Derivatives of Formulas of Trigonometric Functions

$(\sin x)'=\cos x$	$(\cos x)'=-\sin x$
$(\tan x)'=\sec^2 x$	$(\cot x)'=-\csc^2 x$
$(\sec x)'=\tan x\sec x$	$(\csc x)'=-\cot x\csc x$

Example 2.14 Find the derivatives of the following:

(1) $\sin(ax+b)$; (2) $\sin(ax-b)$; (3) $\sqrt{\tan 2x}$;

(4) $x^2\sec(ax-b)$; (5) $\dfrac{1+\sin x}{1-\sin x}$; (6) $\dfrac{1-\tan x}{\sec x}$.

Solution

(1) $[\sin(ax+b)]'=\cos(ax+b)\cdot(ax+b)'=a\cos(ax+b)$.

(2) $[\sin(ax^2-b)]'=\cos(ax^2-b)\cdot(ax^2-b)'=2ax\cos(ax^2-b)$.

(3) $(\sqrt{\tan 2x})'=\dfrac{1}{2\sqrt{\tan 2x}}(\tan 2x)'=\dfrac{\sec^2 2x}{\sqrt{\tan 2x}}$.

(4) $[x^2\sec(ax-b)]'=2x\sec(ax-b)+ax^2\sec(ax-b)\tan(ax-b)$.

(5) $\left(\dfrac{1+\sin x}{1-\sin x}\right)'=\dfrac{\cos x(1-\sin x)-(1+\sin x)(-\cos x)}{(1-\sin x)^2}=\dfrac{2\cos x}{(1-\sin x)^2}$.

(6) $\left(\dfrac{1-\tan x}{\sec x}\right)'=[\cos x(1-\tan x)]'=-\sin x(1-\tan x)+\cos x(-\sec^2 x)$

$$= \sin x \tan x - \sin x - \cos x \sec^2 x.$$

Example 2.15 Find $\frac{dy}{dx}$, when $x - y = \sin xy$.

Solution Now differentiating both sides with respect to x, we get

$$1 - y' = \cos xy(xy)' = \cos xy \cdot (y + xy')$$

$$y' = \frac{1 - y\cos xy}{1 + x\cos xy}$$

Example 2.16 Find $\frac{dy}{dx}$, when $y = 2\theta - \tan\theta$ and $x = \tan\theta$.

Solution We have

$$\frac{dy}{dx} = \frac{y'(\theta)}{x'(\theta)} = \frac{2 - \sec^2\theta}{\sec^2\theta} = 2\cos^2\theta - 1 = \cos 2\theta$$

2.9 Derivatives of Inverse Trigonometric Functions

2.9.1 Derivatives of arcsin$x$ and arccos$x$

Let $y = \arcsin x$, then $x = \sin y$. Differentiating both sides with respect to y, we get

$$\frac{dx}{dy} = \cos y = \sqrt{1 - \sin^2 y} = \sqrt{1 - x^2}$$

$$\frac{dy}{dx} = \frac{1}{\sqrt{1 - x^2}}$$

Similarly, we get

$$\frac{d}{dx}(\arccos x) = -\frac{1}{\sqrt{1 - x^2}}$$

2.9.2 Derivatives of arctan$x$ and arccot$x$

Let $y = \arctan x$, then $x = \tan y$. Differentiating both sides with respect to y, we get

$$\frac{dx}{dy} = \sec^2 y = 1 + \tan^2 y = 1 + x^2$$

$$\frac{dy}{dx} = \frac{1}{1 + x^2}$$

Similarly, we get

$$\frac{d}{dx}(\text{arccot}\, x) = -\frac{1}{1 + x^2}$$

The Derivatives of Formulas of Inverse Trigonometric Functions

$$(\arcsin x)' = \frac{1}{\sqrt{1 - x^2}} \qquad (\arccos x)' = -\frac{1}{\sqrt{1 - x^2}}$$

$$(\arctan x)' = \frac{1}{1 + x^2} \qquad (\text{arccot}\, x)' = -\frac{1}{1 + x^2}$$

Example 2.17 Find $\frac{dy}{dx}$, when $y = x^3 \operatorname{arccot} x$.

Solution We have $y = x^3 \operatorname{arccot} x$. Now differentiating both sides with respect to x, we get

$$\frac{dy}{dx} = 3x^2 \operatorname{arccot} x + x^3\left(-\frac{1}{1+x^2}\right) = 3x^2 \operatorname{arccot} x - \frac{x^3}{1+x^2}$$

Example 2.18 Find $\frac{dy}{dx}$, when $y = \arctan \frac{2x}{1-x^2}$.

Solution

$$y' = \left(\arctan \frac{2x}{1-x^2}\right)' = \frac{1}{1+\left(\frac{2x}{1-x^2}\right)^2}\left(\frac{2x}{1-x^2}\right)'$$

$$= \frac{(1-x^2)^2}{(1+x^2)^2} \frac{[2(1-x^2) - 2x(1-x^2)']}{(1-x^2)^2}$$

$$= \frac{2}{1+x^2}$$

2.10 Derivatives of Logarithmic and Exponential Functions

As we have known, exponential function and logarithmic function are inverse function to each other. So, if $y = f(x) = a^x$, we have $\log_a y = x$.

There is a special type of exponential function e^x, where e is the limiting value of $\lim_{n \to \infty}\left(1+\frac{1}{n}\right)^n$. The value of e lies between 2 and 3 and is approximately 2.718.

The corresponding logarithmic function is called the natural logarithmic function and is denoted by $\ln x$, base e being understood.

Thus
$$e = \lim_{n \to \infty}\left(1+\frac{1}{n}\right)^n$$

Further, we have
$$e = \lim_{x \to \infty}\left(1+\frac{1}{x}\right)^x$$

When $x \to \infty$, $t = \frac{1}{x} \to 0$, so

$$e = \lim_{x \to \infty}\left(1+\frac{1}{x}\right)^x = \lim_{t \to 0}(1+t)^{\frac{1}{t}}$$

We should be familiar with the following properties of the logarithmic functions.

(1) $\log_a xy = \log_a x + \log_a y$	(2) $\log_a \frac{x}{y} = \log_a x - \log_a y$
(3) $\log_a x^n = n \log_a x$	(4) $\log_a a = 1$
(5) $\log_a b = \log_m b / \log_m a = \log_a m \cdot \log_m b$ (changing the base)	(6) $\log_a 1 = 0$

Now, we find the derivatives of the exponential and the logarithmic functions by the definition of derivative using the following limit results.

(1) $\lim\limits_{x\to 0}\dfrac{\ln(1+x)}{x}=1$;　(2) $\lim\limits_{x\to 0}\dfrac{e^x-1}{x}=1$;　(3) $\lim\limits_{x\to 0}\dfrac{a^x-1}{x}=\ln a$.

2.10.1 Derivative of Natural Logarithmic Function

Let $y=\ln x$, then

$$(\ln x)'=\lim_{\Delta x\to 0}\frac{\Delta y}{\Delta x}=\lim_{\Delta x\to 0}\frac{\ln(x+\Delta x)-\ln x}{\Delta x}$$

$$=\lim_{\Delta x\to 0}\frac{\ln\left(1+\frac{\Delta x}{x}\right)}{x\,\frac{\Delta x}{x}}=\frac{1}{x}$$

Example 2.19 Suppose $y=\ln|x|$, find its derivative.

Solution When $x>0$, $y=\ln x\Rightarrow y'=\dfrac{1}{x}$;

When $x<0$, $y=\ln(-x)\Rightarrow y'=\dfrac{1}{-x}(-x)'=\dfrac{1}{x}$;

Hence, $(\ln|x|)'=\dfrac{1}{x}$.

2.10.2 Derivative of e^x

Let $y=e^x$, then

$$(e^x)'=\lim_{\Delta x\to 0}\frac{\Delta y}{\Delta x}=\lim_{\Delta x\to 0}\frac{e^{x+\Delta x}-e^x}{\Delta x}$$

$$=\lim_{\Delta x\to 0}e^x\frac{(e^{\Delta x}-1)}{\Delta x}=e^x$$

2.10.3 Derivative of Logarithmic Function

Let $y=\log_a x$, then

$$y=\log_a x=\ln x/\ln a \text{ (by changing the base from } a \text{ to e)}$$

$$\frac{dy}{dx}=\frac{1}{\ln a}\frac{d(\ln x)}{dx}=\frac{1}{x\ln a}$$

2.10.4 Derivative of a^x

Let $y=a^x$, then

$$(a^x)'=\lim_{\Delta x\to 0}\frac{\Delta y}{\Delta x}=\lim_{\Delta x\to 0}\frac{a^{x+\Delta x}-a^x}{\Delta x}$$

$$=\lim_{\Delta x\to 0}a^x\frac{(a^{\Delta x}-1)}{\Delta x}=a^x\ln a$$

The Derivatives of Formulas of Elementary Functions

1. $(x^{\alpha})'=\alpha x^{\alpha-1}$ $(\alpha\in\mathbf{R})$.
2. $(a^{x})'=a^{x}\ln a$, specially $(e^{x})'=e^{x}$.
3. $(\log_{a}x)'=\dfrac{1}{x\ln a}$, specially $(\ln x)'=\dfrac{1}{x}$.
4. $(\sin x)'=\cos x$, $(\cos x)'=-\sin x$,
   $(\tan x)'=\sec^2 x$, $(\cot x)'=-\csc^2 x$,
   $(\sec x)'=\sec x\tan x$, $(\csc x)'=-\csc x\cot x$.
5. $(\arcsin x)'=\dfrac{1}{\sqrt{1-x^2}}$, $(\arccos x)'=-\dfrac{1}{\sqrt{1-x^2}}$,
   $(\arctan x)'=\dfrac{1}{1+x^2}$, $(\text{arccot}\, x)'=-\dfrac{1}{1+x^2}$.

2.10.5 Derivative of $f(x)^{g(x)}$ $(f(x)>0)$

Definition 2.6 $f(x)^{g(x)}$ $(f(x)>0)$ is called a **power exponential function.**

Method 1 Let $y=f(x)^{g(x)}$, then $\ln y=g(x)\ln f(x)$. Differentiating both sides with respect to x, we get

$$\frac{1}{y}\frac{dy}{dx}=g'(x)\ln f(x)+\frac{g(x)}{f(x)}f'(x)$$

So, $\dfrac{dy}{dx}=y\left[g'(x)\ln f(x)+\dfrac{g(x)}{f(x)}f'(x)\right]=f(x)^{g(x)}\left[g'(x)\ln f(x)+\dfrac{g(x)}{f(x)}f'(x)\right]$.

This method is called the **logarithmic derivative method.**

Method 2 Using the property of exponential function and then differentiating by the chain rule, we get

$$y=f(x)^{g(x)}=e^{\ln f(x)^{g(x)}}=e^{g(x)\ln f(x)}$$

So,
$$\frac{dy}{dx}=f(x)^{g(x)}\left[g'(x)\ln f(x)+\frac{g(x)}{f(x)}f'(x)\right]$$

Example 2.20 Find the derivative of $y=\ln(5x^2+6)$.

Solution $\dfrac{dy}{dx}=[\ln(5x^2+6)]'=\dfrac{10x}{5x^2+6}$.

Example 2.21 Prove $(x^{\alpha})'=\alpha x^{\alpha-1}$ $(x>0)$, where α be any real number.

Proof For
$$x^{\alpha}=e^{\alpha\ln x}$$
so
$$(x^{\alpha})'=(e^{\alpha\ln x})'=e^{\alpha\ln x}\alpha\frac{1}{x}=\alpha x^{\alpha-1}$$

Example 2.22 Find the derivative of $y=\sin 3x\ln(ax+b)$.

Solution $\dfrac{dy}{dx}=[\sin 3x\ln(ax+b)]'=3\cos 3x\ln(ax+b)+\dfrac{a\sin 3x}{ax+b}$.

Example 2.23 Find the derivative of $y=\dfrac{\sin x}{\ln(3x+4)}$.

Solution $\dfrac{dy}{dx}=\left[\dfrac{\sin x}{\ln(3x+4)}\right]'=\dfrac{\cos x\ln(3x+4)-\dfrac{3\sin x}{3x+4}}{[\ln(3x+4)]^2}$

$=\dfrac{(3x+4)\cos x\ln(3x+4)-3\sin x}{(3x+4)[\ln(3x+4)]^2}$

Example 2.24 Find the derivative of $y=e^{5x+4}\sin 6x$.

Solution $\dfrac{dy}{dx}=(e^{5x+4}\sin 6x)'=5e^{5x+4}\sin 6x+6e^{5x+4}\cos 6x=e^{5x+4}(5\sin 6x+6\cos 6x)$.

Example 2.25 Find the derivative of $x^y=y^x\ (x>0, y>0)$.

Solution By the logarithmic derivative method, we have

$$y\ln x=x\ln y$$

Then
$$\frac{dy}{dx}\ln x+\frac{y}{x}=\ln y+\frac{x}{y}\frac{dy}{dx}$$

Hence
$$\frac{dy}{dx}=\frac{\ln y-\dfrac{y}{x}}{\ln x-\dfrac{x}{y}}=\frac{xy\ln y-y^2}{xy\ln x-x^2}$$

Example 2.26 Find the derivative of $y=\sqrt{\dfrac{a-x}{a+x}}$.

Solution Using the properties of logarithmic function we have

$$\ln y=\frac{1}{2}(\ln|a-x|-\ln|a+x|)$$

Differentiating both sides with respect to x, we get

$$\frac{1}{y}y'=\frac{1}{2}\left(\frac{-1}{a-x}-\frac{1}{a+x}\right)=\frac{a}{x^2-a^2}$$

$$y'=\frac{a}{x^2-a^2}\sqrt{\frac{a-x}{a+x}}$$

Exercise 2

1. Find, from definition, the derivatives of the following.

(1) x^2-2; (2) $\dfrac{1}{x}$;

(3) $x^{\frac{1}{2}}$ (4) $\dfrac{2x+3x^{\frac{3}{4}}+x^{\frac{1}{2}}+1}{x^{\frac{1}{4}}}$;

(5) $(3x^4+5)(4x^5-3)$; (6) $\dfrac{x^2-2x}{x+1}$.

2. Use the chain rule to calculate $\dfrac{dy}{dx}$.

(1) $y = 2u^2 - 3u + 1$ and $u = 2x^2$; (2) $y = (2u^2 + 3)^{\frac{1}{3}}$ and $u = \sqrt{2x+1}$;

(3) $(3x^2 + 2x - 1)^4$; (4) $\sqrt{8 - 5x}$;

(5) $\dfrac{1}{\sqrt{ax^2 + bx + c}}$; (6) $\sqrt{\dfrac{x^2 + a^2}{x^2 - a^2}}$.

3. Find the second derivatives of the following.

(1) $y = 7x^2 + 6x - 5$; (2) $y = 3x^4 - x^2 + 1$;

(3) $y = \dfrac{3}{x^2}$; (4) $\dfrac{1}{2x+1}$.

4. Use the implicit differentiation to obtain in the following.

(1) $x^2 + y^2 = 16$; (2) $\dfrac{x^2}{a^2} - \dfrac{y^2}{b^2} = 1$;

(3) $x^2 + 2x^2 y = y^3$; (4) $x^3 y^6 = (x + y)^9$.

5. Find the derivatives of the following function.

(1) $\cos^2 2x$; (2) $\sqrt{\sin 2x}$;

(3) $\tan(5x^2 + 6)$; (4) $\sec^2 \dfrac{1}{x}$;

(5) $\tan(\cos 5x)$; (6) $\csc^3(\cot 4x)$;

(7) $\sec^2(\tan\sqrt{x})$; (8) $\sin^2(\cos 6x)$.

6. Find the derivatives of the following.

(1) $(x^2 + 3x)\sin 5x$; (2) $(x + \sin 2x)\sec 3x^2$;

(3) $\dfrac{1}{\sqrt{x}}\sin\sqrt{x}$; (4) $\dfrac{\sec nx}{ax - b}$;

(5) $\sin 2mx \sin 2nx$; (6) $\sin 3x \cos 5x$.

7. Find the derivatives of the following.

(1) $\dfrac{1 - 2\sin^2 \dfrac{x}{2}}{\cos^2 x}$; (2) $\dfrac{\sin 2nx}{\cos 2nx}$;

(3) $\dfrac{1 - \cos x}{1 + \cos x}$; (4) $\sqrt{\dfrac{1 - \sin x}{1 + \sin x}}$;

(5) $\dfrac{\cos 2x}{1 - \sin 2x}$; (6) $\dfrac{1 + \cos x}{1 - \cos x}$;

(7) $\dfrac{\sec x + \tan x}{\sec x - \tan x}$; (8) $\dfrac{1 + \tan x}{1 - \tan x}$.

8. Find the derivatives of following.

(1) $\arcsin(3x - 4)$; (2) $\arccos \dfrac{3x^2 - 2}{2}$;

(3) $\arccos \dfrac{1 - x^2}{1 + x^2}$; (4) $\arctan \dfrac{1}{1 - x^2}$;

(5) $\arctan \dfrac{\sin 2x}{1+\cos 2x}$; (6) $\operatorname{arcsec} \dfrac{1}{\sqrt{1-x^2}}$.

9. Find $\dfrac{dy}{dx}$.

(1) $x+y=\cos(x-y)$; (2) $x^2+y^2=\sin xy$;

(3) $x^2y^2=\tan(ax+by)$; (4) $xy=\tan(x^2+y^2)$.

10. Find $\dfrac{dy}{dx}$.

(1) $x=a\cos^2\theta, y=b\sin^2\theta$; (2) $x=2a\sin t\cos t, y=b\cos 2t$;

(3) $x=2a\tan\theta, y=a\sec^2\theta$; (4) $x=a(\cos t+t\sin t), y=a(\sin t-t\cos t)$;

(5) $x=a(\tan t-t\sec^2 t), y=a\sec^2 t$.

11. Find the derivatives of the following.

(1) $y=\ln\sin x$; (2) $y=\ln(x+\tan x)$;

(3) $y=\ln(1+e^{5x})$; (4) $y=\ln\ln x$;

(5) $y=\ln\sec x$; (6) $y=\ln(1+\sin^2 x)$;

(7) $y=\ln(e^{ax}+e^{-ax})$; (8) $y=\ln(\sqrt{a^2+x^2}+b)$.

12. Find the derivatives of the following.

(1) $y=e^{\sqrt{\cos x}}$; (2) $y=e^{(1+\ln x)}$;

(3) $y=e^{\sin(\ln x)}$; (4) $y=\tan(\ln x)$;

(5) $y=\cos(\ln\sec x)$; (6) $y=\sec(\ln\tan x)$.

13. Find $\dfrac{dy}{dx}$.

(1) $y=x^y$; (2) $e^{\sin x}+e^{\sin y}=1$;

(3) $x^y \cdot y^x=1$; (4) $x^{\sin y}=y^{\sin x}$.

Chapter 3 Application of Derivatives

The application of derivatives plays a very important role in the field of science, engineering, economics, commerce and so on. For instance, with the help of derivative, we can study the properties of a function, find the equation of a tangent line, solve the problem of instantaneous velocity and maximum revenue and minimum cost etc.

3.1 Monotonicity of Functions

A function can be presented in various forms. One of the most important forms is the graph. To know the nature of the graph of the function, we must have the idea of increasing and decreasing tendency of the function on an interval. Later on, the increasing and decreasing nature of the function are also used in finding the maximum and the minimum values of the function.

Here, in this section as an application, we use derivatives to examine the increasing and decreasing tendency of the function on an interval.

The figure 3 - 1 given below is the continuous curve represented by the given function $y=f(x)$. Consider the points A, B, C, and D on the curve where $OP=x_1$, $OQ=x_2$, $OR=x_3$, $OS=x_4$, $PA=f(x_1)$, $QB=f(x_2)$, $RC=f(x_3)$ and $SD=f(x_4)$.

From the figure, we see that the curve (i. e., the function) increases from A to B then decreases from B to C and again increases from C to D.

That is, in the part AB, $x_2>x_1 \Rightarrow f(x_2)>f(x_1)$

in the part BC, $x_3>x_2 \Rightarrow f(x_3)<f(x_2)$

and in the part CD, $x_4>x_3 \Rightarrow f(x_4)>f(x_3)$

Now, we have the following definitions of monotone increasing and monotone decreasing functions.

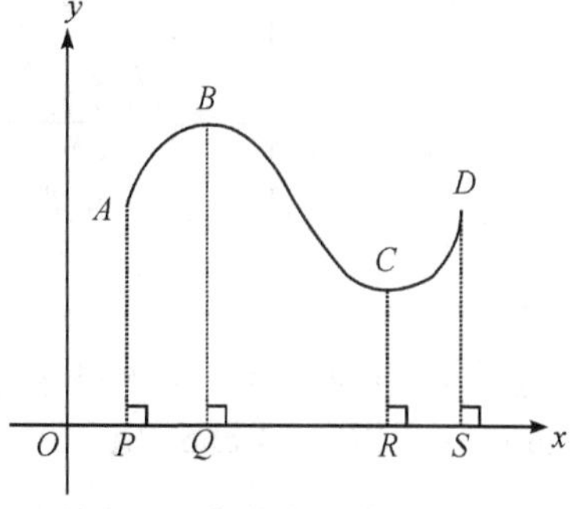

Figure 3 - 1

3.1.1 Monotone Increasing Function

Definition 3.1 A function $y=f(x)$ is said to be increasing on the interval (a,b) if for every $x_1, x_2 \in (a,b)$

$$x_1<x_2 \Rightarrow f(x_1)<f(x_2)$$

This means that as x increases, y, i. e., $f(x)$ also increases. So, the slope of the tangent at any point of such a curve is positive, see figure 3 - 2. Thus, $y=f(x)$ is an increasing function, if $\frac{\mathrm{d}y}{\mathrm{d}x}=f'(x)>0$.

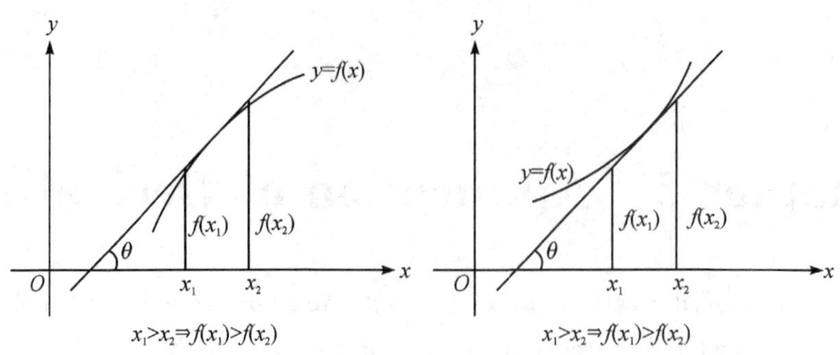

Figure 3 - 2

3.1.2 Monotone Decreasing Function

Definition 3.2 A function $y = f(x)$ is said to be decreasing on the interval (a, b) if for every $x_1, x_2 \in (a, b)$

$$x_1 < x_2 \Rightarrow f(x_1) > f(x_2)$$

This result shows that as x increases, y, i. e. $f(x)$, decreases. So, the slope of the tangent at any point of such a curve is negative, see figure 3 - 3. Thus, $y = f(x)$ is a decreasing function if $\frac{\mathrm{d}y}{\mathrm{d}x} = f'(x) < 0$.

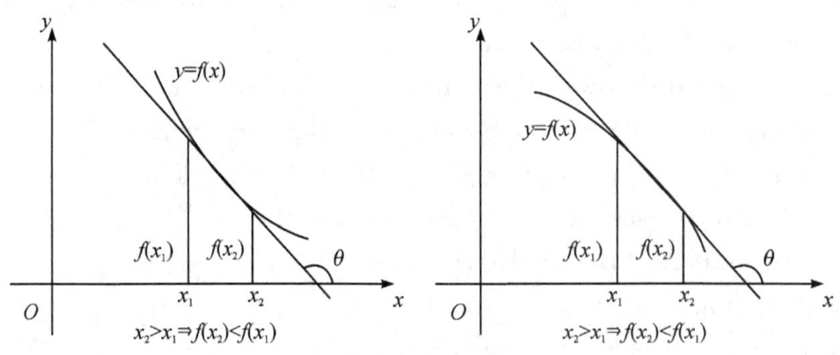

Figure 3 - 3

The following shows the sign of the derivative of the given function $y = f(x)$ defined on an interval (a, b) and the nature of the curve represented by the given function.

Using the Sign of Derivative to Determine the Monotonicity

$f'(x) > 0 \Rightarrow$ function is increasing

$f'(x) < 0 \Rightarrow$ function is decreasing

Example 3.1 Show that the function $f(x) = \frac{1}{2}x^2 - 3x$ increases on the interval $(3, +\infty)$ and decreases on the interval $(-\infty, 3)$.

Proof We have
$$f(x)=\frac{1}{2}x^2-3x$$
$$\Rightarrow \quad f'(x)=x-3$$

For $x>3$, $f'(x)>0$, and for $x<3$, $f'(x)<0$, so $f(x)$ is increasing function for $x>3$, i.e., in the interval $(3,+\infty)$, and $f(x)$ decreases for $x<3$, i.e., on the interval $(-\infty,3)$.

Example 3.2 Find the interval on which the function $f(x)=2x^3-15x^2+36x+1$ is increasing or decreasing.

Solution
$$f(x)=2x^3-15x^2+36x+1$$
Then,
$$\begin{aligned}f'(x)&=6x^2-30x+36\\&=6(x^2-5x+6)\\&=6(x-2)(x-3)\end{aligned}$$

Let $f'(x)=0$, gives $x=2$ and 3.

For $x>3$, $f'(x)>0$, $f(x)$ is increasing for $x>3$, i.e., on the interval $(3,\infty)$;

Again for $x<2$, $f'(x)>0$, $f(x)$ is increasing for $x<2$, i.e., on the interval $(-\infty,2)$;

For $2<x<3$, $f'(x)<0$, $f(x)$ is decreasing on $2<x<3$.

So, $f(x)$ is increasing on $(-\infty,2)\cup(3,+\infty)$ and decreasing on $x\in(2,3)$.

Example 3.3 Show that the function $f(x)=x-\frac{1}{x}$ is increasing for all $x\in\mathbf{R}\ (x\neq 0)$.

Proof $f(x)=x-\frac{1}{x}$. then, $f'(x)=1+\frac{1}{x^2}$ which is positive for $x\in\mathbf{R}\ (x\neq 0)$.

So, $f(x)$ is increasing for all $x\in\mathbf{R}\ (x\neq 0)$, that is for
$$x\in(-\infty,0)\cup(0,\infty)$$

3.2 Maximum and Minimum

3.2.1 Local Maximum and Local Minimum

Definition 3.3 If $f(x_0)>f(x_0\pm\Delta x)$, then $f(x)$ is said to have the local maximum value at $x=x_0$. If $f(x_0)<f(x_0\pm\Delta x)$, then $f(x)$ is said to have the local minimum value at $x=x_0$.

To understand the problem properly, it is necessary to know the nature of the curve in the neighbourhood of the point where the local maximum or the local minimum value of the function occurs. The nature of the curve at a particular point is known by studying the slopes of the tangents at the points in its neighbourhood.

The slope or the gradient of a tangent line at a point of the graph of a function f may be positive, negative or zero. In figure 3-4, $x\in(a,c)$, the slope of the curve is positive, i.e., $f'(x)>0$, the inclination of the tangent line is acute angle, and the tangent line slopes upwards (or goes up-hill) from left to right. In this case, the function is increasing on the interval (a,c). Within the interval (c,b), where the slope of the tangent line is negative, i. e., $f'(x)<0$, $x\in(c,b)$, the inclination of the tangent line is obtuse angle, and the tangent line slopes downwards (or goes down-hill) from left to right. At $x=c$, $f'(x)=0$, the tangent line neither goes "uphill" nor goes "downhill". The points at which $f'(x)=0$ is called **stationary points.** A stationary point is on the graph of the function $y=f(x)$ where the tangent is parallel to the x-axis. The function is monotone increasing on the left side of the stationary point, and the function is monotone decreasing on the right side of the stationary point, so the function takes a local maximum at this point.

The another situation is shown in figure 3 - 5. The function $f(x)$ have local minimum at $x=c'$.

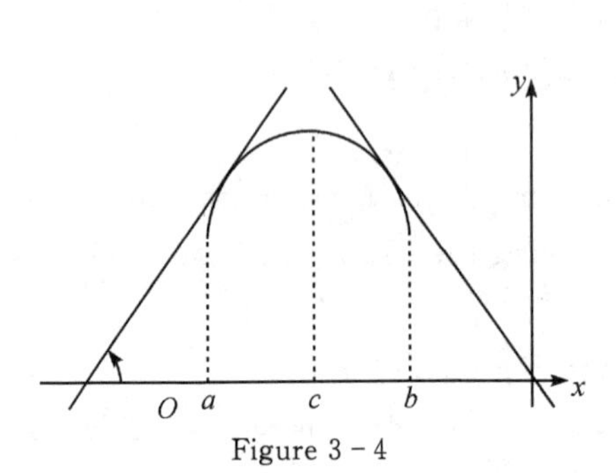

Figure 3 - 4

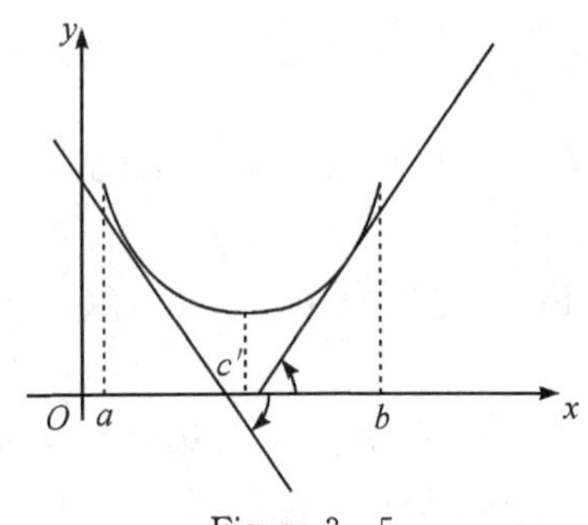

Figure 3 - 5

$f''(x)$ is the derivative of $f'(x)$. So, if $f''(x)>0$, then$f'(x)$ is increasing. In figure 3 - 6(a), the slope increases from a larger negative value to a smaller negative value. In figure 3 - 6(b), the slope increase from smaller positive value to a larger positive value. The curves in figure 3 - 6(c) is just a combination of the curves in figure 3 - 6(a) and 3 - 6(b). These curves are said to be concave. So the condition for a graph to be concave is that the second derivative of the function must be positive.

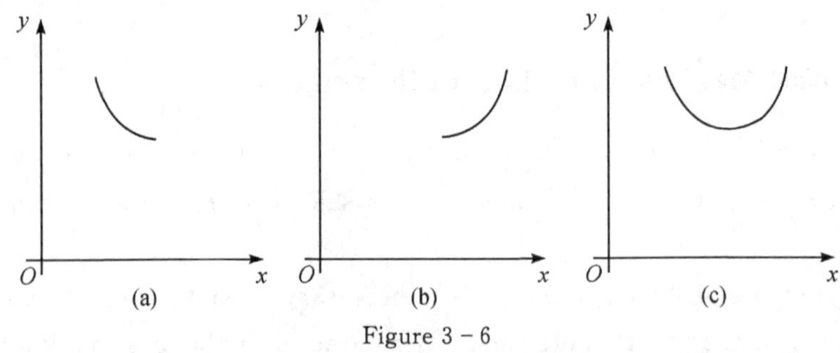

Figure 3 - 6

Similarly, if $f''(x)<0$, then $f'(x)$ is decreasing. In figure 3 - 7(a), the slope decreases from a larger positive value to a smaller positive value. In figure 3 - 7(b), the slope decreases from smaller negative value to a larger negative value. The curve in figure 3 - 7(c) is just a combination of the curves in figure 3 - 7(a) and 3 - 7(b). These curves are said to be convex. So the condition for a graph to be convex is that the second derivative of the function must be negative.

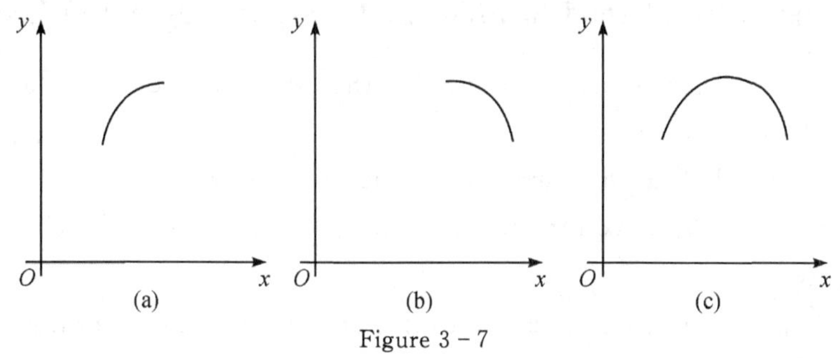

Figure 3 - 7

Again consider the graph of a function f in the following figure 3 - 8. On the interval (a, c), the portion of the graph is concave and so $f''(x) > 0$, while the portion of the graph on the interval (c, b) is convex and so $f''(x) < 0$. So naturally there must exist a point on the interval at which $f''(x)=0$. This point is of considerable importance. It separates the portion of the graph which is convex from the portion which is concave. Such a point of the graph is called a **point of inflection.**

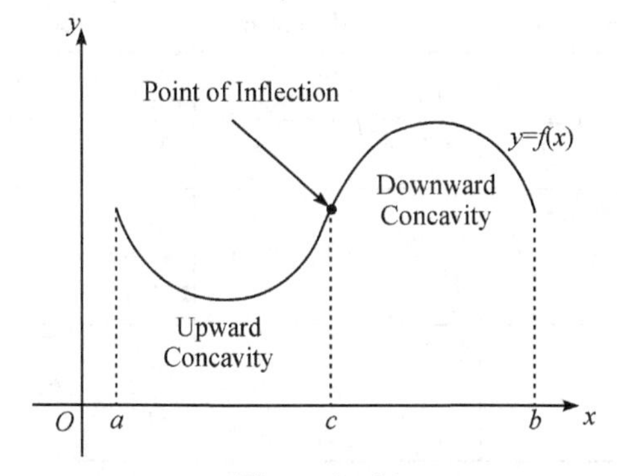

Figure 3 - 8

3.2.2 Procedure to Find the Local Maximum and Local Minimum

To find the maximum or minimum value of the function $y=f(x)$, we use the following steps:

(1) Find $f'(x)$ or $\frac{\mathrm{d}y}{\mathrm{d}x}$.

(2) Making $f'(x)=0$, solve for the stationary points x_i $(i=1,2,\cdots,k)$.

(3) Note the sign of $\frac{\mathrm{d}y}{\mathrm{d}x}$ when x changes its value from $x_i-\Delta x$ to $x_i+\Delta x$.

(4) If $\frac{\mathrm{d}y}{\mathrm{d}x}$ changes its sign from + to −, then $f(x)$ has the maximum value at $x=x_i$. If $\frac{\mathrm{d}y}{\mathrm{d}x}$ changes its sign from − to +, $f(x)$ has minimum value at $x=x_i$. If $\frac{\mathrm{d}y}{\mathrm{d}x}$ does not

change its sign, then $f(x)$ has no local maximum or minimum value.

3.2.3 Alternative Method to Find the Local Maximum and Local Minimum

The following steps are to be used in finding the local maximum and the local minimum of the function $f(x)$ at a point.

(1) Find $f'(x)$ and $f''(x)$ of the given function $y=f(x)$.

(2) Making $f'(x)=0$, solve for the stationary points $x_i\ (i=1,2,\cdots,k)$.

(3) Find $f''(x_i)$.

If $f''(x_i)<0$, then $f(x)$ has maximum value at $x=x_i$ and the maximum value $=f(x_i)$.

If $f''(x_i)>0$, then $f(x)$ has minimum value at $x=x_i$ and the minimum value $=f(x_i)$.

If $f''(x_i)=0$ and $f'''(x_i)\neq 0$, then $f(x)$ has no maximum and no minimum value at $x=x_i$.

The condition for the function $y=f(x)$ to have the maximum, minimum or no maximum, no minimum at the point $x=x_i$ are given below.

Conditions	For the Function $y=f(x)$			
	Local Maximum	Local Minimum	No Local Max. or No Local Min.	
First Order Derivative	$\frac{dy}{dx}=f'(x)=0$	$\frac{dy}{dx}=f'(x)=0$	$\frac{dy}{dx}=f'(x)=0$	$\frac{dy}{dx}=f'(x)\neq 0$
Second Order Derivative	$\frac{d^2y}{dx^2}=f''(x)<0$	$\frac{d^2y}{dx^2}=f''(x)>0$	$\frac{d^2y}{dx^2}=f''(x)=0$	
Third Order Derivative			$\frac{d^3y}{dx^3}=f'''(x)\neq 0$	

Example 3.4 Find the local maximum and the local minimum values of

$$f(x)=2x^3-3x^2-36x$$

Solution We have $f(x)=2x^3-3x^2-36x$, then $f'(x)=6x^2-6x-36$, $f''(x)=12x-6$.

For the maximum or minimum value of $f(x)$,

$$f'(x)=0$$

Then

$$6x^2-6x-36=0$$

$$x^2-x-6=0$$

or

$$(x-3)(x+2)=0$$

(1) When $x=3$, $f''(x)=36-6=30$, which is positive. So $f(x)$ has a minimum value at $x=3$ and the minimum value is $54-27-108=-81$.

(2) When $x=-2$, $f''(x)=-24-6=-30$, which is negative. So $f(x)$ has a maximum value at $x=-2$ and the maximum value is $-16-12+72=44$.

Example 3.5 Show that $f(x)=x^3-3x^2+6x+4$ has neither a maximum nor a minimum value.

Proof

Here,
$$f(x)=x^3-3x^2+6x+4$$
$$f'(x)=3x^2-6x+6=3(x^2-2x+2)=3[(x-1)^2+1]$$
which is always positive for all real values of x and can never be zero.

So, $f(x)$ has neither a maximum nor a minimum value.

3.2.4 Global Maximum and Global Minimum

A function $y=f(x)$ is said to have the global maximum value at $x=x_0$ if $f(x_0)$ is the greatest of all its values for all x belonging to the domain of the function, i. e., $f(x)\geqslant f(x_0)$ for any $x\in D(f)$.

Figure 3 - 9 shows that the local maximum and the local minimum is not necessarily the global maximum and the global minimum, and is just considered in the neighbour hood of the point. A function may have several local maximums. Local maximum may not be greater than the local minimum. Maximum value may be obtained at the maximum point or endpoint, and the minimum value may be obtained at minimum point or endpoint.

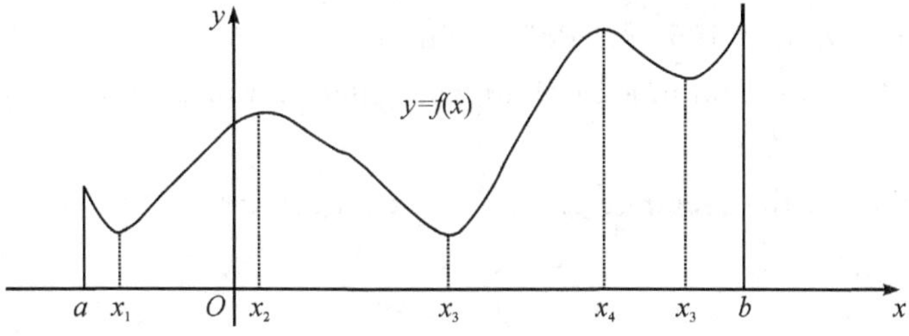

Figure 3 - 9

Sometimes, the global maximum and the global minimum may be taken at the endpoint of the interval as shown in figure 3 - 10.

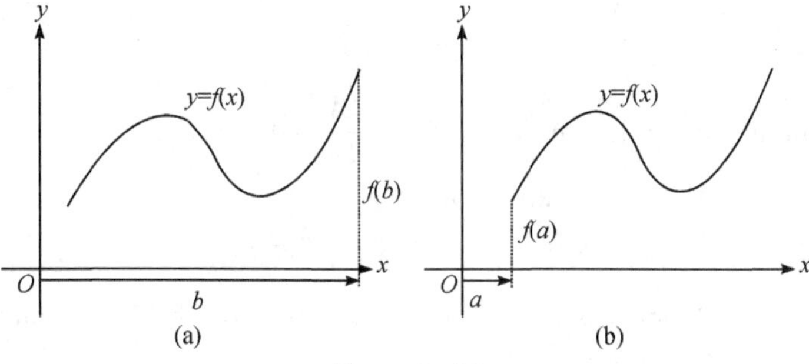

Figure 3 - 10

In figure 3 - 10 (a), $f(b)$ is the global maximum of the function $f(x)$. In figure 3 - 10 (b), $f(a)$ is the global minimum of the function $f(x)$.

3.2.5 Procedure to Find the Global Maximum and Global Minimum

Let $y = f(x)$ be the given function defined on interval $[a, b]$.

(1) Find the first derivative of $f(x)$, i. e. , find $f'(x)$.

(2) Making $f'(x) = 0$, solve for x to get the stationary points. Let the values of x be $x = x_i (i = 1, 2, \cdots, k)$.

(3) Find the values of $f(x)$ at $x = x_i (i = 1, 2, \cdots, k)$, a and b.

Hence, $\max\limits_{a \leqslant x \leqslant b} f(x) = \max\{f(x_i), f(a), f(b)\}$, $\min\limits_{a \leqslant x \leqslant b} f(x) = \min\{f(x_i), f(a), f(b)\}$.

Example 3.6 Find the global maximum (greatest value) and the global minimum value (least value) of the function $f(x) = 2x^3 - 9x^2 + 12x + 20$ defined on interval $[-1, 5]$.

Solution We have $f(x) = 2x^3 - 9x^2 + 12x + 20$, then

$$f'(x) = 6x^2 - 18x + 12$$

$$f'(x) = 0 \text{ gives}$$

$$6x^2 - 18x + 12 = 0 \Rightarrow (x - 1)(x - 2) = 0 \Rightarrow x = 1, 2$$

When $x = -1$, $f(x) = 2(-1)^3 - 9(-1)^2 + 12 \times (-1) + 20 = -3$

$$x = 5, \ f(x) = 2 \times 5^3 - 9 \times 5^2 + 12 \times 5 + 20 = 105$$

$$x = 1, \ f(x) = 2 \times 1^3 - 9 \times 1^2 + 12 \times 1 + 20 = 25$$

$$x = 2, \ f(x) = 2 \times 2^3 - 9 \times 2^2 + 12 \times 2 + 20 = 24$$

So, global max. value $= 105$ and global min. value $= -3$.

Example 3.7 Find the maximum area of a rectangular plot of land which can be enclosed by a rope of 60 meters.

Solution Let the sides of the rectangular plot of land be x and y. So

$$2x + 2y = 60$$

or

$$x + y = 30$$

The area of the land is

$$A = xy = x(30 - x) = 30x - x^2$$

So

$$\frac{dA}{dx} = 30 - 2x$$

and

$$\frac{d^2A}{dx^2} = -2$$

For the maximum or minimum value of A, $\frac{dA}{dx} = 0$

$$30 - 2x = 0 \Rightarrow x = 15$$

and

$$y = 30 - x = 30 - 15 = 15$$

Since $\frac{d^2A}{dx^2} = -2 < 0$, so A has maximum value when $x = 15$, the maximum area $A = 15 \text{ m} \times 15 \text{ m} = 225 \text{ m}^2$.

3.2.6 Procedure to Find the Concave and Convex of the Graph of the Function

Let $y = f(x)$ be the given function. To fine the concave or convex of the graph of the function on an interval we can use the following steps:

(1) Find the second derivative of $f(x)$, i. e. , find $f''(x)$.

(2) Find the interval within which $f''(x) > 0$. Then we conclude that the graph of $y = f(x)$ will be concave on the interval.

(3) Again find the interval within which $f''(x) < 0$. Then we conclude that the graph of $y = f(x)$ will be convex on that interval.

Example 3.8 Let $f(x) = 2x^3 - 6x^2 + 5$. Find where the graph is concave and where it is convex.

Solution We have $f(x) = 2x^3 - 6x^2 + 5$, then

$$f'(x) = 6x^2 - 12x$$

$$f''(x) = 12x - 12 = 12(x - 1)$$

$$f''(x) = 0 \text{ gives } x = 1$$

For $x > 1$, $f''(x) > 0$ and for $x < 1$, $f''(x) < 0$.

Hence the graph is concave if $x > 1$ and is convex if $x < 1$. So the point $(1,1)$ is the point of inflection.

3.3 Derivative as the Rate Measure

Let $y = f(x)$ be the continuous function. By the definition of a function, y changes while x changing. If Δx and Δy be the small changes of x and y respectively, then

$$\frac{\Delta y}{\Delta x} = \frac{f(x + \Delta x) - f(x)}{\Delta x}$$

is the change of y per unit change of x and hence is the average rate of change of y with respect to x on the interval $[x, x + \Delta x]$. The average rate of change becomes the instantaneous rate of change when $\Delta x \to 0$, provided that the limit exists.

Thus,

$$\lim_{\Delta x \to 0} \frac{\Delta y}{\Delta x}\bigg|_{x=x_0} = \lim_{\Delta x \to 0} \frac{f(x_0 + \Delta x) - f(x_0)}{\Delta x}$$

i. e. ,

$$\frac{dy}{dx}\bigg|_{x=x_0} = f'(x_0) = \lim_{\Delta x \to 0} \frac{f(x_0 + \Delta x) - f(x_0)}{\Delta x}$$

is the instantaneous rate of change of y with respect to x at $x = x_0$.

The most appropriate examples of the average rate of change and the instantaneous rate of change occur in the cases of average velocity and the instantaneous velocity of the particle. If Δs be the change in distance made by the particle in time Δt, then $\frac{\Delta s}{\Delta t}$ gives the average rate of change of the distance s, and $\lim_{\Delta t \to 0} \frac{\Delta s}{\Delta t} = \frac{ds}{dt}$ is the instantaneous rate of change of s which gives the velocity of the particle in time t.

In the same way, if v be the velocity of the particle in time t, then $\frac{dv}{dt}$ is the rate of change of velocity known as the acceleration of the particle at time t.

Example 3.9 A particle moves in a straight line so that its distance in meters from a given point in the line after t seconds is $s=3+5t+t^3$. Find,

(1) The velocity at the end of $2\frac{1}{4}$ seconds;

(2) The acceleration at the end of $3\frac{2}{3}$ seconds;

(3) The average velocity during the 5th second.

Solution

$$s=3+5t+t^3$$

$$\frac{ds}{dt}=5+3t^2$$

$$\frac{d^2s}{dt^2}=6t$$

(1) When $t=2\frac{1}{4}$ s, $\frac{ds}{dt}=\text{velocity}=5+3\times\left(\frac{9}{4}\right)^2=20\frac{3}{16}$ m/s;

(2) When $t=3\frac{2}{3}$ s, $\frac{d^2s}{dt^2}=\text{acceleration}=6\times\frac{11}{3}=22$ m/s^2;

(3) Average velocity $=\frac{\Delta s}{\Delta t}=\frac{s(5)-s(4)}{5-4}=\frac{153-87}{1}=66$ m/s.

In business and economics, the rates at which certain quantities are changing often provide useful insight for various economic systems. A manufacture for example, is interested not only in the total cost $C(x)$ at certain production levels, but also in the rate of change of costs at various production level.

In economics, the word marginal refers to a rate of change, that is, to a derivatives. Thus, if

$C(x)=$ total cost of producing x items

then $C'(x)=$ marginal cost

$=$ instantaneous rate of change of total cost with respect to the number of items produced at a production level of items

Example 3.10(Marginal Cost) Suppose that the total cost $C(x)$ (in thousands of yuans) for manufacturing x sailboats per year is given by the function

$$C(x)=575+25x-0.25x^2$$

(1) Find the marginal cost at a production level of boats x per year.

(2) Find the marginal cost at production level of 40 boats per year, and interpret the results.

Solution (1) $C'(x)=25-0.5x$.

(2) $C'(40)=25-0.5\times40=5$, that is 5000 yuan per boats.

At a production level of 40 boats per year, the total cost is increasing at the rate of 5000 yuan per boat.

Example 3.11 (Sales Analysis) The total sales S (in thousands of games) of a home video game t months after the game is introduced are given by

$$S(t)=\frac{125t^2}{t^2+100}$$

(1) Find $S'(t)$.

(2) Find $S(10)$ and $S'(10)$. Write a brief interpretation of these results.

(3) Use the results from (2) to estimate the total sales after 11 months.

Solution

(1) $S'(t)=\dfrac{(t^2+100)(125t^2)'-125t^2(t^2+100)'}{(t^2+100)^2}=\dfrac{25000t}{(t^2+100)^2}$.

(2) $S(10)=62.5$, $S'(10)=6.25$.

The total sales after 10 months are 62500 games, and sales are increasing at the rate of 6250 games per month.

(3) The total sales will increase by approximately 6250 games during the next month, thus, the estimated total sales after 11 months are 62500+6250=68750 games.

Example 3.12 A circular plate of metal expands by heat and its radius increases at the rate of 0.25 cm/s. Find the rate at which the surface-area is increasing when the radius is 7 cm.

Solution Let r and s be the radius and the surface-area of the circular plate in time t respectively. Then,

$$s=\pi r^2$$

$$\frac{ds}{dt}=\frac{d}{dt}(\pi r^2)=2\pi r\frac{dr}{dt}$$

When $r=7$ cm,

$$\frac{ds}{dt}=2\times\pi\times 7\times 0.25=3.5\pi\ \text{cm}^2/\text{s}$$

Example 3.13 Two aeroplanes in flight cross above town at 2:00 PM. One plane travels east at 300 km/h, the other north at 400 km/h. At what rate does the distance between the planes change at 4:00 PM?

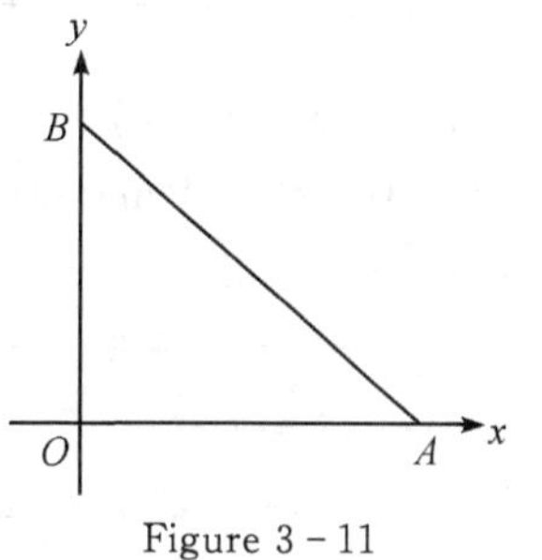

Figure 3 - 11

Solution See figure 3 - 11.

Let A and B be the positions of the planes in t hours.

Let $AB=x$ be the distance between the positions of two planes in t hours.

Then $OA=300t$ and $OB=400t$

Since $$AB^2=OA^2+OB^2$$

We have

$$x^2=(300t)^2+(400t)^2\Rightarrow x=500t\Rightarrow\frac{dx}{dt}=500$$

At $t=2$ hours, $\dfrac{dx}{dt}=500$, the two planes are changing their positions at the rate of 500 km/

h.

Example 3.14 Water flows into an inverted conical tank at the rate of $24\text{cm}^3/\text{min}$. When the depth of water is 9 cm, how fast is the level rising? Assume that the height of the tank is 15 cm and the radius at the top is 5 cm.

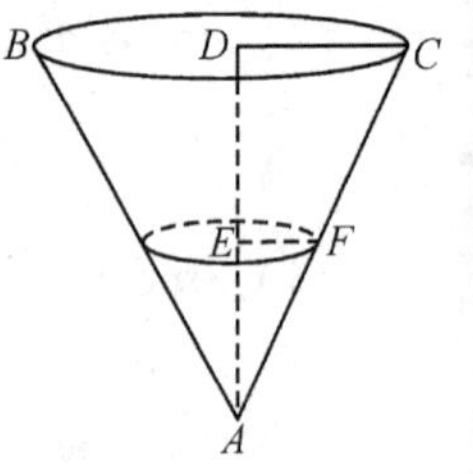

Figure 3 - 12

Solution See figure 3 - 12.

Let cone ABC be the conical water tank into which water is flowing at the rate of $24\text{cm}^3/\text{min}$. At a certain time t, let h be the height (AE) of the water and r the radius (EF) of the water surface. Now $\triangle ACD$, $\triangle AFE$ are similar. So

$$\frac{AE}{AD}=\frac{EF}{DC} \quad \text{or} \quad \frac{h}{15}=\frac{r}{5} \quad \Rightarrow \quad r=\frac{1}{3}h$$

Let V be the volume of water in the tank, then

$$V=\frac{1}{3}\pi r^2 h=\frac{1}{3}\pi\frac{1}{9}h^2 h=\frac{1}{27}\pi h^3$$

$$\frac{\mathrm{d}V}{\mathrm{d}t}=\frac{1}{27}\pi\cdot 3h^2\frac{\mathrm{d}h}{\mathrm{d}t}$$

$$\frac{\mathrm{d}h}{\mathrm{d}t}=\frac{9}{\pi h^2}\cdot\frac{\mathrm{d}V}{\mathrm{d}t}=\frac{9}{\pi h^2}\times 24$$

When $h=9$ cm, $\frac{\mathrm{d}h}{\mathrm{d}t}=\frac{9}{81\pi}\times 24=\frac{8}{3\pi}\text{cm/min}$.

Exercise 3

1. Find the intervals in which the following functions are increasing or decreasing.

   (1) $f(x)=3x^2-6x+5$; (2) $f(x)=x^4-\frac{1}{3}x^3$;

   (3) $f(x)=6+12x+3x^2-2x^3$; (4) $f(x)=x^3-12x$, defined on $[-3,5]$.

2. Find the global maximum, the global minimum values of the following function on the given intervals.

   (1) $f(x)=x^3-6x^2+9x$ on $[0,5]$;

   (2) $f(x)=2x^3-15x^2+36x+10$ on $[1,4]$.

3. Find the local maximum, minimum and points of inflection.

   (1) $f(x)=3x^2-6x+3$; (2) $f(x)=2x^3-9x^2-24x+3$;

   (3) $f(x)=4x^3-15x^2+12x+7$; (4) $f(x)=x+\frac{100}{x}-5$.

4. Show that the following functions have neither maximum nor minimum value.

   (1) $f(x)=x^3-6x^2+24x+4$; (2) $f(x)=x^3-6x^2+12x-3$.

5. Determine where the graph is concave and where it is convex of the following functions:

   (1) $f(x)=x^4-2x^3+5$; (2) $y=3x^5+10x^3+15x$.

6. A man who has 144 meters of fencing material wishes to enclose a rectangular garden.

Find the maximum area he can enclose.

7. Show that the rectangle of largest possible area, for a given perimeter, is a square.
8. A closed cylindrical can is made of volume of 52cm^3. Find its dimensions if the surface is to be minimum.
9. Find two numbers whose sum is 10 and the sum of their squares is minimum.
10. A particle moves in a straight line. The distance s (measured in metre) covered by the particle in time t (in second) is given by

$$s = 2t^2 + 5t - 4$$

Find the velocity and the acceleration of the particle in 6 s.

11. (**Sales Analysis**) The total S (in the thousands of CDs) for a compact disk are given by

$$S(t) = \frac{90t^2}{t^2 + 50}$$

Where t is the number of months since the release of the CD.

(1) Find $S'(t)$.

(2) Find $S(10)$ and $S'(10)$. Write a brief interpretation of these results.

(3) Use the results of part (2) to estimate the total sales after 11 months.

12. (**Medicine**) A drug is injected into the bloodstream of patient through her right arm. The concentration of a drug (in milligrams per cubic centimeter) in the bloodstream of the left arm t hours after the injection is given by

$$C(t) = \frac{0.14t}{t^2 + 1}$$

(1) Find $C'(t)$.

(2) Find $C'(0.5)$ and $C'(3)$, and interpret the results.

13. Suppose the fixed cost of producing a certain product be 2000 yuan, and the variable cost of producing x products be $0.01x^2 + 10x$ yuan. If the selling price of the product is 30 yuan, try to find the output when the marginal cost, marginal profit and marginal profit are zero.
14. Suppose the quantity demanded of a product be $x = 1000 - 10p$, where p is the price, and calculate the marginal revenue function and the marginal revenue as $x = 100, 200, 500, 600$, and interpret the result when $x = 500$.
15. (1) From a cylindrical drum containing oil and kept vertical, the oil is leaking so that the level of the oil is decreasing at the rate of 2 cm/min. If the radius and height of the drum are 10.5 cm and 40 cm respectively, find the rate at which the volume of the oil is decreasing.

(2) Water is poured into a right circular cylinder of radius 8 cm at the rate of 18 cm^3/min. Find the rate at which the surface of water is rising in the cylinder.

(3) Gasoline is pumped into a vertical cylindrical tank at the rate of 24 cm^3/min. The radius of the tank is 9 cm. How fast is the surface rising?

16. (1) A spherical balloon is inflated at the rate of 18 cm^3/min. At what rate is the radius

increasing when the radius is 8 cm?

(2) A spherical ball of salt is dissolving in water in such a way that the rate of decrease in volume at any instant is proportional to the surface area. Prove that the radius is decreasing at the constant rate.

17. A man of height 1.5 m walks away from a lamp post of height 4.5 m at the rate of 20 cm/s. How fast is the shadow lengthening when the man is 42 cm from the post?

18. (1) A kite is 24 m high and there are 25 m of cord out. If the kite moves horizontally at the rate of 36 km/h directly away from the person who is fling it, how fast is the cord out?

(2) A 2.5 m ladder leans against a vertical wall. If the top slides downwards at the rate of 12 cm/s, find the speed of the lower end when it is 2 m from the wall.

Chapter 4 Indefinite Integral

In this chapter, we will introduce the concept of the indefinite integral and its computation, so that we can complete the solution of computational problem of definite integrals in next chapter.

4.1 Antiderivatives and Indefinite Integrals

4.1.1 Definitions of Antiderivative and Indefinite Integral

Let f be a continuous function defined on an open interval (a, b). Then a function F is said to be an **antiderivative** of f on the interval, if the derivative of F is equal to f on the interval, i. e., if

$$\frac{\mathrm{d}F(x)}{\mathrm{d}x}=f(x), \quad x \in (a, b)$$

As the derivative of a constant C is zero, $F(x)+C$ is also an antiderivative of f.

The converse is also true: Any two antiderivatives of a function differ by a constant. Let F and G be antiderivatives of a function f, then

$$\frac{\mathrm{d}[F(x)-G(x)]}{\mathrm{d}x}=\frac{\mathrm{d}F(x)}{\mathrm{d}x}-\frac{\mathrm{d}G(x)}{\mathrm{d}x}=f(x)-f(x)=0$$

From this it follows that there exists a constant C such that

$$F(x)-G(x)=C$$

All these go to establish the fact that if F is an antiderivative of f, $F(x)+C$ gives all the possible antiderivatives of f, when C runs through all real numbers.

Now it is desirable to have a general form of all antiderivatives of f. This general form, which we call **indefinite integral** of f, is denoted by

$$\int f(x)\mathrm{d}x$$

Where, $\int$ is called integration notation, $f(x)$ is called integrand, and x is called integration variable.

If F is an antiderivative of f, we have

$$\int f(x)\mathrm{d}x=F(x)+C$$

4.1.2 Properties of Indefinite Integral

Suppose f and g are both continuous on (a, b), k_1 and k_2 are constants, then

(1) $\int[k_1 f(x)+k_2 g(x)]\mathrm{d}x = k_1\int f(x)\mathrm{d}x + k_2\int g(x)\mathrm{d}x$;

(2) $\left[\int f(x)\mathrm{d}x\right]' = f(x)$ or $\mathrm{d}\left[\int f(x)\mathrm{d}x\right] = f(x)$;

(3) $\int f'(x)\mathrm{d}x = f(x)+C$ or $\int \mathrm{d}f(x) = f(x)+C$.

4.2 Techniques of Integration

4.2.1 Formulas

Using the derivatives of formulas of elementary functions in Chapter 2, we can obtain

The Indefinite Integrals of Formulas of Elementary Functions

$$\int x^{\alpha}\mathrm{d}x = \frac{x^{\alpha+1}}{\alpha+1}+C\ (\alpha\neq -1).$$

$\int \frac{1}{x\ln a}\mathrm{d}x = \log_a x + C$, $\quad \int \frac{1}{x}\mathrm{d}x = \ln|x|+C$.

$\int a^x\mathrm{d}x = \frac{a^x}{\ln a}+C$, $\quad \int \mathrm{e}^x\mathrm{d}x = \mathrm{e}^x + C$.

$\int \cos x\,\mathrm{d}x = \sin x + C$, $\quad \int \sin x\,\mathrm{d}x = -\cos x + C$.

$\int \sec^2 x\,\mathrm{d}x = \tan x + C$, $\quad \int \csc^2 x\,\mathrm{d}x = -\cot x + C$.

$\int \sec x\tan x\,\mathrm{d}x = \sec x + C$, $\quad \int \csc x\cot x\,\mathrm{d}x = -\csc x + C$.

Example 4.1 Calculate

(1) $\int\left(4x^{1/3}+5x^{2/3}+\frac{1}{x^2}\right)\mathrm{d}x$; $\quad$ (2) $\int \frac{2x+\sqrt{x}+1}{x}\mathrm{d}x$;

(3) $\int \frac{1+2x^2}{x^2(1+x^2)}\mathrm{d}x$; $\quad$ (4) $\int \frac{1}{\sin^2 x\cos^2 x}\mathrm{d}x$.

Solution

(1)
$$\begin{aligned}\int\left(4x^{1/3}+5x^{2/3}+\frac{1}{x^2}\right)\mathrm{d}x &= \int 4x^{1/3}\mathrm{d}x + \int 5x^{2/3}\mathrm{d}x + \int x^{-2}\mathrm{d}x\\ &= 4\times\frac{x^{4/3}}{\frac{4}{3}} + 5\times\frac{x^{5/3}}{\frac{5}{3}} + \frac{x^{-1}}{-1} + C\\ &= 3x^{4/3}+3x^{5/3}-\frac{1}{x}+C.\end{aligned}$$

(2) $\int \frac{2x+\sqrt{x}+1}{x}\mathrm{d}x = 2\int \mathrm{d}x + \int \frac{\mathrm{d}x}{\sqrt{x}} + \int \frac{1}{x}\mathrm{d}x = 2x + 2\sqrt{x} + \ln|x| + C.$

(3) $\int \frac{1+2x^2}{x^2(1+x^2)}\mathrm{d}x = \int \frac{1}{1+x^2}\mathrm{d}x + \int \frac{1}{x^2}\mathrm{d}x = \arctan x - \frac{1}{x} + C.$

(4) $\int \frac{1}{\sin^2 x\cos^2 x}\mathrm{d}x = \int \frac{\sin^2 x + \cos^2 x}{\sin^2 x\cos^2 x}\mathrm{d}x = \int \sec^2 x\,\mathrm{d}x + \int \csc^2 x\,\mathrm{d}x$

$= \tan x - \cot x + C.$

4.2.2 Integration by Substitution Method

In the preceding section we have seen that integrals can be integrated easily when they are in standard forms. A given integral may not always be in a standard form. So in order to turn it to a standard form we have to change the variable into a new one by a suitable substitution. There are two types of integration by substitution.

Integration by the First Type of Substitutions

Suppose f and φ' both be continuous, and $F'(u) = f(u)$, then

$$\int f[\varphi(x)]\varphi'(x)\mathrm{d}x = \left(\int f(u)\mathrm{d}u\right)_{u=\varphi(x)} = F(u)_{u=\varphi(x)} + C = F[\varphi(x)] + C$$

Example 4.2 Calculate:

(1) $\int \cos(3x+1)\mathrm{d}x$;

(2) $\int (kx+b)^{\alpha}\mathrm{d}x\ (\alpha \neq 0, \alpha \neq -1)$;

(3) $\int \frac{\mathrm{d}x}{\sqrt{a^2-x^2}}\ (a>0)$;

(4) $\int \frac{\mathrm{d}x}{a^2+x^2}\ (a>0)$;

(5) $\int \mathrm{e}^{\sin x}\cos x\,\mathrm{d}x$;

(6) $\int \frac{\mathrm{d}x}{a^2-x^2}\ (a>0)$;

(7) $\int (2x+3)(4x+5)^4\mathrm{d}x$;

(8) $\int \frac{1-\mathrm{e}^{3x}}{\mathrm{e}^{5x}}\mathrm{d}x$;

Solution

(1) $\int \cos(3x+1)\mathrm{d}x = \frac{1}{3}\int \cos(3x+1)\mathrm{d}(3x+1) = \frac{1}{3}\int \cos u\,\mathrm{d}u\,\big|_{u=3x+1}$

$= \frac{1}{3}\sin u\,\big|_{u=3x+1} + C = \frac{1}{3}\sin(3x+1) + C$

(2) $\int (kx+b)^{\alpha}\mathrm{d}x = \frac{1}{k}\int (kx+b)^{\alpha}\mathrm{d}(kx+b) = \frac{1}{k}\int u^{\alpha}\mathrm{d}u\,\big|_{u=kx+b}$

$= \frac{1}{k(\alpha+1)}u^{\alpha+1}\,\big|_{u=kx+b} + C$

$= \frac{1}{k(\alpha+1)}(kx+b)^{\alpha+1} + C$

(3) $\int \frac{dx}{\sqrt{a^2-x^2}} = \frac{1}{a}\int \frac{dx}{\sqrt{1-\left(\frac{x}{a}\right)^2}} = \int \frac{d\frac{x}{a}}{\sqrt{1-\left(\frac{x}{a}\right)^2}} = \int \frac{du}{\sqrt{1-u^2}}\Big|_{u=\frac{x}{a}}$

$= \arcsin u\Big|_{u=\frac{x}{a}} + C$

$= \arcsin\frac{x}{a} + C$

(4) When we become familiar with this method, it is sometimes possible to avoid writing down the intermediate variable u.

$$\int \frac{dx}{a^2+x^2} = \frac{1}{a^2}\int \frac{dx}{1+\left(\frac{x}{a}\right)^2} = \frac{1}{a}\int \frac{d\frac{x}{a}}{1+\left(\frac{x}{a}\right)^2} = \frac{1}{a}\arctan\frac{x}{a} + C$$

(5) $\int e^{\sin x}\cos x\,dx = \int e^{\sin x}\,d\sin x = e^{\sin x} + C.$

(6) $\int \frac{dx}{a^2-x^2} = \frac{1}{2a}\int\left(\frac{1}{a-x}+\frac{1}{a+x}\right)dx$

$= \frac{1}{2a}\left[-\int \frac{1}{a-x}d(a-x) + \int \frac{1}{a+x}d(a+x)\right]$

$= \frac{1}{2a}[-\ln|a-x| + \ln|a+x|] + C$

$= \frac{1}{2a}\ln\left|\frac{a+x}{a-x}\right| + C.$

(7) $\int (2x+3)(4x+5)^4 dx = \frac{1}{2}\int (4x+6)(4x+5)^4 dx$

$= \frac{1}{2}\int (4x+5+1)(4x+5)^4 dx$

$= \frac{1}{2}\int [(4x+5)^5 + (4x+5)^4]\,dx$

$= \frac{1}{8}\left[\frac{(4x+5)^6}{6} + \frac{(4x+5)^5}{5}\right] + C.$

(8) $\int \frac{1-e^{3x}}{e^{5x}}dx = \int (e^{-5x} - e^{-2x})\,dx = -\frac{1}{5e^{5x}} + \frac{1}{2e^{2x}} + C.$

Example 4.3 Evaluate:

(1) $\int \tan x\,dx$;　　(2) $\int \cot x\,dx$;

(3) $\int \sqrt{1-\sin 2x}\,dx$, $x \in \left(\frac{\pi}{2}, \pi\right)$;　　(4) $\int \frac{dx}{1-\cos x}$;

(5) $\int \sin^3 x\,dx$;　　(6) $\int \sin^2 x\,dx$;

(7) $\int \frac{dx}{1-\sin x}$;　　(8) $\int \sin 6x \cdot \cos 3x\,dx$.

(9) $\int \sec x\,dx$; (10) $\int \csc x\,dx$.

Solution

(1) $\int \tan x\,dx = \int \frac{\sin x}{\cos x}dx = -\int \frac{1}{\cos x}d\cos x = -\ln|\cos x| + C = \ln|\sec x| + C$.

(2) By the same procedure we have

$$\int \csc x\,dx = \ln|\sin x| + C$$

(3) $$\int \sqrt{1-\sin 2x}\,dx = \int \sqrt{(\sin^2 x + \cos^2 x - 2\sin x\cos x)}\,dx$$
$$= \int \sqrt{(\sin x - \cos x)^2}\,dx$$
$$= \int |\sin x - \cos x|\,dx = -\cos x - \sin x + C.$$

(4) $$\int \frac{dx}{1-\cos x} = \int \frac{dx}{2\sin^2 \frac{x}{2}} = \frac{1}{2}\int \csc^2 \frac{x}{2}dx = \frac{1}{2}\left(-\frac{\cot \frac{x}{2}}{\frac{1}{2}}\right) + C = -\cot \frac{x}{2} + C.$$

(5) $$\int \sin^3 x\,dx = \int (1-\cos^2 x)\sin x\,dx = -\int (1-\cos^2 x)d\cos x = -\cos x + \frac{1}{3}\cos^3 x + C.$$

(6) $$\int \sin^2 x\,dx = \frac{1}{2}\int (1-\cos 2x)dx = \frac{1}{2}\left(\int 1dx - \int \cos 2x\,dx\right) = \frac{1}{2}\left(x - \frac{\sin 2x}{2}\right) + C.$$

(7) $$\int \frac{dx}{1-\sin x} = \int \frac{1+\sin x}{1-\sin^2 x}dx = \int \frac{1+\sin x}{\cos^2 x}dx$$
$$= \int \sec^2 x\,dx + \int \tan x \cdot \sec x\,dx = \tan x + \sec x + C.$$

(8) $$\int \sin 6x \cdot \cos 3x\,dx = \frac{1}{2}\int (\sin 9x + \sin 3x)\,dx = -\frac{1}{18}(\cos 9x + 3\cos 3x) + C.$$

(9) $$\int \sec x\,dx = \int \frac{\sec x(\sec x + \tan x)}{\sec x + \tan x}dx = \int \frac{d(\sec x + \tan x)}{\sec x + \tan x}$$
$$= \ln|\sec x + \tan x| + C.$$

(10) $$\int \csc x\,dx = \int \frac{\csc x(\csc x + \cot x)}{\csc x + \cot x}dx = -\int \frac{d(\csc x + \cot x)}{\csc x + \cot x}$$
$$= -\ln|\csc x + \cot x| + C.$$

Example 4.4 Compute:

(1) $\int \frac{x\,dx}{2x^2+3}$; (2) $\int \frac{\sqrt{\arctan x}\,dx}{1+x^2}$;

(3) $\int \frac{(2ax+b)}{(ax^2+bx+c)^{1/2}}dx$; (4) $\int \frac{dx}{x(1+\ln x)}$;

(5) $\int x\cos(ax^2+b)dx$; (6) $\int \sin^3 x\cos^3 x\,dx$;

(7) $\int e^{\cos^2 x}\sin x\cos x\,dx$; (8) $\int \frac{1}{1+e^x}dx$.

Solution

(1) $\int \frac{x\mathrm{d}x}{2x^2+3}=\frac{1}{2}\int \frac{\mathrm{d}x^2}{2x^2+3}=\frac{1}{4}\int \frac{\mathrm{d}2x^2+3}{(2x^2+3)}=\frac{1}{4}\ln(2x^2+3)+C.$

(2) $\int \frac{\sqrt{\arctan x}\,\mathrm{d}x}{1+x^2}=\int \sqrt{\arctan x}\,\mathrm{d}\arctan x=\frac{2}{3}(\arctan x)^{\frac{3}{2}}+C.$

(3) $\int \frac{(2ax+b)}{(ax^2+bx+c)^{1/2}}\mathrm{d}x=\int \frac{\mathrm{d}(ax^2+bx+c)}{(ax^2+bx+c)^{1/2}}=2\sqrt{ax^2+bx+c}+C.$

(4) $\int \frac{\mathrm{d}x}{x(1+\ln x)}=\int \frac{\mathrm{d}\ln x}{1+\ln x}=\int \frac{\mathrm{d}(1+\ln x)}{1+\ln x}=\ln|1+\ln x|+C.$

(5) $\int x\cos(ax^2+b)\mathrm{d}x=\frac{1}{2}\int \cos(ax^2+b)\mathrm{d}x^2$

$$=\frac{1}{2a}\int \cos(ax^2+b)\mathrm{d}(ax^2+b)=\frac{1}{2a}\sin(ax^2+b)+C.$$

(6) $\int \sin^3 x\cos^3 x\,\mathrm{d}x=\int \sin^3 x\cos^2 x\,\mathrm{d}\sin x$

$$=\int \sin^3 x(1-\sin^2 x)\,\mathrm{d}\sin x=\frac{1}{4}\sin^4 x-\frac{1}{6}\sin^6 x+C.$$

(7) $\int \mathrm{e}^{\cos^2 x}\sin x\cos x\,\mathrm{d}x=-\int \mathrm{e}^{\cos^2 x}\cos x\,\mathrm{d}\cos x=-\frac{1}{2}\int \mathrm{e}^{\cos^2 x}\mathrm{d}\cos^2 x=-\frac{1}{2}\mathrm{e}^{\cos^2 x}+C.$

(8) $\int \frac{1}{1+\mathrm{e}^x}\mathrm{d}x=\int \frac{\mathrm{e}^{-x}}{1+\mathrm{e}^{-x}}\mathrm{d}x=-\int \frac{\mathrm{d}\mathrm{e}^{-x}}{1+\mathrm{e}^{x}}=-\ln(1+\mathrm{e}^{-x})+C.$

If $\int f(x)\mathrm{d}x$ is difficult to compute directly, we may substitute x with $x=\varphi(t)$, then we can get the antiderivative function with respect to t, and again express the function in terms of x.

Integration by the Second Type of Substitutions

Suppose $f(x)$ and $\varphi'(t)$ both be continuous and $\varphi'(t)>0$ (or <0), then

$$\int f(x)\mathrm{d}x=\int f(x)\mathrm{d}x\Big|_{x=\varphi(t)}=\left\{\int f[\varphi(t)]\varphi'(t)\mathrm{d}t\right\}\Big|_{t=\varphi^{-1}(x)}$$

Example 4.5 Calculate:

(1) $\int x\sqrt{x-4}\,\mathrm{d}x$; (2) $\int \sqrt{a^2-x^2}\,\mathrm{d}x$; (3) $\int \frac{\mathrm{d}x}{\sqrt{x^2-a^2}}$;

(4) $\int \frac{\mathrm{d}x}{(a^2+x^2)^2}$; (5) $\int \sqrt{\frac{a-x}{x}}\,\mathrm{d}x$; (6) $\int \frac{\mathrm{d}x}{1+\sqrt[3]{x+2}}$;

(7) $\int \frac{\mathrm{d}x}{\sqrt{x^2-a^2}}$.

Solution

(1) Let $x=t^2+4$, then $\mathrm{d}x=2t\mathrm{d}t$, so

$$\int x\sqrt{x-4}\,\mathrm{d}x=\int (t^2+4)\,t\cdot 2t\mathrm{d}t=2\int (t^4+4t^2)\,\mathrm{d}t$$

$$=\left(\frac{2}{5}t^5+\frac{8}{3}t^3\right)_{t=\sqrt{x-4}}+C$$

$$=\frac{2}{5}(x-4)^{\frac{5}{2}}+\frac{8}{3}(x-4)^{\frac{3}{2}}+C$$

(2) Let $x=a\sin t$, then $\mathrm{d}x=a\cos t\mathrm{d}t$, so that

$$\int\sqrt{a^2-x^2}\,\mathrm{d}x=a^2\int\cos^2 t\mathrm{d}t=\frac{a^2}{2}\int(1+\cos 2t)\,\mathrm{d}t$$

$$=\frac{a^2}{2}\left(t+\frac{\sin 2t}{2}\right)+C$$

Since $x=a\sin t$, in the right triangle (Figure 4 - 1(a)), sides $AC=a$, $BC=x$, so that $AB=\sqrt{a^2-x^2}$.

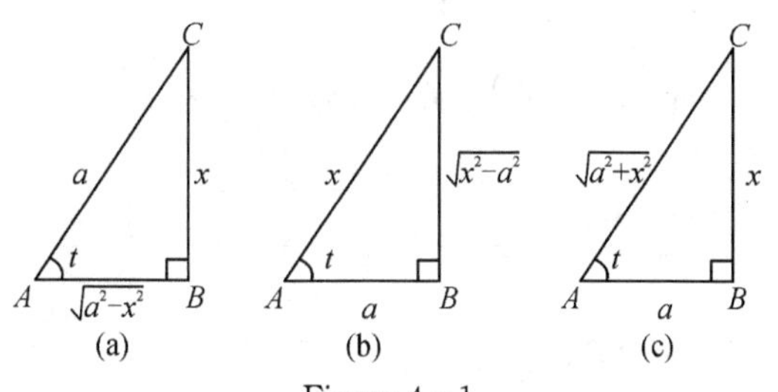

Figure 4 - 1

Hence, $\cos t=\frac{\sqrt{a^2-x^2}}{a}$, $\sin 2t=2\sin t\cos t=2\,\frac{x}{a}\,\frac{\sqrt{a^2-x^2}}{a}$, so that

$$\int\sqrt{a^2-x^2}\,\mathrm{d}x=a^2\int\cos^2 t\mathrm{d}t=\frac{a^2}{2}\int(1+\cos 2t)\,\mathrm{d}t$$

$$=\frac{a^2}{2}\arcsin\frac{x}{a}+\frac{1}{2}x\sqrt{a^2-x^2}+C.$$

(3) Let $x=a\sec t$, then $\mathrm{d}x=a\sec t\tan t\mathrm{d}t$, so that

$$\int\frac{\mathrm{d}x}{x\sqrt{x^2-a^2}}=\int\frac{a\sec t\tan t\mathrm{d}t}{a\sec t\sqrt{a^2\sec^2 t-a^2}}=\int\frac{a\sec t\tan t\mathrm{d}t}{a\sec t\cdot a\tan t}$$

$$=\frac{1}{a}\int\mathrm{d}t=\frac{1}{a}t+C$$

For $x=a\sec t$, $BC=\sqrt{x^2-a^2}$, so $\sec t=\frac{x}{a}$, from figure 4 - 1(b) we know that, $\cos t=\frac{a}{x}$, thus $t=\arccos\frac{a}{x}$, then

$$\int\frac{\mathrm{d}x}{x\sqrt{x^2-a^2}}=\frac{1}{a}\arccos\frac{a}{x}+C$$

(4) Let $x=a\tan t$, then $\mathrm{d}x=a\sec^2 t\mathrm{d}t$, $a^2+x^2=a^2+a^2\tan^2 t=a^2\sec^2 t$, so that

$$\int\frac{\mathrm{d}x}{(a^2+x^2)^2}=\frac{1}{a^3}\int\cos^2 t\mathrm{d}t\ \ \frac{1}{2a^3}\left(t+\frac{\sin 2t}{2}\right)+C$$

Since $x = a\tan t$, $\tan t = \dfrac{x}{a}$, from figure 4-1(c), we know that $AC = \sqrt{a^2 + x^2}$, thus

$$\sin 2t = 2\sin t\cos t = 2\,\frac{x}{\sqrt{a^2+x^2}}\,\frac{a}{\sqrt{a^2+x^2}}$$

$$\int \frac{\mathrm{d}x}{(a^2+x^2)^2} = \frac{1}{2a^3}\left[\arctan\frac{x}{a} + \frac{x}{\sqrt{a^2+x^2}}\,\frac{a}{\sqrt{a^2+x^2}}\right] + C$$

$$= \frac{1}{2a^3}\left[\arctan\frac{x}{a} + \frac{ax}{a^2+x^2}\right] + C$$

(5) Let $x = a\sin^2 t$, then $\mathrm{d}x = a \cdot 2\sin t\cos t\,\mathrm{d}t$, so

$$\int\sqrt{\frac{a-x}{x}}\,\mathrm{d}x = \int\sqrt{\frac{a - a\sin^2 t}{a\sin^2 t}}\cdot 2a\sin t\cos t\,\mathrm{d}t = a\int 2\cos^2 t\,\mathrm{d}t$$

$$= a\int(1+\cos 2t)\,\mathrm{d}t$$

$$= a\left(t + \frac{\sin 2t}{2}\right) + C = a(t + \sin t\cos t) + C$$

$$= a\left(\arcsin\sqrt{\frac{x}{a}} + \sqrt{\frac{x}{a}}\cdot\sqrt{1-\frac{x}{a}}\right) + C$$

$$= a\arcsin\sqrt{\frac{x}{a}} + \sqrt{ax - x^2} + C$$

(6) Let $x = t^3 - 2$, then $\sqrt[3]{x+2} = t$, $\mathrm{d}x = 3t^2\,\mathrm{d}t$, thus

$$\int\frac{\mathrm{d}x}{1+\sqrt[3]{x+2}} = \int\frac{3t^2}{1+t}\,\mathrm{d}t = 3\int\left(t - 1 + \frac{1}{t+1}\right)\mathrm{d}t$$

$$= 3\left[\frac{t^2}{2} - t + \ln(1+t)\right] + C$$

$$= 3\left[\frac{1}{2}\sqrt[3]{(x+2)^2} - \sqrt[3]{x+2} + \ln(1+\sqrt[3]{x+2})\right] + C$$

(7) Let $x = a\sec t$, then $\mathrm{d}x = a\sec t\tan t\,\mathrm{d}t$, so that

$$\int\frac{\mathrm{d}x}{\sqrt{x^2-a^2}} = \int\frac{a\sec t\tan t\,\mathrm{d}t}{\sqrt{a^2\sec^2 t - a^2}} = \int\frac{a\sec t\tan t\,\mathrm{d}t}{a\tan t}$$

$$= \int\sec t\,\mathrm{d}t \quad \text{(by example 4.3(9))}$$

$$= \ln|\sec t + \tan t| + C$$

Since $x = a\sec t$, $\sec t = \dfrac{x}{a}$, from figure 4-1(b), we know that $BC = \sqrt{x^2 - a^2}$, thus $\tan t = \dfrac{\sqrt{x^2-a^2}}{a}$, so that

$$\int\frac{\mathrm{d}x}{\sqrt{x^2-a^2}} = \ln\left|\frac{x}{a} + \frac{\sqrt{x^2-a^2}}{a}\right| + C = \ln\left|x + \sqrt{x^2-a^2}\right| + C' \quad (C' = C - \ln a)$$

To summarize the substitutions used in the examples above, we have the following substitutions to remove the radical in the integrands.

(1) $\sqrt{a^2-x^2}$, let $x=a\sin t$;	(2) $\sqrt{x^2+a^2}$, let $x=a\tan t$;
(3) $\sqrt{x^2-a^2}$, let $x=a\sec t$;	(4) $\sqrt[n]{ax+b}$, let $\sqrt[n]{ax+b}=t$.

4.2.3 Integration by Parts

If a given function to be integrated is in the product form and it cannot be integrated either by reducing the integrand into the standard form nor by substitution, we can use the following rule known as the integration by parts.

Let $u(x)$ and $v(x)$ be two differential functions of x, then using the product rule of differentiation, we have

$$(uv)'=u'v+uv'$$

or

$$uv'=(uv)'-vu'$$

Integrating both sides with respect to x, we have

The Formula of Integration by Parts

$$\int uv'\mathrm{d}x=\int (uv)'\mathrm{d}x-\int vu'\mathrm{d}x$$

or

$$\int u\,\mathrm{d}v=uv-\int v\,\mathrm{d}u$$

The successfulness of the use of the above formula depends upon the proper choice of $u(x)$ and $v(x)$ such that the second integral should be easily integrated.

It should be noted that the constant comes from the first term on the right side of the formula $\int uv'\mathrm{d}x=\int (uv)'\mathrm{d}x-\int vu'\mathrm{d}x$, can be kept in the integral $\int vu'\mathrm{d}x$.

Example 4.6 Calculate the following integrals.

(1) $\int x\mathrm{e}^x\,\mathrm{d}x$; (2) $\int x\cos x\,\mathrm{d}x$; (3) $\int \ln x\,\mathrm{d}x$;

(4) $\int x\arctan x\,\mathrm{d}x$; (5) $\int x\ln x\,\mathrm{d}x$; (6) $\int \mathrm{e}^x\sin x\,\mathrm{d}x$.

Solution

(1) Combining the two factors e^x and $\mathrm{d}x$ into $\mathrm{d}\mathrm{e}^x$, by the formula of integration by parts

$$\int x\mathrm{e}^x\,\mathrm{d}x=\int x\,\mathrm{d}\mathrm{e}^x=x\mathrm{e}^x-\int \mathrm{e}^x\,\mathrm{d}x=x\mathrm{e}^x-\mathrm{e}^x+C$$

Note: if we combine x with $\mathrm{d}x$, then

$$\int x\mathrm{e}^x\,\mathrm{d}x=\frac{1}{2}\int \mathrm{e}^x\,\mathrm{d}x^2=\frac{1}{2}\left(x^2\mathrm{e}^x-\int x^2\,\mathrm{d}\mathrm{e}^x\right)=\frac{1}{2}\left(x^2\mathrm{e}^x-\int x^2\mathrm{e}^x\,\mathrm{d}x\right)$$

As we see the integral becomes more complicated. Therefore, proper choice of the $u(x)$ and $v(x)$ is the key for application of the formula.

(2) $\int x\cos x\,\mathrm{d}x=\int x\,\mathrm{d}\sin x=x\sin x-\int \sin x\,\mathrm{d}x=x\sin x+\cos x+C$.

(3) $\int \ln x \mathrm{d}x = \ln x \cdot x - \int x \mathrm{d}\ln x = x\ln x - \int x \frac{1}{x}\mathrm{d}x = x\ln x - x + C.$

(4) $\int x\arctan x \mathrm{d}x = \frac{1}{2}\int \arctan x \mathrm{d}x^2 = \frac{1}{2}\left(x^2\arctan x - \int x^2 \mathrm{d}\arctan x\right)$

$= \frac{1}{2}\left(x^2\arctan x - \int \frac{x^2}{1+x^2}\mathrm{d}x\right)$

$= \frac{1}{2}(x^2\arctan x - x + \arctan x) + C.$

(5) $\int x\ln x \mathrm{d}x = \frac{1}{2}\int \ln x \mathrm{d}x^2 = \frac{1}{2}\left(x^2\ln x - \int x^2 \frac{1}{x}\mathrm{d}x\right) = \frac{1}{2}\left(x^2\ln x - \frac{1}{2}x^2\right) + C.$

(6) Let $I = \int \mathrm{e}^x \cos x \mathrm{d}x$, then

$I = \int \mathrm{e}^x \cos x \mathrm{d}x = \int \cos x \mathrm{d}\mathrm{e}^x = \mathrm{e}^x\cos x - \int \mathrm{e}^x \mathrm{d}\cos x = \mathrm{e}^x\cos x + \int \mathrm{e}^x \sin x \mathrm{d}x$

$= \mathrm{e}^x\cos x + \int \sin x \mathrm{d}\mathrm{e}^x = \mathrm{e}^x\cos x + \mathrm{e}^x\sin x - \int \mathrm{e}^x\cos x \mathrm{d}x$

$= \mathrm{e}^x\cos x + \mathrm{e}^x\sin x - I$

Then move the last integral I to the left side, and the constant C to the right side. Transposition of terms gives

$$\int \mathrm{e}^x\cos x \mathrm{d}x = \frac{1}{2}\mathrm{e}^x(\sin x + \cos x) + C$$

Exercise 4

1. Find the following indefinite integral.

(1) $\int (2x+1)(3x+2)\mathrm{d}x$;

(2) $\int \left(x^2 - \frac{1}{x^2}\right)\mathrm{d}x$;

(3) $\int \left(\sqrt{x} - \frac{1}{\sqrt{x}}\right)\mathrm{d}x$;

(4) $\int \frac{3x^2 - 5x + 2}{x}\mathrm{d}x$;

(5) $\int (x^2 + 3x + 5)x^{-1/3}\mathrm{d}x$;

(6) $\int (a - bx)^5\mathrm{d}x$;

(7) $\int \frac{\mathrm{d}x}{\sqrt{2x+7}}$;

(8) $\int \frac{3x-1}{x-2}\mathrm{d}x$;

(9) $\int \frac{\mathrm{d}x}{\sqrt{x+a} - \sqrt{x-a}}$;

(10) $\int \frac{3x+2}{\sqrt{5x+3}}\mathrm{d}x$;

(11) $\int \left[x + \frac{1}{(x+3)^2}\right]\mathrm{d}x$;

(12) $\int \frac{x^2+3x+3}{x+1}\mathrm{d}x$;

(13) $\int (\mathrm{e}^{px} + \mathrm{e}^{-qx})\mathrm{d}x$;

(14) $\int \mathrm{e}^x(\mathrm{e}^{2x}+1)\mathrm{d}x$.

2. Calculate the following integrals:

(1) $\int \cos(a^2x + b)\mathrm{d}x$;

(2) $\int \sec^2(2x+3)\mathrm{d}x$;

(3) $\int \sin^2 ax \, dx$;

(4) $\int \tan^2 ax \, dx$;

(5) $\int \sin^4 x \, dx$;

(6) $\int \frac{1}{\cos^2 x \sin^2 x} dx$;

(7) $\int \frac{1}{\sec^2 x \tan^2 x} dx$;

(8) $\int \sqrt{1 + \sin 2ax} \, dx$;

(9) $\int \frac{dx}{1 - \sin ax}$;

(10) $\int \sin 7x \sin 5x \, dx$.

3. Integrate the following:

(1) $\int 3x^2 (x^3 + 1)^3 dx$;

(2) $\int \frac{2x + 3}{(3x^2 + 9x + 5)^3} dx$;

(3) $\int \frac{(x^2 + 1) dx}{\sqrt{x^3 + 3x + 4}}$;

(4) $\int \frac{1}{x} \ln x \, dx$;

(5) $\int \cos^5 x \sin^3 x \, dx$;

(6) $\int (a \sin x - b)^3 \cos x \, dx$;

(7) $\int \cot x (\ln \sin x)^3 dx$;

(8) $\int \tan^2 \theta \sec^4 \theta \, d\theta$;

(9) $\int \tan^3 x \sec^4 x \, dx$;

(10) $\int \tan^3 x \, dx$;

(11) $\int e^{\sin x \cos x} \cos 2x \, dx$;

(12) $\int \left(1 - \frac{1}{x^2}\right) e^{x+1/x} dx$;

(13) $\int \frac{\sin \sqrt{x}}{\sqrt{x}} dx$;

(14) $\int \frac{e^{2x}}{1 + e^x} dx$;

(15) $\int \frac{e^x - 1}{e^x + 1} dx$.

4. Compute:

(1) $\int x \sqrt{x + 1} \, dx$;

(2) $\int (x + 2) \sqrt{3x + 2} \, dx$;

(3) $\int \frac{dx}{\sqrt{(a^2 - x^2)^{3/2}}}$;

(4) $\int \frac{x^2 dx}{\sqrt{a^2 - x^2}}$;

(5) $\int \frac{dx}{\sqrt{x^2 - 4}}$;

(6) $\int \frac{dx}{x^2 \sqrt{x^2 + 1}}$;

(7) $\int \sqrt{\frac{a + x}{a - x}} dx$;

(8) $\int \sqrt{\frac{x}{a - x}} dx$.

5. Find the following indefinite integrals:

(1) $\int x \ln x \, dx$;

(2) $\int x e^{5x} dx$;

(3) $\int x \sec^2 x \, dx$;

(4) $\int x \sin x \, dx$;

(5) $\int \sec^3 x \, dx$;

(6) $\int \arcsin x \, dx$;

(7) $\int x \sec x \tan x \, dx$;

(8) $\int x \sin^2 x \, dx$.

Chapter 5 Definite Integral and its Application

Unlike the previous chapter where the indefinite integral of an expression resulted in a new expression, when finding the definite integral we produce a numerical value. Definite integrals are important because they can be used to find different types of measures, for example, area, volume and so on.

5.1 Definite Integral

5.1.1 Calculation of Area Under a Parabola

Computation of the area under the curve of a function (also called the area of the trapezoid with the curved top) was indeed a big challenge for mathematicians in the early days. As an illustration, let us consider finding the area between the curve $y = x^2$, the x-axis, $x = 0$ and $x = a$.

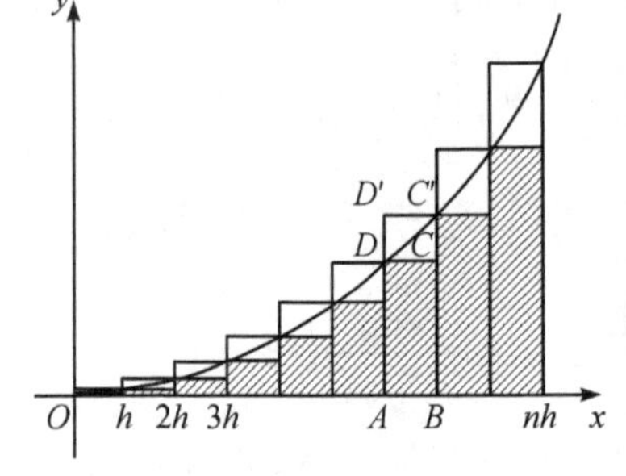

Figure 5 - 1

Draw the parabola $y = x^2$ as shown in figure 5 - 1 and divide the interval $(0, a)$ into n sub-intervals of equal length $h = a/n$. Draw n rectangles as shown in figure 5 - 1. All the shaded rectangles such as $ABCD$ lie under the curve and constitute a set of rectangles inscribed in the parabolic region whose area we have to find out. Let us denote the sum of the areas of these inscribed rectangles by s_n.

So as $y = x^2$ gives $y = h^2$, when $x = h$, we have

$$\begin{aligned} s_n &= 0 + h \cdot h^2 + h \cdot (2h)^2 + h \cdot (3h)^2 + \cdots + h \cdot [(n-1)h]^2 \\ &= h^3 [1^2 + 2^2 + \cdots + (n-1)^2] \\ &= \frac{a^3}{n^3}\left[\frac{1}{6}(n-1) \cdot n \cdot (2n-1)\right] \\ &= \frac{1}{6}a^3\left(1 - \frac{1}{n}\right)\left(2 - \frac{1}{n}\right) \end{aligned}$$

Therefore,

$$\lim_{n \to \infty} s_n = \frac{1}{3}a^3$$

Let us consider the rectangles of the type $ABC'D'$. They constitute a set of rectangles circumscribing the parabolic region. Let us denote the sum of the areas of these circumscribing rectangles by S_n. Therefore,

$$\begin{aligned} S_n &= h \cdot h^2 + h \cdot (2h)^2 + h \cdot (3h)^2 + \cdots + h \cdot (nh)^2 \\ &= h^3 (1^2 + 2^2 + \cdots + n^2) \end{aligned}$$

$$=\frac{a^3}{n^3}\frac{1}{6}n(n+1)(2n+1)$$

$$=\frac{1}{6}a^3\left(1+\frac{1}{n}\right)\left(2+\frac{1}{n}\right)$$

Therefore,
$$\lim_{n\to\infty}S_n=\frac{1}{3}a^3$$

If we denote the area of the parabolic region by A, obviously we have

$$\lim_{n\to\infty}s_n \leqslant A \leqslant \lim_{n\to\infty}S_n$$

$$\frac{1}{3}a^3 \leqslant A \leqslant \frac{1}{3}a^3$$

Therefore,
$$A=\frac{1}{3}a^3$$

With this method, we can easily find the areas of plane regions.

5.1.2 Calculation of Area Under a Curve

If $f(x)>0$ be a function continuous on interval $[a,b]$ and interval $[a,b]$ be divided into n equal parts with each of length h, so that $h=\frac{b-a}{n}$, then the area between $x=a$, $x=b$, $y=f(x)$ and x-axis can be calculated by

$$\lim_{h\to 0}h\left[f(a)+f(a+h)+f(a+2h)+\cdots+f(a+(n-1)h)\right] \tag{5.1}$$

or

$$\lim_{h\to 0}h\left[f(a+h)+f(a+2h)+f(a+3h)+\cdots+f(a+nh)\right] \tag{5.2}$$

Example 5.1 Find the area bounded by the curve $y=2x^2-3$, x-axis, $x=0$, $x=a$. (Use the limit of a sum)

Solution Here $f(x)=2x^2-3$, Then using (5.2)

$$A=\lim_{h\to 0}h\left[f(h)+f(2h)+f(3h)+\cdots+f(nh)\right]$$

We have,

$$A=\lim_{h\to 0}h\{(2h^2-3)+[2(2h)^2-3]+[2(3h)^2-3]+\cdots+[2(nh)^2-3]\}$$

$$=\lim_{h\to 0}h[2h^2(1^2+2^2+3^2+\cdots+n^2)-3n]$$

$$=\lim_{h\to 0}\left[2h^3\cdot\frac{n(n+1)(2n+1)}{6}-3nh\right]$$

$$=\lim_{h\to 0}\left[\frac{1}{3}\cdot nh(nh+h)(2nh+h)-3nh\right]\quad\left(h=\frac{a}{n}\right)$$

$$=\frac{1}{3}\cdot a(a+0)(2a+0)-3a$$

$$=\frac{2}{3}a^3-3a$$

This is known as **the definite integral** of $f(x)$ with respect to x, and the range is from a to b and can be written as $\int_a^b f(x)\mathrm{d}x$.

Hence,

$$\int_a^b f(x)\mathrm{d}x = \lim_{h\to 0} h\,[f(a)+f(a+h)+f(a+2h)+\cdots+f(a+(n-1)h)]$$

where $h=\dfrac{b-a}{n}$. The above equation can also be written as

$$\int_a^b f(x)\mathrm{d}x = \lim_{h\to 0} h\,[f(a+h)+f(a+2h)+f(a+3h)+\cdots+f(a+nh)]$$

In particular, if $a=0$, then

$$\int_0^b f(x)\mathrm{d}x = \lim_{h\to 0} h\,[f(h)+f(2h)+f(3h)+\cdots+f(nh)]$$

where $h=\dfrac{b-0}{n}=\dfrac{b}{n}$.

5.1.3 Riemann Sums and Definite Integrals

Now we shall generalize the idea developed in the last section. For this, the function, we consider, must be continuous. Let f be a function defined and continuous on interval $[a, b]$. See figure 5 - 2. Consider a finite set of points $x_0, x_1, \cdots, x_n$ such that

$$a=x_0<x_1<\cdots<x_{n-1}<x_n=b$$

This set of points is called a partition of the interval $[a, b]$. This set divides the interval $[a, b]$ into n sub-intervals whose lengths can be denoted by

$$\Delta x_i = x_i - x_{i-1},\ i=1,2,3,\cdots,n$$

The longest sub-interval is called the norm of the partition, that is, norm $=\max\Delta x_i$, $i=1,2,\cdots,n$.

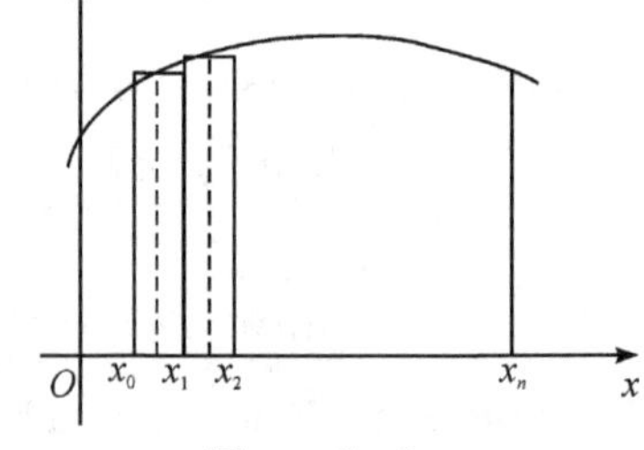

Figure 5 - 2

Take any point t_i of the interval $[x_{i-1}, x_i]$. Then the sum $\sum_{i=1}^{n} f(t_i)\Delta x_i$ is called a Riemann sum of the function on interval $[a, b]$. Now as n tends to infinity, $\Delta x_i \to 0$. In this case, as we have seen earlier, the Riemann sum tends to a limit which is equal to the area under the curve of the continuous function f.

Definition 5.1 If the Riemann sum $\sum_{i=1}^{n} f(t_i)\Delta x_i$ tends to a limit I, as the norm approaches zero for every choice of the point t_i in $[x_{i-1}, x_i]$, then I is called **the definite integral** of f on a to b. We denote this limit by

$$I=\int_a^b f(x)\mathrm{d}x = \lim_{\text{norm}\to 0}\sum_{i=1}^{n} f(t_i)\Delta x_i$$

Here a and b are called the lower and upper limits of the integral respectively.

The existence of definite integral for some functions is guaranteed by the theorem below.

Theorem 5.1 If f is continuous on interval $[a, b]$, then $\int_a^b f(x)\mathrm{d}x$ exists.

The proof of this theorem is beyond the scope of this textbook, so we will not prove it.

1. Geometric Interpretation of Definite Integral

If $f(x) > 0$ is continuous on $[a, b]$, then the area bounded by the curve $y = f(x)$, the x-axis and $x = a$, $x = b$ is equal to $\int_a^b f(x)\mathrm{d}x$.

Using the geometric interpretation of definite integral, it is easy to find the value of $\int_0^a \sqrt{x^2 - a^2}\,\mathrm{d}x$ be equal to the area of a quarter-circle, that is $\frac{\pi}{4}a^2$.

2. Properties of Definite Integral

(1) $\int_a^a f(x)\mathrm{d}x = 0$;

(2) $\int_a^b f(x)\mathrm{d}x = -\int_b^a f(x)\mathrm{d}x$;

(3) $\int_a^b f(x)\mathrm{d}x = \int_b^c f(x)\mathrm{d}x + \int_c^b f(x)\mathrm{d}x$ (where c is between a and b);

(4) $\int_a^b kf(x)\mathrm{d}x = k\int_a^b f(x)\mathrm{d}x$;

(5) $\int_a^b [\alpha f(x) + \beta g(x)]\mathrm{d}x = \alpha\int_a^b f(x)\mathrm{d}x + \beta\int_a^b g(x)\mathrm{d}x$.

5.2 Two Fundamental Theorems of Calculus

Let $P(x, y)$ and $Q(x + \Delta x, y + \Delta y)$ be two adjacent points on the curve, $f(x) \geqslant 0$ and be continuous on $[a, b]$. See figure 5-3. We consider the areas of rectangles $PMNR$ and $SMNQ$.

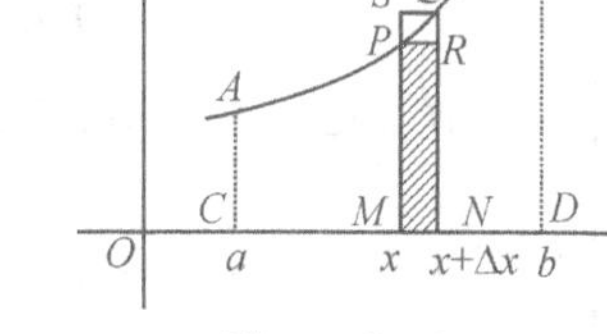

Figure 5-3

Let $A(x) =$ area of $ACMP = \int_a^x f(x)\mathrm{d}x$, then

$$A(x + \Delta x) = \text{area of } ACNQ = \int_a^{x+\Delta x} f(x)\mathrm{d}x$$

Thus,

$$A(x + \Delta x) - A(x) = \text{area of } PMNQ$$

As shown in figure 5-3, we know,

area of $PMNR <$ area of $PMNQ <$ area of $SMNQ$

$$\Rightarrow f(x) \cdot \Delta x < A(x + \Delta x) - A(x) < f(x + \Delta x) \cdot \Delta x$$

$$\Rightarrow f(x) < \frac{A(x + \Delta x) - A(x)}{\Delta x} < f(x + \Delta x)$$

$$\Rightarrow \lim_{\Delta x \to 0} f(x) < \lim_{\Delta x \to 0} \frac{A(x + \Delta x) - A(x)}{\Delta x} < \lim_{\Delta x \to 0} f(x + \Delta x)$$

$$\Rightarrow f(x) < A'(x) < \lim_{\Delta x \to 0} f(x + \Delta x)$$

Since $y = f(x)$ is continuous, then $\lim_{\Delta x \to 0} f(x + \Delta x) = f(x)$

so

$$A'(x) = f(x)$$

That is, $A(x)$ is the antiderivative of $f(x)$.

Note: In the proof of the above result, we have assumed that $f(x) \geqslant 0$ (i. e. the curve lies above the x-axis).

Based on the analysis above, we can obtain a very important theorem of calculus.

The First Fundamental Theorem of Calculus

If f is a continuous function and $\Phi(x) = \int_a^x f(t)\mathrm{d}t$, then

$$\frac{\mathrm{d}}{\mathrm{d}x}\Phi(x) = f(x)$$

Remark:

(1) The independent variable of $\Phi(x)$ is the upper limit of the definite integral $\int_a^x f(t)\mathrm{d}t$ and it ranges from a to b ;

(2) The independent variable of $\Phi(x)$ and $f(x)$ are different.

The first fundamental theorem of integral calculus establishes the relationship between the two basic concepts of the derivative and the definite integral. Now with the help of this fundamental theorem, we can prove the following theorem.

The Second Fundamental Theorem of Calculus

If f is continuous on $[a,b]$ and $F(x)$ is an antiderivative of f, then

$$\int_a^b f(x)\mathrm{d}x = F(x)\Big|_a^b = F(b) - F(a)$$

Poof Let $\Phi(x) = \int_a^x f(t)\mathrm{d}t$.

Obviously, we get $\Phi(a) = 0$. Now, as Φ and F are antiderivatives of the same function f, they differ only by a constant. So $\Phi(x) = F(x) + C$, for some constant C, and $\Phi(a) = F(a) + C$, or $0 = F(a) + C$, or $F(a) = -C$, then we have

$$\Phi(x) = F(x) - F(a)$$

and

$$\Phi(b) = F(b) - F(a)$$

Again,

$$\Phi(b) = \int_a^b f(t)\mathrm{d}t$$

so

$$\int_a^b f(t)\mathrm{d}t = F(x)\Big|_a^b = F(b) - F(a) \tag{5.3}$$

Formula (5.3) is also called **Newton-Leibniz formula (N-L formula)**.

5.3 Some Special Cases for Finding Area

Case 1 If the curve $y=f(x)$ lies under the x-axis (i. e. $f(x)\leqslant 0$), as shown in figure 5-4, the area bounded by the curve $y=f(x)$, x-axis and the two lines $x=a$, $x=b$ is given by

$$\int_a^b -f(x)\,dx = -\int_a^b f(x)\,dx$$

Case 2 If the curve $x=f(y)$ lies on the right of y-axis, as shown in figure 5-5, the area bounded by the curve $x=f(y)(f(y)\geqslant 0)$, the y-axis and the two lines $y=a$ and $y=b$ is equal to $\int_a^b x\,dy$.

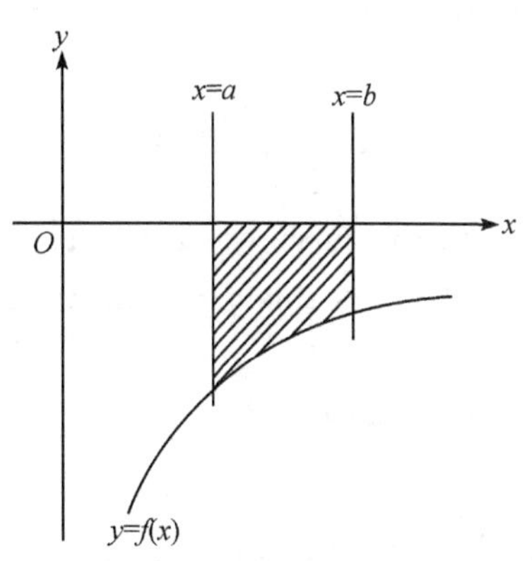

Figure 5-4

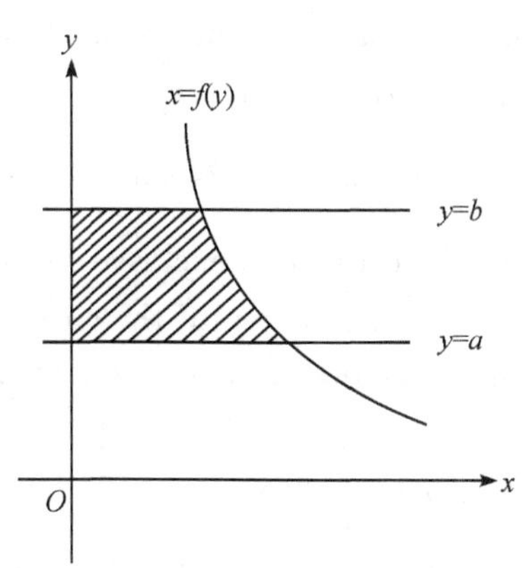

Figure 5-5

Case 3 If the curve $x=f(y)$ lies on the left of y-axis(i. e. $f(y)\leqslant 0$) as shown in figure 5-6, then the area bounded by the curve $x=f(y)$, y-axis and the lines $y=a$ and $y=b$ is equal to

$$\int_a^b (-x)\,dy = -\int_a^b f(y)\,dy$$

Case 4 Area between two curves. Now let us find out the area enclosed by the curves represented by two given functions f_1 and f_2 and the given lines $x=a$ and $x=b$, see figure 5-7.

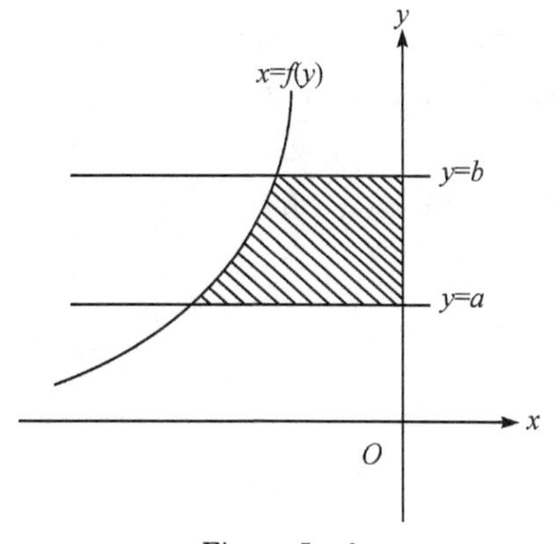

Figure 5-6

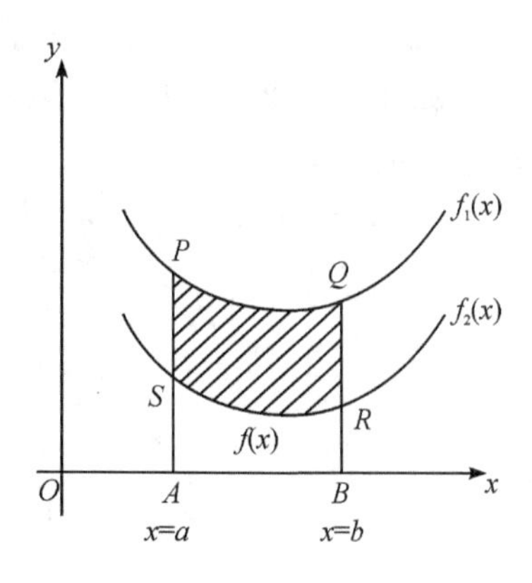

Figure 5-7

$$\text{Area of } PQRS = \int_a^b f_1(x)\,\mathrm{d}x - \int_a^b f_2(x)\,\mathrm{d}x$$
$$= \int_a^b [f_1(x) - f_2(x)]\,\mathrm{d}x$$

Example 5.2 Find the area enclosed by $y = 3x$, the x-axis and lines $x = 0$, $x = 4$.

Solution The required area $= \int_0^4 3x\,\mathrm{d}y = \frac{3x^2}{2}\Big|_0^4 = \frac{3 \times 4^2}{2} = 24.$

Example 5.3 Find the area bounded by the curve $y^2 = 4ax$, the x-axis and the line which cuts the curve at the point $(a, 2a)$.

Solution The point $(0,0)$ satisfies the equation of the curve. So the curve passes through the origin. Hence, we have to find the area bounded by the curve $y^2 = 4ax$, the x-axis and the lines $x = 0$ and $x = a$. Therefore,

$$\text{The required area} = \int_0^a 2\sqrt{ax}\,\mathrm{d}x = 2\sqrt{a}\int_0^a x^{1/2}\,\mathrm{d}x = 2\sqrt{a}\,\frac{x^{3/2}}{3/2}\Big|_0^a = \frac{4}{3}a^2$$

Example 5.4 Find the area enclosed by the x-axis and the curve $y = x^2 - 4x + 3$.

Solution The given equation of the curve is $y = x^2 - 4x + 3$, this curve meets the x-axis at the point which has the ordinates of $y = 0$, then

$$x^2 - 4x + 3 = 0 \Rightarrow (x-1)(x-3) = 0$$

Either $x = 1$ or $x = 3$, then

$$\text{the required area} = -\int_1^3 (x^2 - 4x + 3)\,\mathrm{d}x = -\left(\frac{x^3}{3} - 2x^2 + 3x\right)\Big|_1^3 = \frac{4}{3}$$

Example 5.5 Find the area bounded by the y-axis, the curve $x^2 = 4(y-2)$ and the line $y = 11$.

Solution See figure 5 - 8. The curve $x^2 = 4(y-2)$ meets the y-axis at the point where $x = 0$, $y = 2$. Then,

$$\text{the required area} = \int_2^{11} 2\sqrt{y-2}\,\mathrm{d}y = \left[\frac{4}{3}(y-2)^{3/2}\right]\Big|_2^{11} = 36$$

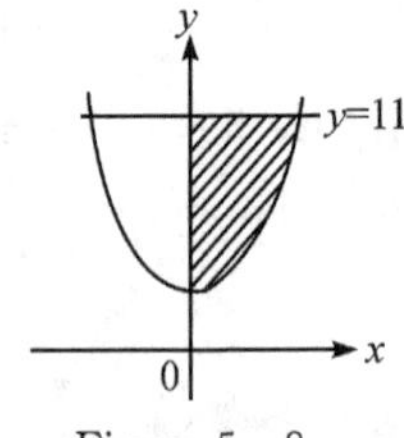

Figure 5 - 8

Example 5.6 Find the area bounded by the curves $y = x^2$ and $y = 2x$.

Solution Eliminating y with the given equations, we get

$$x^2 = 2x \quad \text{or} \quad x(x-2) = 0$$

Hence, $x = 0$ and $x = 2$ are the horizontal ordinates of the points at which the given curves intersect.

Therefore, the required area as shown in figure 5 - 9 is given by

$$\int_0^2 (2x - x^2)\,\mathrm{d}x = \left(x^2 - \frac{1}{3}x^3\right)\Big|_0^2$$
$$= 4 - \frac{8}{3} = \frac{4}{3}$$
$$= 1\frac{1}{3}$$

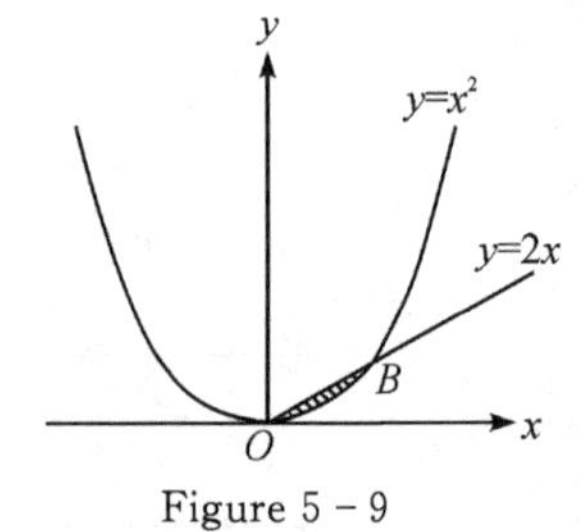

Figure 5 - 9

5.4 Calculation of Definite Integrals

5.4.1 N-L Formula for Definite Integrals

In evaluating a definite integral $\int_a^b f(x)\mathrm{d}x$, we use the following steps:

(1) Find an antiderivative of $f(x)$ denoted by $F(x)$;

(2) Substitute $x=b$ in $F(x)$ to get $F(b)$;

(3) Substitute $x=a$ in $F(x)$ to get $F(a)$;

(4) Then subtract $F(b)$ from $F(a)$, we will get the value of the definite integral $\int_a^b f(x)\mathrm{d}x$.

Example 5.7 Evaluate the following definite integrals.

(1) $\int_0^3 x^5\mathrm{d}x$; (2) $\int_{-1}^2 (x^2+x+1)\mathrm{d}x$;

(3) $\int_0^1 \left(\mathrm{e}^{2x}+\frac{3}{x+1}\right)\mathrm{d}x$; (4) $\int_{-2}^0 (x-\mathrm{e}^{-x})\mathrm{d}x$;

(5) $\int_{\frac{\pi}{6}}^{\frac{\pi}{2}} \sin 3x\mathrm{d}x$. (6) $\int_5^2 (3x-4)^4\mathrm{d}x$.

Solution

(1) $\int_0^3 x^5\mathrm{d}x=\left(\frac{1}{6}x^6\right)\Big|_0^3=\frac{1}{6}(3^6-0)=\frac{243}{2}$.

(2) $\int_{-1}^2 (x^2+x+1)\mathrm{d}x=\left(\frac{1}{3}x^3+\frac{1}{2}x^2+x\right)\Big|_{-1}^2=\frac{15}{2}$.

(3) $\int_0^1 \left(\mathrm{e}^{2x}+\frac{3}{x+1}\right)\mathrm{d}x=\left(\frac{1}{2}\mathrm{e}^{2x}+3\ln(x+1)\right)\Big|_0^1=\frac{1}{2}\mathrm{e}^2+3\ln 2-\frac{1}{2}$.

(4) $\int_{-2}^0 (x-\mathrm{e}^{-x})\mathrm{d}x=\left(\frac{1}{2}x^2+\mathrm{e}^{-x}\right)\Big|_{-2}^0=-1-\mathrm{e}^2$.

(5) $\int_{\frac{\pi}{6}}^{\frac{\pi}{2}} \sin 3x\mathrm{d}x=-\frac{1}{3}\cos 3x\Big|_{\frac{\pi}{6}}^{\frac{\pi}{2}}=0$.

(6) $\int_5^2 (3x-4)^4\mathrm{d}x=\frac{1}{15}(3x-4)^5\Big|_5^2=-\frac{161019}{15}$.

5.4.2 Substitution for Definite Integrals

> Suppose that the function f is continuous and $x=\varphi(t)$ is continuously differentiable on interval $[\alpha,\beta]$, $\varphi(\alpha)=a$, $\varphi(\beta)=b$, then
>
> $$\int_a^b f(x)\mathrm{d}x=\int_\alpha^\beta f[\varphi(t)]\varphi'(t)\mathrm{d}t$$

Example 5.8 Find the following definite integrals.

(1) $\int_0^4 \frac{dx}{1+\sqrt{x}}$; (2) $\int_0^{\pi/6} \frac{\cos\theta d\theta}{\sqrt{1-\sin\theta}}$;

(3) $\int_0^1 \sqrt{1-x^2}\,dx$; (4) $\int_{\sqrt{2}}^2 \frac{dx}{\sqrt{x^2-1}}$.

Solution

(1) Let $x=t^2$, then $dx=2t\,dt$ and when $x=0, t=0$; $x=4, t=2$. So we have

$$\int_0^4 \frac{dx}{1+\sqrt{x}} = \int_0^2 \frac{2t\,dt}{1+t} = 2[t-\ln(1+t)]\Big|_0^2 = 4-2\ln 3$$

(2) Let $t=\sin\theta$, then $dt=\cos\theta d\theta$ and when $\theta=0, t=0$; $\theta=\frac{\pi}{6}, t=\frac{1}{2}$. So we have

$$\int_0^{\pi/6} \frac{\cos\theta d\theta}{\sqrt{1-\sin\theta}} = \int_0^{\frac{1}{2}} \frac{dt}{\sqrt{1-t}} = -2\sqrt{1-t}\Big|_0^{\frac{1}{2}} = 2-\sqrt{2}$$

(3) Let $x=\sin t$, then $dx=\cos t\,dt$ and when $x=0$, $t=0$; $x=1$, $t=\frac{\pi}{2}$. So we have

$$\int_0^1 \sqrt{1-x^2}\,dx = \int_0^{\frac{\pi}{2}} \sqrt{1-\sin^2 t}\,d\sin t = \int_0^{\frac{\pi}{2}} \cos^2 t\,dt = \frac{1}{2}\left(t+\frac{1}{2}\sin 2t\right)\Big|_0^{\frac{\pi}{2}} = \frac{\pi}{4}$$

(4) Let $x=\sec t$, then $dx=\sec t\tan t\,dt$ and when $x=\sqrt{2}$, $t=\frac{\pi}{4}$; $x=2$, $t=\frac{\pi}{3}$. So we have

$$\int_{\sqrt{2}}^2 \frac{dx}{\sqrt{x^2-1}} = \int_{\frac{\pi}{4}}^{\frac{\pi}{3}} \sec t\,dt = \ln(\sec t+\tan t)\Big|_{\frac{\pi}{4}}^{\frac{\pi}{3}} = \ln\frac{2+\frac{\sqrt{3}}{3}}{1+\sqrt{2}}$$

5.4.3 Integration by Parts for Definite Integrals

Suppose that $u(x)$, $v(x)$ are continuously differentiable on the interval $[a,b]$, then

$$\int_a^b u\,dv = uv\Big|_a^b - \int_a^b v\,du$$

Example 5.9 Compute the following definite integrals.

(1) $\int_1^e \ln x\,dx$; (2) $\int_0^4 e^{\sqrt{x}}\,dx$; (3) $\int_0^{\pi/2} x\sin x\,dx$.

Solution

(1) $\int_1^e \ln x\,dx = x\ln x\Big|_1^e - \int_1^e x\frac{1}{x}dx = 1.$

(2) Let $\sqrt{x}=t$, then $x=t^2$, $dx=2dt$, thus

$$\int_0^4 e^{\sqrt{x}}\,dx = \int_0^2 e^t \cdot 2t\,dt = 2\int_0^2 t\,de^t = 2\left(te^t\Big|_0^2 - \int_0^2 e^t dt\right) = 2(e^2+1)$$

(3) $\int_0^{\pi/2} x\sin x\,dx = -x\cos x\Big|_0^{\pi/2} + \int_0^{\pi/2}\cos x\,dx = -x\cos x\Big|_0^{\pi/2} + \sin x\Big|_0^{\pi/2} = 1.$

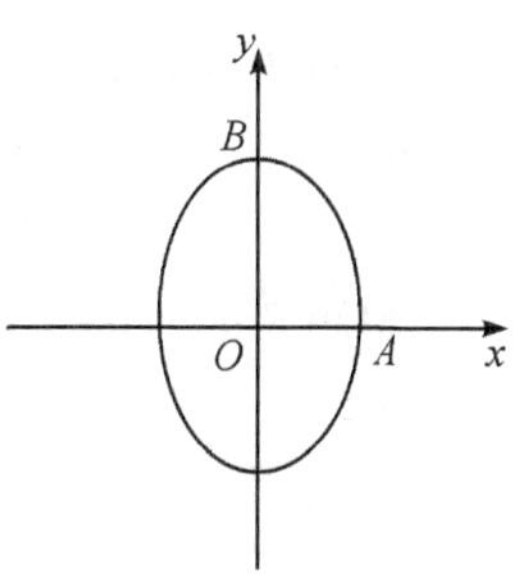

Figure 5 - 10

Example 5.10 Find the area of the ellipse $\frac{x^2}{9}+\frac{y^2}{16}=1$.

Solution See figure 5 - 10. The curve is symmetrical about x- and y-axis. So to find the area of the whole ellipse, we can first find the area of the portion lying in the first quadrant and then multiply it by 4.

Here $OA=3$, $OB=4$. The area of the portion lying in the first quadrant is bounded by the curve $y=\frac{4}{3}\sqrt{9-x^2}$, x-axis and the lines $x=0$, $x=3$. So its area is

$$A=\int_0^3 \frac{4}{3}\sqrt{9-x^2}\,dx$$

Let $x=3\sin\theta$, then $dx=3\cos\theta d\theta$, and

$$\sqrt{9-x^2}=\sqrt{9-9\sin^2\theta}=3\cos\theta$$

When $x=0$, $\theta=0$; when $x=3$, $\theta=\frac{1}{2}\pi$, then

$$A=\int_0^{\pi/2}\frac{4}{3}\cdot 3\cos\theta\cdot 3\cos\theta d\theta=12\int_0^{\pi/2}\cos^2\theta d\theta$$

$$=12\int_0^{\pi/2}\frac{1+\cos 2\theta}{2}d\theta=6\left(\theta+\frac{\sin 2\theta}{2}\right)\Big|_0^{\frac{\pi}{2}}=3\pi$$

Therefore, the whole area $=4A=4\times 3\pi=12\pi$.

5.5 Calculation of Volume

A solid of revolution is formed by revolving a plane region about a line which is called the axis of revolution. In this section we will only using the x-axis or the y-axis.

5.5.1 Disk Method

To find the volume of solid of revolution, we can use the same approach as finding the area of a region enclosed by a curve, the x-axis and the lines $x=a$ and $x=b$. See figure 5 - 11.

Then, the volume of such a solid can be cut into a large number of slices (i. e., disks), each having a thickness of Δx_i and radius of $f(x_i)$. The volume produced is then the sum of the volumes of these disks, i. e.,

$$V\approx\sum_{i=1}^{n-1}\pi[f(x_i)]^2\Delta x_i,\ \Delta x_i=\frac{b-a}{n}$$

So, as $n\to\infty$, $\Delta x_i\to 0$ then,

$$V=\lim_{\Delta x\to 0}\sum_{i=1}^{n-1}\pi[f(x_i)]^2\Delta x_i=\int_a^b\pi[f(x)]^2dx$$

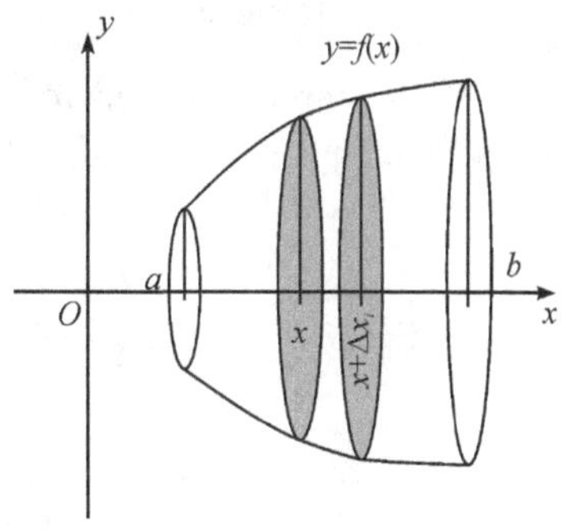

Figure 5 - 11

This process can be simpified as two steps:

Step 1 Approximate: Slice the solid into thin pieces and label a typical piece to give the element of the volume

$$\Delta V \approx \pi[f(x)]^2 \Delta x$$

Step 2 Precise: Integrate the element of the volume on the given interval to get the volume

$$V = \int_a^b \pi[f(x)]^2 \mathrm{d}x$$

Step 1 and Step 2 is called the **method of elements of integration.**

Therefore, we have

> The volume of a solid of revolution is given by
>
> $$V = \int_a^b \pi[f(x)]^2 \mathrm{d}x$$
>
> When a plane region enclosed by the curve $y = f(x)$ and the lines $x = a$ and $x = b$ is revolved about the x-axis.
>
> And $$V = \int_c^d \pi[f^{-1}(y)]^2 \mathrm{d}y$$
>
> when a plane region enclosed by the curve $x = f^{-1}(y)$ and the lines $y = c$ and $y = d$ is revolved about the y-axis.

Example 5.11 The curve $y = \sqrt{x-1}\,(1 \leqslant x \leqslant 5)$ is rotated about x-axis to form a solid of revolution. Sketch its solid and find its volume. If the same curve is rotated about the y-axis, a different solid is formed. Sketch the second solid and find its volume.

Solution See figure 5 - 12, the curve has a restricted domain and is rotated about the x-axis.

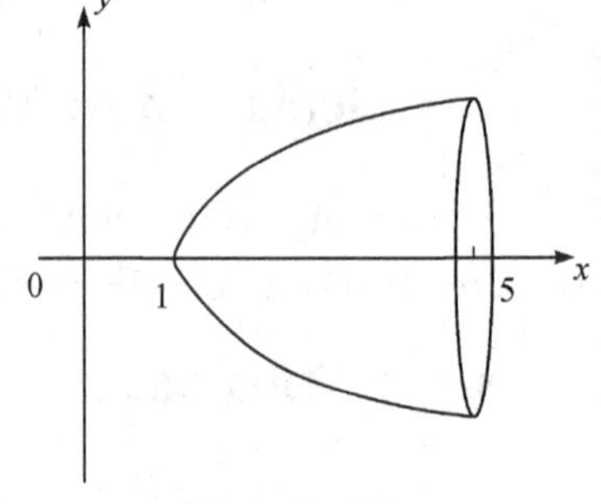

Figure 5 - 12

So, the solid formed has a volume given by

$$V = \int_1^5 \pi(\sqrt{x-1})^2 \mathrm{d}x = \pi\int_1^5 (x-1)\,\mathrm{d}x = 8\pi$$

If the same curve is rotated about the y-axis, the solid formed like this as shown in figure 5 - 13.

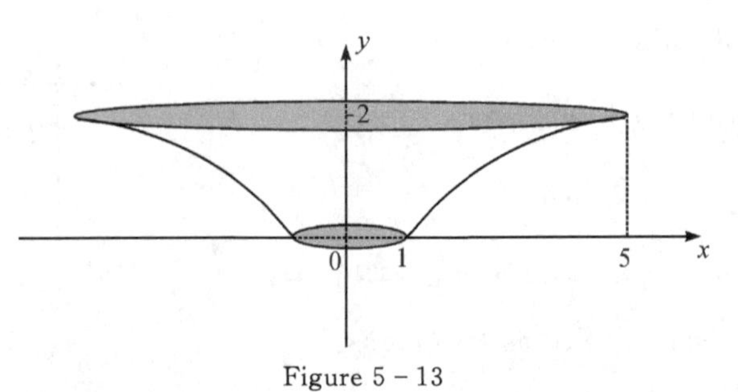

Figure 5 - 13

It is important to realize that the integral limits are in terms of the variable y and so are 0 and 2.

Also, x must be made the subject of the rule for the curve

$$y=\sqrt{x-1}\Rightarrow y^2=x-1\Rightarrow x=y^2+1$$

When $x=1$, $y=0$ and when $x=5$, $y=2$, entering these values into the formula

$$V=\pi\int_0^2(y^2+1)^2\mathrm{d}y=13\frac{11}{15}\pi$$

Example 5.12 Find the volume of a circular cone whose semi-angle is α and has a height of h.

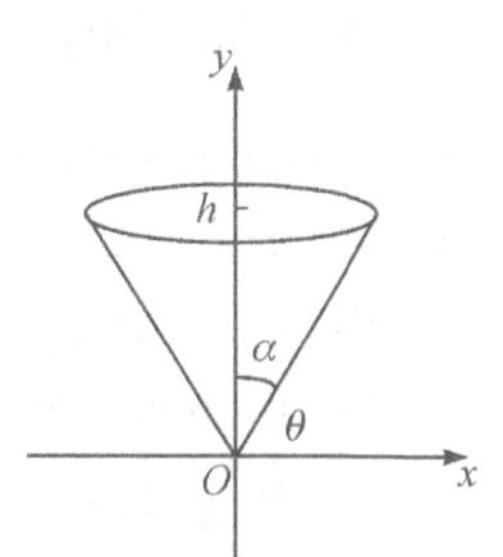

Figure 5 - 14

Solution A circular cone of height h generated by revolving the region bounded by an oblique straight line about the y-axis from $y=0$ to $y=h$.

First, to find the equation of the straight line. As shown in figure 5 - 14, the straight line has equation of

$$y=kx$$

To find k, we note that $k=\tan\theta$, where θ is the angle between the line and the positive x-axis, therefore we have $\theta=\frac{\pi}{2}-\alpha$, and so $k=\tan\left(\frac{\pi}{2}-\alpha\right)=\cot\alpha$.

Therefore, the equation of the straight line is

$$y=\cot\alpha\cdot x$$

Next, as we are revolving the plane about the y-axis, we use the method of elements of integration to have

$$\Delta V\approx\pi x^2\Delta y$$

With $a=0$, $b=h$, and from $y=x\cot\alpha$, we have $x=y\tan\alpha$, that is,

$$V=\pi\int_0^h(y\tan\alpha)^2\mathrm{d}y=\frac{1}{3}\pi\tan^2\alpha\cdot h^3$$

We may skip the step 1 of the method of element of integration when we are familiar with the process.

Example 5.13 Find the volume of the solid formed by revolving the region enclosed with the curve $f(x)=\sqrt{25-x^2}$ and the line $g(x)=3$ about the x-axis.

Solution We start by drawing a diagram with intersections, as shown in figure 5 - 15. Then we determine the coordinates of intersections.

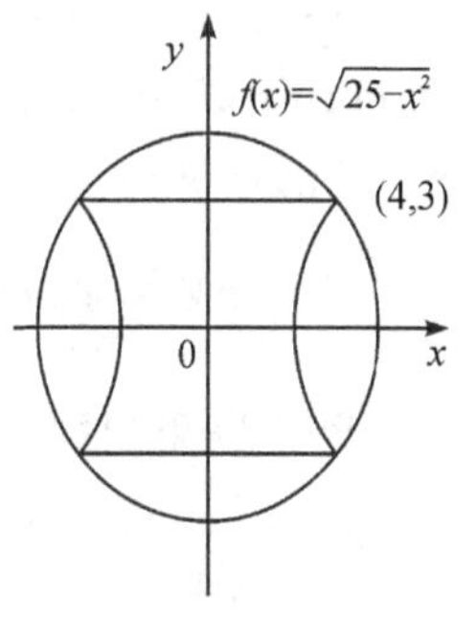

Figure 5 - 15

Let $f(x)=g(x)$ we have, $\sqrt{25-x^2}=3\Rightarrow x=\pm4$. The solid is hollow, i. e., $-3\leqslant y\leqslant3$.

Next, we find the difference between the two volumes generated by revolving the two curves(like finding the area between two curves)

$$V=\pi\int_{-4}^{4}[f(x)]^2\mathrm{d}x-\pi\int_{-4}^{4}[g(x)]^2\mathrm{d}x$$
$$=\pi\int_{-4}^{4}[f^2(x)-g^2(x)]\mathrm{d}x$$
$$=2\pi\int_{0}^{4}\left[(\sqrt{25-x^2})^2-3^2\right]\mathrm{d}x \quad \text{(by symmetry)}$$
$$=\frac{256}{3}\pi$$

5.5.2 Shell Method

There is another method for finding the volume of a solid of revolution: The method of cylindrical shells. For many problems, it is easier to apply than the methods of disks.

Now, consider a region of the type shown in figure 5 - 16(a). Slice it vertically and revolve it about y-axis. It will generate a solid of revolution, and each slice will generate a piece that is approximately a cylindrical shell, see figure 5 - 16(b). To get the volume of this solid, we calculate the volume $\Delta V\approx 2\pi x f(x)\Delta x$ of a typical shell as shown in figure 5 - 16 (c), and then integrate it to obtain the whole volume on a give interval.

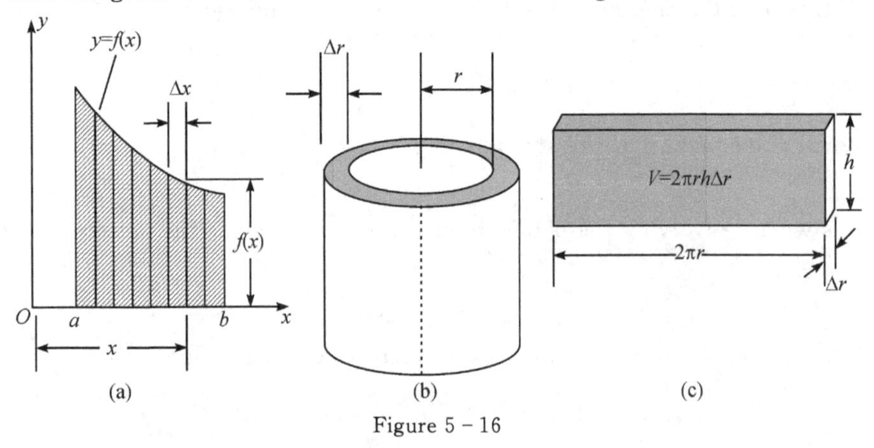

Figure 5 - 16

Example 5.14 The reigon bounded by $y=\frac{1}{\sqrt{x}}$, the x-axis, $x=1$ and $x=9$ is revolved about the y-axis. Find the volume of the resulting solid.

Solution From figure 5 - 17 we see that the volume of the shell generated by the slice is

$$\Delta V\approx 2\pi x f(x)\Delta x=2\pi x\frac{1}{\sqrt{x}}\Delta x$$

The volume is then found by integrating.

$$V=2\pi\int_{1}^{9}x\frac{1}{\sqrt{x}}\mathrm{d}x=2\pi\int_{1}^{9}\sqrt{x}\,\mathrm{d}x=2\pi\left(\frac{2}{3}x^{3/2}\right)\Big|_{1}^{9}=\frac{104}{3}\pi$$

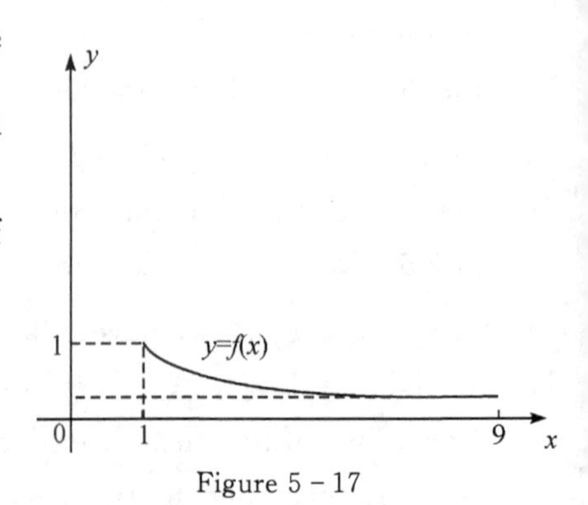

Figure 5 - 17

Example 5.15 The region bounded by the line $y=\frac{r}{h}x$, $x=h$, the x-axis is revolved about the x-axis, thereby generating a cone(assume that $xr>0, h>0$). Find its volume by the disk method and by the shell method.

Solution **Disk Method**: See figure 5-18.

Follow the steps suggested by the method of element of integration.

$$\Delta V \approx \pi[f(x)]^2\Delta x=\pi\left(\frac{r}{h}x\right)^2\Delta x$$

Then

$$V=\int_0^h \pi\left(\frac{r}{h}x\right)^2 dx=\frac{1}{3}\pi r^2 h$$

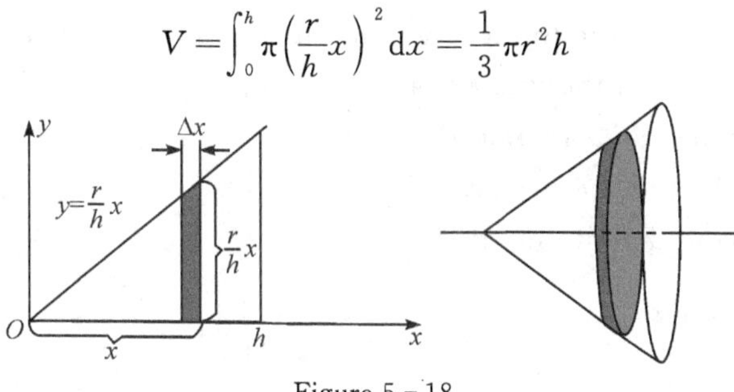

Figure 5-18

Shell Method: See figure 5-19.

$$\Delta V \approx 2\pi y\left(h-\frac{h}{r}y\right)^2\Delta y$$

Then

$$V=\int_0^r 2\pi y\left(h-\frac{h}{r}y\right)^2 dy=\frac{1}{3}\pi r^2 h$$

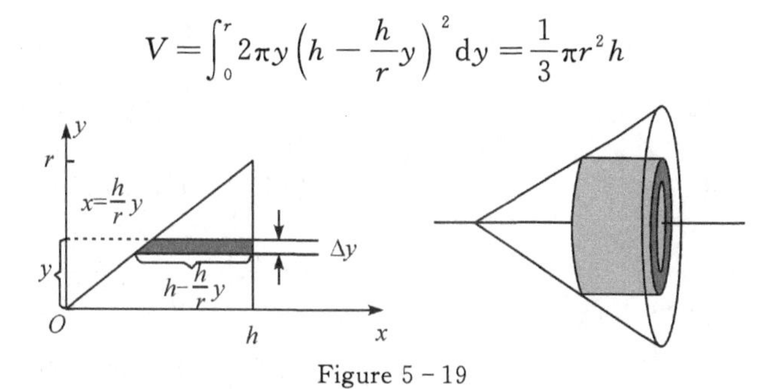

Figure 5-19

Exercise 5

1. Find the area of the plane region using the limit of a sum. (The x-axis is included in each of the following question.)

   (1) $y=1-x$, $x=0$, $x=b$ $(0<b<1)$; (2) $y=4x^2$, $x=0$, $x=c$ $(c>0)$;

   (3) $y=e^x$, $x=0$, $x=a$ $(a>0)$.

2. Find the area bounded by the x-axis and the following curves and lines:

   (1) $x^2=4by$, $x=a$, $x=b$ $(a, b>0)$; (2) $y=4x^3$, $x=2$, $x=4$;

(3) $y = 3x^2 - 2$, $x = 1$, $x = 4$; (4) $y^2 - x - 4 = 0$, $x = 2$, $x = 5$;

(5) $y = e^{ax}$, $x = b$, $x = c$ $(0 < b < c)$; (6) $y = \ln(1 + x)$, $x = 0$, $x = 1$.

3. Find the areas in the first quadrant bounded by the x-axis and the following given curves and lines.

(1) $y^2 = 8ax$ and the ordinate at the point $(4a, 0)$;

(2) $y^2 = 4a(x - a)$ and the ordinate at the point $(h, 0)$;

(3) $x^2 = 4by$ and the ordinate at the point $(b, 0)$.

4. Find the areas of the regions between following curves and lines.

(1) The curve $y^2 = 16x$ and the line $y = 2x$;

(2) The curve $y = x^3$ and the line $x = y$ lying in the first quadrant;

(3) The curve $y^2 = x^3$ and the line $x = 4$;

(4) The curve $y^2 = 4ax$ and the line $x^2 = 4ay$;

(5) $y = x^2 - 8x + 15$ and the x-axis.

5. Evaluate the following definite integrals.

(1) $\int_1^2 (2x^2 + 3x + 4)\mathrm{d}x$; (2) $\int_0^{-1} \frac{\mathrm{d}x}{x + 2}$;

(3) $\int_0^1 x^3 \sqrt{1 + 2x^4}\,\mathrm{d}x$; (4) $\int_1^2 e^{2x^2 - 1} x\,\mathrm{d}x$;

(5) $\int_0^a \frac{x\,\mathrm{d}x}{(a^2 + x^2)^{3/2}}$; (6) $\int_0^1 \frac{2x\,\mathrm{d}x}{x^2 + 3}$;

(7) $\int_0^{\pi/4} \tan^2\theta\,\mathrm{d}\theta$; (8) $\int_0^1 \cos^2 \pi x\,\mathrm{d}x$;

(9) $\int_0^{\pi/2} \sin^3 x\,\mathrm{d}x$; (10) $\int_0^{\pi/4} \frac{\mathrm{d}x}{1 - \sin x}$;

(11) $\int_0^{\pi/2} \sqrt{1 + \sin x}\,\mathrm{d}x$; (12) $\int_0^{\pi/2} \cos 3x \cos 2x\,\mathrm{d}x$;

(13) $\int_0^{\pi/4} \cos^3 x \sin^2 x\,\mathrm{d}x$; (14) $\int_0^{\pi/4} \tan^3 x\,\mathrm{d}x$;

(15) $\int_0^{\pi/4} \tan^2 x \sec^4 x\,\mathrm{d}x$; (16) $\int_0^{-1} \frac{\mathrm{d}x}{\sqrt{4 - x^2}}$;

(17) $\int_\pi^{\pi/2} x \cos x\,\mathrm{d}x$; (18) $\int_1^e x \ln x\,\mathrm{d}x$.

6. Find the volume of the solid of revolution that is produced by rotating the region about x-axis and y-axis respectively, and the region is enclosed by $y = \sin x$ $(0 \leqslant x \leqslant \pi)$ and x-axis.

7. The curve $y = \frac{1}{x}$ between $x = \frac{1}{5}$ and $x = 1$ is rotated about the y-axis. Find the volume of the solid of revolution formed in this way.

8. Find the equation of the straight line that passes through the origin and the point (r, h). Hence use calculus to prove that the volume of a right circular with base radius r and height h is given by $V = \frac{1}{3}\pi r^2 h$.

9. Find the equation of circle with center (0,0) and radius r. Use calculus to prove that the volume of a sphere is given by the formula $V=\frac{4}{3}\pi r^3$.

10. Find the volume of the solid of revolution that is formed by rotating the region bounded by the curves $y=\sqrt{x}$ and $y=\sqrt{x^3}$, and the axis of revolution is:
(1) The x-axis; (2) The y-axis.

11. The region is enclosed by the curve $y=\frac{1}{x}$, the line $x=1$ and $x=2$, find the volume of the region rotating about the line $x=1$.

Chapter 6 Differentiation of Functions with Several Variables

A function of a single variable $y=f(x)$ is interpreted graphically as a planar curve. In this section we generalize the concept to functions of more than one variable. We shall see that a function of two variables $z=f(x,y)$ can be interpreted as a surface. Functions with two or more variables often arise in engineering and science, it is important to be able to deal with such functions with confidence and skill. In this section we will learn how to sketch simple surfaces. In later sections we shall examine how to determine the rate of change of $f(x,y)$ with respect to x and y, also, how to obtain the optimum values of functions of several variables.

6.1 Functions of Several Variables

6.1.1 Recognize Functions of Several Variables

We know that $f(x)$ is used to represent a function of one variable: The input variable is x and the output is the value $f(x)$. Here x is the **independent variable** and $y=f(x)$ is the **dependent variable.**

The domain of functions of two variables, $z=f(x,y)$, are regions of two dimensional space and consist of all the coordinate pairs, (x,y), that we could plug into the function and get back a real number.

Consider a function with two independent input variables x and y, for example

$$f(x,y)=x+2y+3$$

Its domain is xOy plane.

If we specify values for x and y then we will get a single value $f(x,y)$. For example, if $x=3$ and $y=1$ then $f(x,y)=3+2+3=8$, it can be denoted as $f(3,1)=8$.

Example 6.1 Determine the domain of each of the following and sketch the graph of the integral domain.

(1) $f(x,y)=\sqrt{x+y}$; (2) $f(x,y)=\sqrt{x}+\sqrt{y}$; (3) $f(x,y)=\ln(9-x^2-9y^2)$.

Solution

(1) In this case we can't take the square root of a negative number, so we have $x+y\geqslant 0$;

(2) Similarly, $x\geqslant 0$ and $y\geqslant 0$;

(3) In this case we can't take the logarithm of a negative number or zero. Therefore, $9-x^2-9y^2>0\Rightarrow x^2+9y^2<9$.

Figure 6 - 1 indicates the integral domain of the functions above.

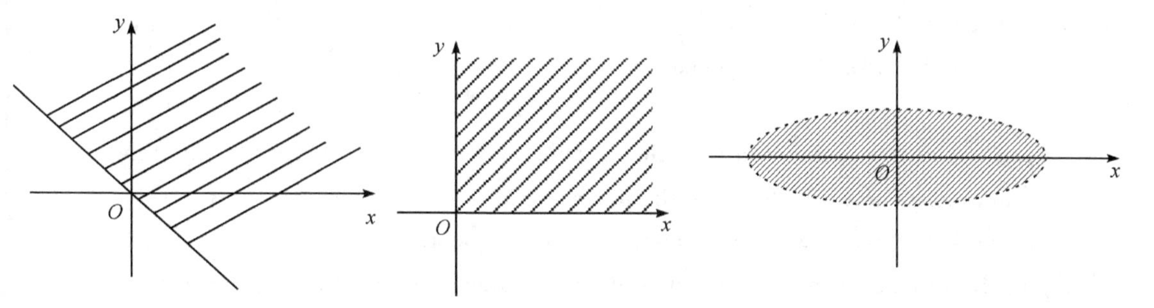

Figure 6 - 1

Example 6.2 Find the values of $f(2,1), f(-1,-3)$ and $f(0,0)$ for the following functions.

(1) $f(x,y)=x^2+y^2+1$; (2) $f(x,y)=2x+xy+y^3$.

Solution (1) $f(2,1)=2^2+1^2+1=6$;

$f(-1,-3)=(-1)^2+(-3)^2+1=11$;

$f(0,0)=1$.

(2) $f(2,1)=4+2+1=7$;

$f(-1,-3)=-2+3-27=-26$;

$f(0,0)=0$.

In a similar way we can define a function of three independent variables. Let these variables be x, y and z, the function be $u=f(x,y,z)$.

6.1.2 Geometrical Interpretation of Functions with Two Variables

Let us begin with looking back upon the simplest case—**plane.** Planes are particularly simple examples of **surfaces.**

Now let us recall the general equation of a plane

$$Ax+By+Cz=D$$

where A, B, C, D are constants. If $C\neq 0$, we can rewrite this equation in the form of $z=\frac{D}{C}-\frac{A}{C}x-\frac{B}{C}y$.

Generally a surface is defined by a relation of the form $z=f(x,y)$, where the expression on the right side is any relation involving two variables x, y.

6.1.3 Sketching Surfaces

A plane is relatively easy to sketch since it is flat, all we need to know about it is the intersections with three coordinate axes. For more general surfaces, we will sketch curves which lie on the surface. If we draw enough of these curves, our "eye" will naturally interpret the shape of the surface.

Let us see, for example, how we sketch $z=x^2+y^2$.

Step 1 Fix x at value x_0.

In this case the relation becomes

$$z=x_0^2+y^2$$

Since z is now a function of a **single** variable y, with x_0^2 held constant, this relation $z=x_0^2+y^2$ defines a **curve** which **lies in the plane** $x=x_0$.

In figure 6 - 2 we have drawn this curve (a **parabola**). Now by changing the value chosen for x_0 we will obtain a sequence of curves, each a parabola, lying in a different plane, and each being a part of the surface we are trying to sketch.

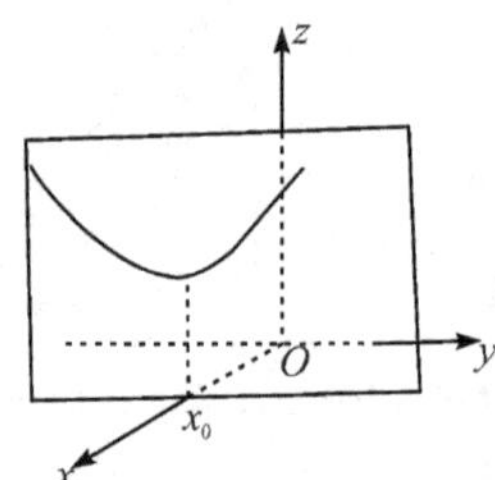

Figure 6 - 2

What we need to do is to **slice** the surface by planes parallel to the yOz plane. Each slice intersects the surface in a curve. In this case, we have not yet plotted enough curves to accurately visualize the surface, so we need to draw other surface curves.

Step 2 Fix y at value y_0.

Here $y = y_0$ (the equation of a plane parallel to the xOz plane). In this case the surface becomes $z=x^2+y_0^2$.

Again z is a function of single variable x (since y_0 is fixed) and describes a curve(a **parabola**), but this time residing on the plane $y=y_0$, see figure 6 - 3. For each different y_0 a different parabola is obtained: Each lying on the surface $z=x^2+y^2$.

Figure 6 - 3

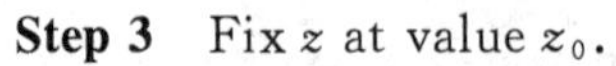

Step 3 Fix z at value z_0.

We have $z = z_0$ (the equation of a plane parallel to the xOy plane). In this case the surface becomes $z_0=x^2+y^2$.

But this is the equation of a **circle** centered on $x=0$, $y=0$, of radius $\sqrt{z_0}$, see figure 6 - 4. (Clearly we **must** choose $z_0 \geqslant 0$, because $x^2 + y^2$ cannot be negative.) As varying z_0 we obtain different circles, each lying on a different plane $z=z_0$.

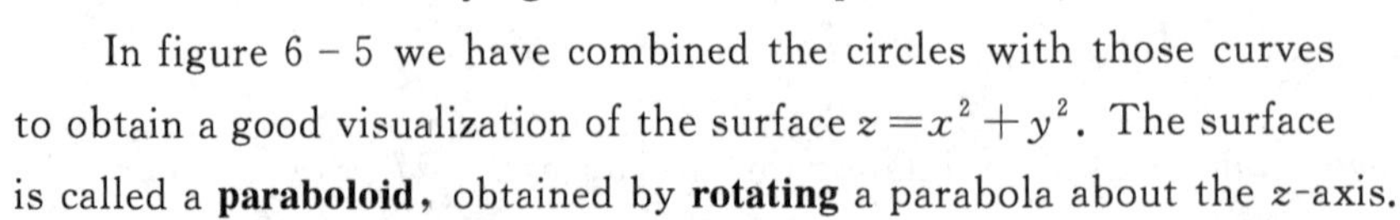

Figure 6 - 4

In figure 6 - 5 we have combined the circles with those curves to obtain a good visualization of the surface $z=x^2+y^2$. The surface is called a **paraboloid**, obtained by **rotating** a parabola about the z-axis.

With the wide availability of sophisticated graphics packages, the need to be able to sketch a surface is not as important as once it was. However, we urge readers to attempt simple surface sketching in the initial stages of this study as it will enhance understanding of functions with two variables.

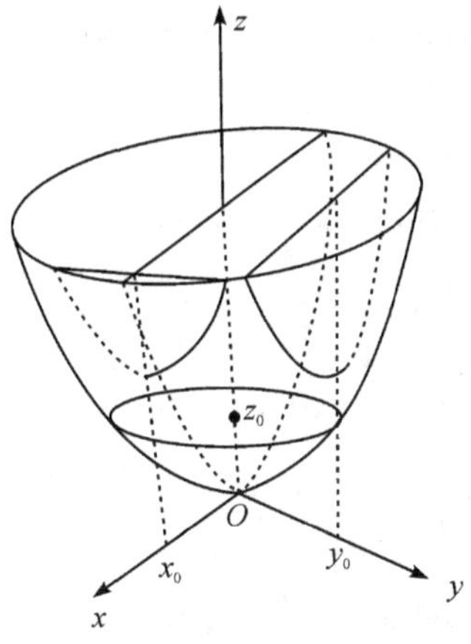

Figure 6 - 5

6.2 Limit of Function with Two Variables

In this section we will take a look at limits involving functions with more than one variable. Before getting into this, let's briefly recall how limits of functions with one variable work. We say that,

$$\lim_{x\to x_0} f(x) = a$$

provided

$$\lim_{x\to x_0^-} f(x) = \lim_{x\to x_0^+} f(x) = a$$

In other words, we will have: $\lim\limits_{x\to x_0} f(x) = a$ provided $f(x)$ approaches a as x moving towards x_0 from both sides.

Now, notice that in this case there are only two paths that x can take as moving towards x_0. We can either moving from the left or the right. Then for the limit of a function with one variable to exist, the function $f(x)$ must approach the same value a as x taking each of these paths towards x_0.

For functions with two variables we will have to do something similar. We will be asking to take the limit of the function $f(x, y)$ as x approaches x_0 and y approaches y_0. This can be written in the following ways

$$\lim_{\substack{x\to x_0 \\ y\to y_0}} f(x, y) \quad \text{or} \quad \lim_{(x,y)\to(x_0,y_0)} f(x, y)$$

Just like the limits of functions with one variable, for this limit to exist, the function must approach the same value regardless of the path that takes as moving towards (x_0, y_0). Because the point (x_0, y_0) is on a plane, there are literally an infinite number of paths that can be took as moving towards (x_0, y_0).

In other words, to show that a limit exists we would technically need to check an infinite number of paths and verify that the function is approaching the same value regardless of the path taking to approach the point. On the other hand, If we can find two paths upon which the function approaches different values as getting near the point, then we will know that the limit doesn't exist.

Let's take a look at some examples.

Example 6.3 Determine if the following limits exist.

(1) $\lim\limits_{\substack{x\to 0\\ y\to 0}} \frac{x-y}{x+y}$;   (2) $\lim\limits_{\substack{x\to 0\\ y\to 0}} \frac{x^3 y}{x^6+y^2}$.

Solution (1) Let the point (x,y) tend to $(0,0)$ along the straight line $y=kx$. Then we have

$$\lim_{\substack{x\to 0\\ y\to 0}} \frac{x-y}{x+y}=\lim_{\substack{x\to 0\\ y=kx}} \frac{x-kx}{x+kx}=\frac{1-k}{1+k}$$

It can be seen that the limit value $\frac{1-k}{1+k}$ depends on k. It means that when (x,y) tends to $(0,0)$ along different lines $y=kx$, the function $f(x,y)$ approaches different values. Hence, $\lim\limits_{\substack{x\to 0\\ y\to 0}} \frac{x-y}{x+y}$ does not exist.

(2) Let the point (x,y) tend to $(0,0)$ along the line $y=x$. Then we have

$$\lim_{\substack{x\to 0\\ y\to 0}} \frac{x^3 y}{x^6+y^2}=\lim_{\substack{x\to 0\\ y=x}} \frac{x^4}{x^6+x^2}=\lim_{\substack{x\to 0\\ y=x}} \frac{x^2}{x^4+1}=0$$

Let the point (x,y) tend to $(0,0)$ along curve $y=x^3$. Then we have

$$\lim_{\substack{x\to 0\\ y\to 0}} \frac{x^3 y}{x^6+y^2}=\lim_{\substack{x\to 0\\ y=x^3}} \frac{x^6}{x^6+x^6}=\frac{1}{2}$$

It can be seen that the function $f(x,y)$ approaches different values when (x,y) tends to $(0,0)$ along different paths. Hence, $\lim\limits_{\substack{x\to 0\\ y\to 0}} \frac{x^3 y}{x^6+y^2}$ does not exist.

How to find the limit of function with two variables? We can use one of the main ideas from limits of function with only one variable to help us get limits here. That is continuity.

6.3 Continuity of Function with Two Variables

From a graphical standpoint this definition means the same thing as it did when we first saw continuity in Chapter 1. A function will be continuous at a point if the graph doesn't have any holes or breaks at that point.

Definition 6.1

If $\lim\limits_{\substack{x\to x_0\\ y\to y_0}} f(x,y)=f(x_0,y_0)$, then the function $f(x,y)$ is called **continuous** at the point (x_0,y_0).

This definition provide a way to find the limit of a continuous function.

Example 6.4 Determine if the following limits exist or not. If they do exist, give the value of the limit.

(1) $\lim\limits_{\substack{x\to 5\\ y\to 1}} \dfrac{xy}{x+y}$;   (2) $\lim\limits_{\substack{x\to 5\\ y\to 0}} \dfrac{e^x+e^y}{\cos x-\sin y}$.

Solution Because the given functions are continuous at the given points, all we have to do is plug in the point.

(1) $\lim\limits_{\substack{x\to 5\\ y\to 1}} \dfrac{xy}{x+y}=\dfrac{5}{6}$;

(2) $\lim\limits_{\substack{x\to 5\\ y\to 0}} \dfrac{e^x+e^y}{\cos x-\sin y}=\dfrac{e^5+1}{\cos 5}$.

6.4 Partial Derivatives

When a function with more than one independent input variable changes because of changes in one or more of the input variables, it is important to calculate the change in the function itself. This can be investigated by holding all but one of the variables constant and finding the changing rate of the function with respect to the one remaining variable. This process is called partial differentiation. In this section we will show how to carry out the process.

6.4.1 First-Order Partial Derivatives

1. The Partial Derivative with Respect to x

For a function of a single variable, $y=f(x)$, changing the independent variable x leads to a corresponding change of the dependent variable y. The **rate of change** of y with respect to x is given by the derivative, written as $\dfrac{\mathrm{d}f}{\mathrm{d}x}$. A similar situation occurs with functions of more than one variable. For clarity, we shall concentrate on functions of just two variables.

In the function $z=f(x,y)$ the **independent variables** are x, y and the **dependent variable** is z. We have seen in section 6.1 that as x and y vary the z-value traces out a surface. Now both of the variables x and y may change **simultaneously** inducing a change in z. However, rather than consider this general situation, to begin with we shall hold one of the independent variables **fixed.** This is equivalent to moving along a curve obtained by intersecting the surface by one of the plane parallel to coordinate plane.

Consider $z=x^3+2x^2y+y^2+2x+1$, if we keep y constant and vary x, then what is the rate of change of the function f?

Suppose we hold y at the value 3, then

$$f(x,3)=x^3+6x^2+9+2x+1=x^3+6x^2+2x+10$$

In effect, we now have a function of independent variable x only. If differentiate it with respect to x we obtain the expression

$$f'(x,3)=3x^2+12x+2$$

We say that f has been **partially differentiated** with respect to x. We denote the partial derivative of f with respect to x by $\frac{\partial f}{\partial x}$ (to be read as "partial dee f by dee x"). In this example, when $y=3$

$$\frac{\partial f}{\partial x}=3x^2+12x+2$$

Now if we return to the original formulation

$$f(x,y)=x^3+2x^2y+y^2+2x+1$$

and treat y as a constant then the process of partial differentiation with respect to x gives

$$\begin{aligned}\frac{\partial f}{\partial x}&=3x^2+4xy+0+2+0\\&=3x^2+4xy+2\end{aligned}$$

The Partial Derivative of f with Respect to x

For a function with two variables, $z=f(x,y)$, the partial derivative of f with respect to x is denoted by $\frac{\partial f}{\partial x}$ and is obtained by differentiating $f(x,y)$ with respect to x in the usual way but treating the y-variable as constant.

Alternative notations for $\frac{\partial f}{\partial x}$ are $f_x(x,y)$, f_x or $\frac{\partial z}{\partial x}$.

2. The Partial Derivative with Respect to y

For functions of two variables, $f(x,y)$, the x and y variables are on the same footing, so what we have done for the x-variable can be implemented for the y-variable. We can thus imagine keeping the x-variable fixed and determining the rate of change of f as y changes. This rate of change is denoted by $\frac{\partial f}{\partial y}$.

The Partial Derivative of f with Respect to y

For a function of two variables, $z=f(x,y)$, the partial derivative of f with respect to y is denoted by $\frac{\partial f}{\partial y}$ and is obtained by differentiating $f(x,y)$ with respect to y in the usual way but treating the x-variable as constant.

Alternative notations for $\frac{\partial f}{\partial y}$ are $f_y(x,y)$, f_y or $\frac{\partial z}{\partial y}$.

Returning to $f(x,y)=x^3+2x^2y+y^2+2x+1$ once again, we therefore obtain

$$\frac{\partial f}{\partial y}=0+2x^2+2y+0+0=2x^2+2y$$

Example 6.5 Find $\frac{\partial f}{\partial y}$ for: (1) $f(x,y)=x+\sqrt{y}$; (2) $f(x,y)=\arccos\frac{x}{y}$.

Solution (1) $\frac{\partial f}{\partial y}=0+\frac{1}{2\sqrt{y}}=\frac{1}{2\sqrt{y}}$;

(2) $\frac{\partial f}{\partial y}=-\frac{1}{\sqrt{1-\frac{x^2}{y^2}}}\left(-\frac{x}{y^2}\right)=\frac{x}{y\sqrt{y^2-x^2}}$.

We can calculate the partial derivative of f with respect to x and the value of $\frac{\partial f}{\partial x}$ at a specific point.

Example 6.6 Find $f_x(1,-2)$ and $f_y(-3,2)$ for $f(x,y)=x^2+y^3+2xy$.

Solution **Solution 1**

$$f_x(x,y)=2x+2y,\ \text{so } f_x(1,-2)=2-4=-2$$

$$f_y(x,y)=3y^2+2x,\ \text{so } f_y(-3,2)=12-6=6$$

Solution 2

$$f(x,-2)=x^2-8-4x\Rightarrow f_x(x,-2)=(x^2-8-4x)'_x=2x-4$$

so
$$f_x(1,-2)=(2x-4)_{x=1}=-2$$

Similarly, $f(-3,y)=9+y^3-6y\Rightarrow f_y(-3,y)=(9+y^3-6y)'_y=3y^2-6$

so
$$f_y(-3,2)=(3y^2-6)\big|_{y=2}=12-6=6.$$

3. Geometric Interpretation of Partial Derivatives

By the definition of partial derivatives we know that the partial derivative of the function $z=f(x,y)$ with respect to x at the point (x_0,y_0), $f_x(x_0,y_0)$, is just the derivative of the function of one variable $f_x(x,y_0)$ at the point x_0. That means the plane intersects this surface in the plane curve RPQ, and the value of $f_x(x_0,y_0)$ is the slope of the tangent line to this curve at the point P. See figure 6 - 6. Similarly, the plane intersects the surface in the plane curve LMP, and is the slope of the tangent line to this curve at the point P.

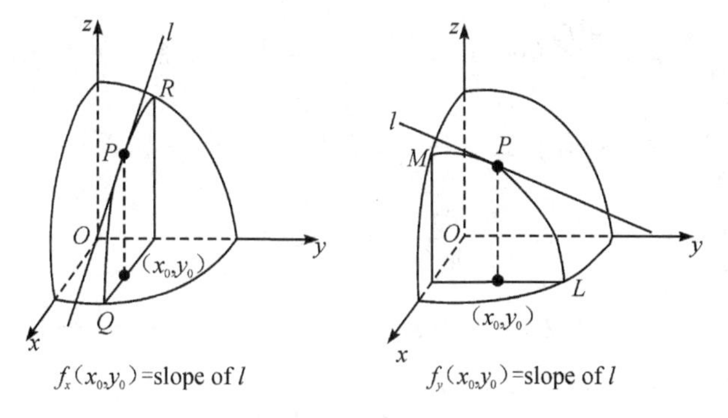

Figure 6 - 6

4. Functions of Several Variables

As we have seen, a function with two variables, $f(x,y)$, has two partial derivatives $\frac{\partial f}{\partial x}$

and $\frac{\partial f}{\partial y}$. In an exactly analogous way, a function with three variables, $f(x,y,z)$, has three partial derivatives $\frac{\partial f}{\partial x}$, $\frac{\partial f}{\partial y}$ and $\frac{\partial f}{\partial z}$, and so on for functions with more than three variables. Each partial derivative is obtained in the same way as stated before.

Example 6.7 Find $\frac{\partial f}{\partial x}$ and $\frac{\partial f}{\partial z}$ for (1) $f = x^2 + yz + z^3$; (2) $f = e^z + \ln(x^2 + y^2 - 1)$.

Solution (1) $\frac{\partial f}{\partial x} = 2x$, $\frac{\partial f}{\partial z} = y + 3z^2$; (2) $\frac{\partial f}{\partial x} = \frac{2y}{x^2 + y^2 - 1}$, $\frac{\partial f}{\partial z} = e^z$.

6.4.2 Second-Order Partial Derivatives

Performing partial derivatives of $f(x,y)$ with respect to x (holding y constant) twice successively is denoted by $\frac{\partial^2 f}{\partial x^2}$ (or $f_{xx}(x,y)$) and is defined by $\frac{\partial^2 f}{\partial x^2} = \frac{\partial}{\partial x}\left(\frac{\partial f}{\partial x}\right)$. Similarly, other second-order partial derivatives can be obtained.

$$\frac{\partial^2 f}{\partial x^2} = \frac{\partial}{\partial x}\left(\frac{\partial f}{\partial x}\right) = f_{xx}(x,y), \qquad \frac{\partial^2 f}{\partial x \partial y} = \frac{\partial}{\partial y}\left(\frac{\partial f}{\partial x}\right) = f_{xy}(x,y)$$

$$\frac{\partial^2 f}{\partial y^2} = \frac{\partial}{\partial y}\left(\frac{\partial f}{\partial y}\right) = f_{yy}(x,y), \qquad \frac{\partial^2 f}{\partial y \partial x} = \frac{\partial}{\partial x}\left(\frac{\partial f}{\partial y}\right) = f_{yx}(x,y)$$

Where, $\frac{\partial^2 f}{\partial x \partial y}$ and $\frac{\partial^2 f}{\partial y \partial x}$ are called the **mixed partial derivatives.** $\frac{\partial^2 f}{\partial x \partial y}$ means "differentiate first with respect to x and then with respect to y" and $\frac{\partial^2 f}{\partial y \partial x}$ means "differentiate first with respect to y and then with respect to x".

Example 6.8 Find $\frac{\partial^2 f}{\partial x^2}$ and $\frac{\partial^2 f}{\partial y^2}$ for $f(x,y) = x^3 + x^2 y^2 + 2y^3 + 2x + y$.

Solution

$$\frac{\partial f}{\partial x} = 3x^2 + 2xy^2 + 0 + 2 + 0 = 3x^2 + 2xy^2 + 2$$

$$\frac{\partial^2 f}{\partial x^2} = 6x + 2y^2$$

$$\frac{\partial f}{\partial y} = 0 + 2x^2 y + 6y^2 + 0 + 1 = 2x^2 y + 6y^2 + 1$$

$$\frac{\partial^2 f}{\partial y^2} = 2x^2 + 12y$$

Example 6.9 Find $f_{xx}(-1,1)$ and $f_{yy}(2,-2)$ for $f(x,y) = x^3 + x^2 y^2 + 2y^3 + 2x + y$.

Solution $f_{xx}(-1,1) = (6x + 2y^2)\Big|_{(-1,1)} = -4$, $f_{yy}(2,-2) = (2x^2 + 12y)\Big|_{(2,-2)} = -16$.

Example 6.10 For $f(x,y)=x^3+2x^2y^2+y^3$, find $\frac{\partial^2 f}{\partial x\partial y}$.

Solution $\frac{\partial f}{\partial x}=3x^2+4xy^2+0$, $\frac{\partial^2 f}{\partial x\partial y}=0+8xy=8xy$.

The remaining possibility is to differentiate first with respect to y and then with respect to x, i.e., $\frac{\partial}{\partial x}\left(\frac{\partial f}{\partial y}\right)$.

For the function in example 6.10, $\frac{\partial f}{\partial x}=4x^2y+3y^2$ and $\frac{\partial^2 f}{\partial y\partial x}=8xy$.

Notice that for this function $\frac{\partial^2 f}{\partial x\partial y}=\frac{\partial^2 f}{\partial y\partial x}$.

This equality of mixed derivatives is true for all functions which you are likely to meet in your studies.

Example 6.11 Find $f_{yx}(1,2)$ for the function $f(x,y)=x^3+2x^2y^2+y^3$.

Solution $\frac{\partial f}{\partial y}=4x^2y+3y^2$ and $f_{yx}=8xy$, so $f_{yx}(1,2)=16$.

6.5 Stationary Points

The calculation of the optimum value of a function with two variables is a common requirement in many areas of engineering, for example in thermo dynamics. Unlike the case of a function of one variable we have to use more complicated criteria to distinguish between the various types of stationary point.

6.5.1 The Stationary Points of a Function with Two Variables

Figure 6-7 shows a computer generated picture of the surface defined by the function $z=x^3+y^3-3x-3y$, where both x and y take values on the interval $[-1.8, 1.8]$. There are four features of particular interest on the surface. At point A there is a **local maximum**, at

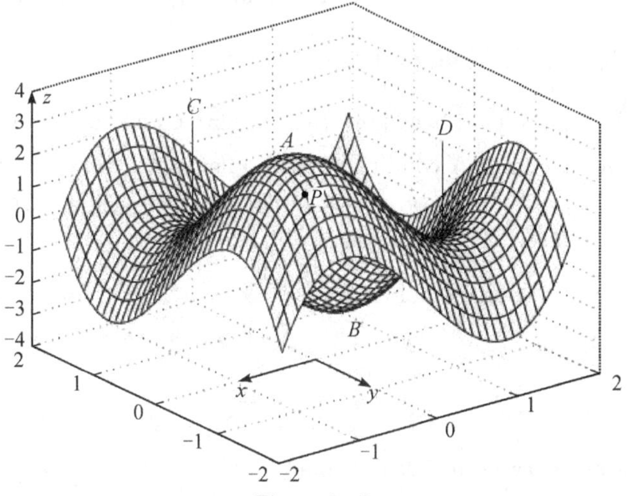

Figure 6-7

B there is a **local minimum**, and at C and D there are what are known as **saddle points.**

At A the surface is at its greatest height in the immediate neighborhood. If we move along the surface from A, we immediately lose height no matter which direction we take. At B the surface is at its least height in the neighborhood. If we move along the surface from B, we immediately gain height, no matter which direction we take.

The features at C and D are quite different. In some directions as we move away from these points along the surface we lose height while in others we gain height. The similarity in shape to a horse's saddle is evident.

At each point P of a smooth surface one can draw a unique plane which touches the surface there. This plane is called the **tangent plane** at P (The tangent plane is a natural generalization of the tangent line which can be drawn at each point of a smooth curve). In figure 6 - 7, at each of the points A, B, C, D, the tangent plane to the surface is horizontal at those points. Such points are thus known as **stationary points** of the function. The next subsections shows how to locate stationary points and how to determine their nature using partial differentiation of the function $f(x,y)$.

Example 6.12 In figure 6 - 8, what are the features at A, B and O?

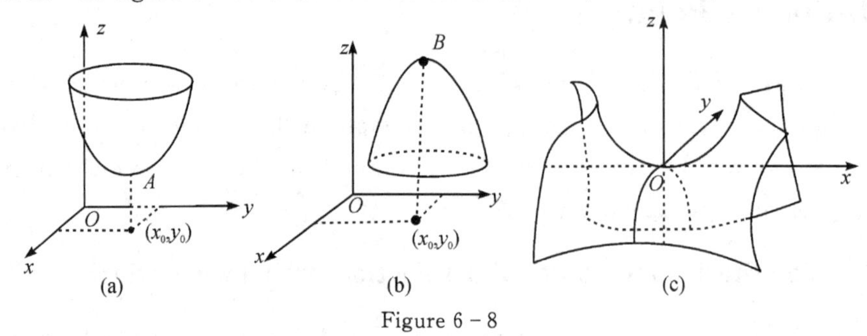

Figure 6 - 8

Solution

A is a local minimum point, B is a local maximum point, O is a saddle point.

6.5.2 Location of Stationary Points

As we said in the previous subsection, the tangent plane to the surface $z=f(x,y)$ is horizontal at a stationary point. A condition which guarantees that the function $f(x,y)$ will have a stationary point at a point (x_0,y_0) is that, at that point both $f_x=0$ and $f_y=0$ simultaneously.

Example 6.13 Find the stationary point of $f(x,y)=8x^2+6y^2-2y^3+5$.

Solution $f_x=16x$ and $f_y=6y(2-y)$. Let $f_x=16x=0$ and $f_y=6y(2-y)=0$, solve the equations, we have

$$x=0, y=0 \text{ and } x=0, y=2$$

Therefore, the stationary points of the function are $(0,0)$ and $(0,2)$.

Example 6.14 Locate the stationary points of $f(x,y)=x^4+y^4-36xy$.

Solution Write the partial derivatives of $f(x,y)$

$$\frac{\partial f}{\partial x}=4x^3-36y,\ \frac{\partial f}{\partial y}=4y^3-36x$$

Then solve the equations

$$\begin{cases}\dfrac{\partial f}{\partial x}=4x^3-36y=0 & (6.1)\\ \dfrac{\partial f}{\partial y}=4y^3-36x=0 & (6.2)\end{cases}$$

From (6.2) we have

$$x=\frac{y^3}{9} \tag{6.3}$$

Now substitute (6.3) into (6.1)

$$\frac{y^9}{9^3}-9y=0\Rightarrow y(y^4-3^4)(y^4+3^4)=0\Rightarrow y=0 \text{ or } y=\pm 3$$

Now, using (6.3), when $y=0, x=0$, when $y=3, x=3$, when $y=-3, x=-3$.
The stationary points are $(0,0)$, $(3,3)$ and $(-3,-3)$.

6.5.3 The Nature of a Stationary Point

We state, without proof, a relatively simple test to determine the nature of a stationary point, once located. If the surface is very flat near the stationary point then the test will not be sensitive enough to determine the nature of the point. The method is dependent upon the values of the second order derivatives: f_{xx}, f_{yy}, f_{xy} and also upon combination of second order derivatives denoted by $AC-B^2$ where

$$A=f_{xx},\ B=f_{xy},\ C=f_{yy}$$

The method is as follows:

Theorem 6.1 Method to Determine the Nature of Stationary Points

(1) Work out the three second order partial derivatives at each stationary point.

(2) Calculate the value of $AC-B^2$ at each stationary point.

(3) Then, test each stationary point in turn:

If $AC-B^2>0$ and $A>0$, then the stationary point is a **local minimum point.**

If $AC-B^2>0$ and $A<0$, then the stationary point is a **local maximum point.**

If $AC-B^2<0$, then the stationary point is a **saddle point.**

If $AC-B^2=0$, then the test is inconclusive (we need an alternative method).

Example 6.15 The function $f(x,y)=x^4+y^4-36xy$ has stationary points at $(0,0)$, $(-3,-3)$, $(3,3)$. Determine the nature of each stationary point.

Solution We have $\dfrac{\partial f}{\partial x}=4x^3-36y$ and $\dfrac{\partial f}{\partial y}=4y^3-36x$, then

$$\frac{\partial^2 f}{\partial x^2}=12x^2,\ \frac{\partial^2 f}{\partial y^2}=12y^2,\ \frac{\partial^2 f}{\partial x\partial y}=-36$$

A tabular presentation is useful for calculating $AC-B^2$.

Term	Stationary Points		
	(0,0)	(−3, −3)	(3,3)
A	0	108	108
C	0	108	108
B	−36	−36	−36
$AC-B^2$	<0	>0	>0

(0,0) is a saddle point, (−3, −3) and (3,3) are both local minimum points.

For most functions the procedures described above enable us to distinguish between the various types of stationary point. However, note the following example, in which these procedures fail.

Given
$$f(x,y)=x^4+y^4+2x^2y^2$$
$$\frac{\partial f}{\partial x}=4x^3+4xy^2,\ \frac{\partial f}{\partial y}=4y^3+4x^2y$$
$$\frac{\partial^2 f}{\partial x^2}=12x^2+4y^2,\ \frac{\partial^2 f}{\partial y^2}=12y^2+4x^2,\ \frac{\partial^2 f}{\partial x\partial y}=8xy$$

The stationary points are located where $\frac{\partial f}{\partial x}=\frac{\partial f}{\partial y}=0$, that is,

$$\begin{cases}4x^3+4xy^2=0\\4y^3+4x^2y=0\end{cases}\Rightarrow x=y=0$$

The only stationary point is (0,0).

Unfortunately, all the second order partial derivatives are zero at (0,0), therefore $AC-B^2=0$, so the method, as described in **theorem 6.1**, fails to give us the necessary information.

However, in this example it is easy to see that the stationary point(0,0) is in fact a local minimum point.

Let us observe $f(x,y)=x^4+y^4+2x^2y^2=(x^2+y^2)^2\geqslant 0$, the only point where $f(x,y)=0$ is the stationary point. This is therefore a local (and global) minimum point.

Example 6.16 (**Profit**) Suppose that a surfboard company has developed the yearly profit equation

$$P(x,y)=-22x^2+22xy-11y^2+110x-44y-23$$

where x is the number (in thousands) of standard surfboards produced every year; y is the number (in thousands) of competition surfboards produced every year; P is profit (in thousands of dollars). How many of each type of surfboard should be produced every year to realize a maximum profit? What is the maximum profit?

Solution

Step 1 Find stationary points

$$P_x(x,y) = -44x + 22y + 110 = 0$$
$$P_y(x,y) = 22x - 22y - 44 = 0$$

Solving this system, we obtain $(3,1)$ as the only stationary point.

Step 2 Compute $A = P_{xx}(3,1), B = P_{xy}(3,1)$ and $C = P_{yy}(3,1)$

$$A = P_{xx}(3,1) = -44,\ B = P_{xy}(3,1) = 22 \text{ and } C = P_{yy}(3,1) = -22$$

Step 3 Evaluate $AC - B^2$ and try to classify the stationary point $(3,1)$, so

$$AC - B^2 = 484 > 0 \text{ and } A = -44 < 0$$

Since the stationary point is unique, $P(3,1) = 120$ is a global maximum point. A maximum profit of 120000 dollars is obtained by producing 3000 standard surfboards and 1000 competition surfboards every year.

Exercise 6

1. Determine the domains and sketch each of the following.

(1) $f(x,y) = \sqrt{2x + 4y - 1}$; (2) $f(x,y) = \ln \dfrac{1}{x - y}$;

(3) $f(x,y) = \sqrt{\dfrac{1}{x^2} - \dfrac{1}{y^2}}$; (4) $f(x,y) = \sqrt{x + y} - \sqrt{x - 3}$.

2. Sketch each of the following quadratic surfaces.

(1) $z = 1 + x^2 + y^2$; (2) $z = \sqrt{x^2 + y^2}$;

(3) $z = 2x^2 + 3y^2$; (4) $4x^2 + 9y^2 + z^2 = 1$.

3. Evaluate each of the following limits.

(1) $\lim\limits_{\substack{x \to \pi \\ y \to 0}} \dfrac{x \sin y}{x - y}$; (2) $\lim\limits_{\substack{x \to 3 \\ y \to -7}} \dfrac{6x - y + xy}{2x^3 + y^3}$;

(3) $\lim\limits_{\substack{x \to 0 \\ y \to 0}} \dfrac{2x^2 + 7y^2}{4y^2 + x^2}$; (4) $\lim\limits_{\substack{x \to 0 \\ y \to 0}} \dfrac{2x^4 y}{x^8 + 6y^2}$.

4. Find all the first partial derivatives of the following functions.

(1) $f(x,y) = xy + \dfrac{x}{y}$; (2) $f(x,y) = \arctan(xy)$;

(3) $f(x,y) = (1 + y^2)^x$; (4) $f(x,y,z) = \ln\sqrt{x^2 + y^2 + z^2}$.

5. Find $f_x(1,1)$, $f_x(-1,-1)$, $f_y(1,2)$, $f_y(2,1)$ of the functions of exercise 4 (1) to (3).

6. Find the slope of the tangent at the point $(3,2,2)$ which is produced by the intersection of the surface $36z = 4x^2 + 9y^2$ and the plane $x = 3$.

7. Find the slope of the tangent at the point $\left(2,1,\dfrac{3}{2}\right)$ which is produced by the intersection of the surface $2z = \sqrt{9x^2 + 9y^2 - 36}$ and the plane $y = 1$.

8. Find $\frac{\partial^2 f}{\partial x^2}, \frac{\partial^2 f}{\partial y^2}, \frac{\partial^2 f}{\partial x \partial y}, \frac{\partial^2 f}{\partial y \partial x}$ of the following functions.

(1) $f(x,y)=x+2y+3$; (2) $f(x,y)=x^2+y^2$;

(3) $f(x,y)=x^3+xy+y^3$; (4) $f(x,y)=x^4+xy^3+2x^3y^2$;

(5) $f(x,y,z)=xy+yz$.

9. Find $f_{xx}(1,-3)$, $f_{yy}(-2,-2)$, $f_{xy}(-1,1)$ of the functions of Exercise 8 (1) to (4).

10. If $f(x,y,z)=3x^2y-xyz+y^2z^2$, find the following values:

(1) $f_x(x,y,z)$; (2) $f_y(0,1,1)$; (3) $f_z(x,y,z)$.

11. Find $\frac{\partial f}{\partial x}$ and $\frac{\partial^2 f}{\partial x \partial t}$ of the following functions.

(1) $f(x,t)=x\sin(xt)+x^2t$; (2) $f(x,t,z)=zxt-\mathrm{e}^{xt}$;

(3) $f(x,t)=3\cos(t+x^2)$.

12. Determine the nature of the stationary points of the function in each case below.

(1) $f(x,y)=8x^2+6y^2-2y^3+5$; (2) $f(x,y)=x^3+15x^2-20y^2+10$;

(3) $f(x,y)=4-x^2-xy-y^2$; (4) $f(x,y)=2x^2+y^2+3xy-3y-5x+8$;

(5) $f(x,y)=(x^2+y^2)^2-2(x^2-y^2)+1$;

(6) $f(x,y)=x^4+y^4+2x^2y^2+2x^2+2y^2+1$.

13. (**Automation-labor mix for minimum cost**) The annual labor and automated equipment cost (in millions of dollars) for a company's production of television sets is given by

$$C(x,y)=2x^2+2xy+3y^2-16x-18y+54$$

where x is the amount spent per year on labor and y is the amount spent per year on automated equipment (both in millions of dollars). Determine how much should be spent on each per year to minimize this cost. What is the minimum cost?

14. (**Package Design**) The packaging department in a company has been asked to design a rectangular box with no top and a partition down the middle, the box must have a volume of 48 cubic centimeters as shown in the figure below. Find the dimensions that will minimize the amount of material used to construct the box.

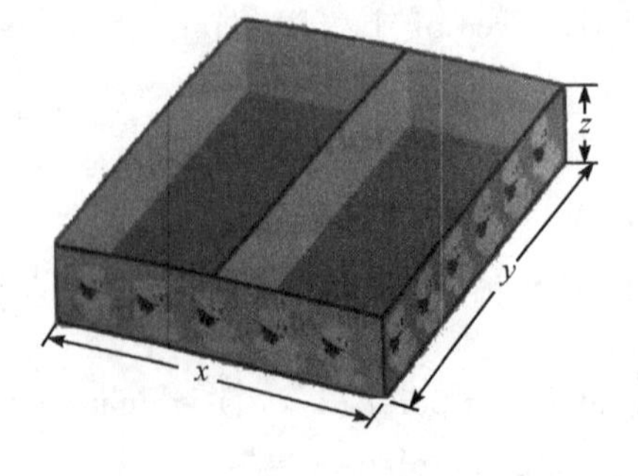

Chapter 7 Multivariable Integrals and Their Applications

Now that we have finished our discussion of derivatives of functions with more than one variable, we need to move on to integrals of functions with two or three variables. However, the functions involving two or three variables will be more complicated.

7.1 Double Integrals

7.1.1 Double Integrals over a Rectangle

Before starting on double integrals let's do a quick review of the definition of definite integrals for functions with single variable. First, when working with the integral $\int_a^b f(x)\mathrm{d}x$, we think of x as coming from the interval $[a, b]$. For these integrals we can say that we are integrating over the interval $[a, b]$. Note that this does assume $a < b$, however, if $b < a$, then we can use the interval $[b, a]$.

When we derived the definition of the definite integral, we first thought of this as an area problem. We first asked what the area under the curve was, and to do this, we divide the interval $[a, b]$ into subintervals of width Δx, and choose a point t_i from each interval as shown in figure 7 - 1.

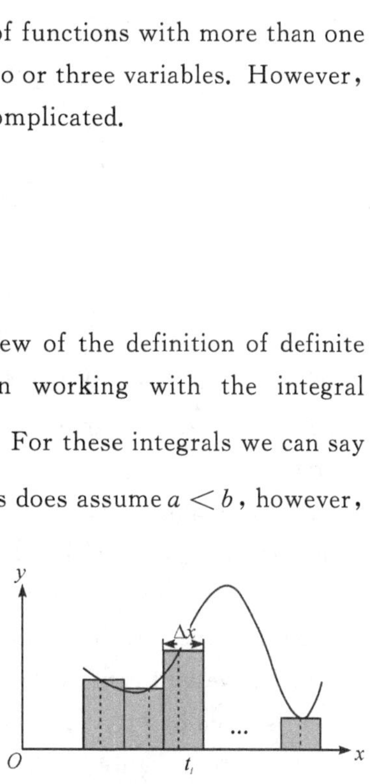

Figure 7 - 1

Each of the rectangles has height of $f(t_i)$ and we could then use the area of each of these rectangles to approximate the area as below.

$$A \approx f(t_1)\Delta x_1 + f(t_2)\Delta x_2 + \cdots + f(t_n)\Delta x_n$$

To get the exact area we then took the limit as n goes to infinity and this was also the definition of the definite integral.

$$\int_a^b f(x)\mathrm{d}x = \lim_{n\to\infty}\sum_{i=1}^{n} f(t_i)\,\Delta x_i$$

In this section we want to integrate a function of two variables, $f(x, y)$. With functions of one variable, we integrated over an interval (i. e. a one-dimensional space) and so it makes some sense then that when integrating a function of two variables we will integrate over a region of $\mathbf{R}^2$ (two-dimensional space).

We will start out by assuming that the region in $\mathbf{R}^2$ is a rectangle which we will denote as follows,

$$D = [a, b] \times [c, d]$$

This means that the ranges for x and y are $a \leqslant x \leqslant b$ and $c \leqslant y \leqslant d$.

Also, we will initially assume that $f(x, y) \geqslant 0$ although this doesn't really have to be the case. Let's start out with the graph of the surface S given by graphing $f(x, y)$ over the rectangle D, which is called a cylinder with a surface roof, as shown in figure 7 - 2.

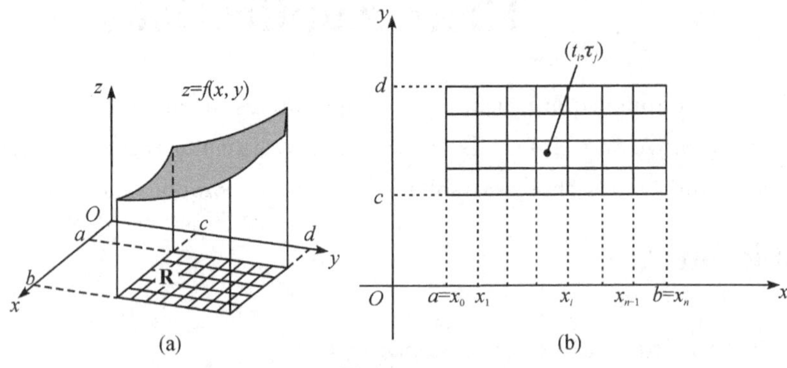

Figure 7 - 2

Now, just like with functions of one variable, let's do not worry about integrals quite yet. Let's first ask what the volume of the region under S (and above the xy-plane of course) is.

We will approximate the volume much as we approximated the area above. We will first divide up $[a, b]$ into n subintervals and $[c, d]$ into m subintervals. This will divide up D into a series of small rectangles and from each of these we will choose a point (t_i, τ_j).

Now, over each of these small rectangles we will construct a box whose height is given by $f(t_i, \tau_j)$. The sketch of that is shown in figure 7 - 3.

Each of the rectangles has a base area of ΔA and a height of $f(t_i, \tau_j)$, so the volume of each of these boxes is $f(t_i, \tau_j)\Delta A$. The volume under the surface S is then approximately

$$V \approx \sum_{i=1}^{n} \sum_{j=1}^{m} f(t_i, \tau_j) \Delta A$$

Figure 7 - 3

We will have a double sum since we need to add up volumes in both the x and y directions.

To get a better estimation of the volume we will take n and m larger and larger, and to get the exact volume, we will need to take the limit as both n and m go to infinity. In other words,

$$V = \lim_{n, m \to \infty} \sum_{i=1}^{n} \sum_{j=1}^{m} f(t_i, \tau_j) \Delta A$$

Now, this looks familiar. It looks like the definition of the integral of a function with single variable a lot. In fact, this is also the definition of a double integral, or more exactly, an integral of a function with two variables over a rectangle. Denote

$$\iint_D f(x,y)\mathrm{d}A = \lim_{n,m\to\infty} \sum_{i=1}^{n} \sum_{j=1}^{m} f(t_i, \tau_j)\Delta A$$

Note the similarities and differences in the notation to single integrals. We have two integrals to denote the fact that we are dealing with a two dimensional region, and we have a differential here as well. Note that the differential is $\mathrm{d}A$ instead of the $\mathrm{d}x$ and $\mathrm{d}y$ that we used to see. Note as well that we don't have limits on the integrals in this notation. Instead we have the D written below the two integrals to denote the region that we are integrating over.

As indicated above, one interpretation of the double integral of $f(x,y)$ over the rectangle D is the volume under the function $f(x,y)$ (and above the xy-plane), or,

$$\text{volume} = \iint_D f(x,y)\mathrm{d}A$$

7.1.2 Iterated Integrals

In the previous section we have the definition of the double integral. However, just like with the definition of a single integral, the definition is very difficult to use in practice, and so we need to start looking into how to actually compute double integrals. We will continue to assume that we are integrating over the rectangle

$$D = [a,b]\times[c,d]$$

We will look at more general regions in the next section.

The following theorem tells us how to compute a double integral over a rectangle.

Theorem 7.1 Computation of double integrals over a rectangle. If $f(x,y)$ is continuous on $D = [a,b]\times[c,d]$ then,

$$\iint_D f(x,y)\mathrm{d}A = \int_a^b \left[\int_c^d f(x,y)\mathrm{d}y\right]\mathrm{d}x = \int_c^d \left[\int_a^b f(x,y)\mathrm{d}x\right]\mathrm{d}y$$

These integrals are called **iterated integrals.**

Now, notice that there are actually two ways to compute double integrals over a rectangular region, first to y and then to x, or first to x and then to y. On some level this is just notation and doesn't really tell us how to compute the double integral. Let's just take the first possibility above and change the notation a little.

$$\iint_D f(x,y)\mathrm{d}A = \int_a^b \left[\int_c^d f(x,y)\mathrm{d}y\right]\mathrm{d}x$$

We will do the double integral by first computing

$$\int_c^d f(x,y)\mathrm{d}y$$

and we compute this by holding x constant and integrating with respect to y as if this were a single integral. This will give a function involving only x's which we can in turn integrate.

We've done a similar process with partial derivatives. To take the derivative of a function with respect to y we treated the x's as constants and differentiated with respect to y as if it was a function of a single variable.

Double integrals workis in the same manner. We think of all the x as constants and integrate with respect to y or we think of all y's as constants and integrate with respect to x.

7.1.3 Properties of Double Integration

Here are some properties of the double integral that we should be introduced before we actually do some examples. Note that all three of these properties are really just extensions of properties of single integrals that have been extended to double integrals.

(1) $\iint_D [f(x,y)+g(x,y)]\mathrm{d}A = \iint_D f(x,y)\mathrm{d}A + \iint_D g(x,y)\mathrm{d}A$.

(2) $\iint_D cf(x,y)\mathrm{d}A = c\iint_D f(x,y)\mathrm{d}A$, where c is any constant.

(3) If the region D can be split into two separate regions D_1 and D_2 then the integral can be written as

$$\iint_D f(x,y)\mathrm{d}A = \iint_{D_1} f(x,y)\mathrm{d}A + \iint_{D_2} f(x,y)\mathrm{d}A$$

Example 7.1 Compute each of the following double integrals over the indicated rectangles.

(1) $\iint_D 6xy^2\mathrm{d}A$, $D=[2,4]\times[1,2]$;

(2) $\iint_D (2x-4y^3)\mathrm{d}A$, $D=[-5,4]\times[0,3]$;

(3) $\iint_D [x^2y^2+\cos(\pi x)+\sin(\pi y)]\mathrm{d}A$, $D=[-2,-1]\times[0,1]$;

(4) $\iint_D \frac{1}{(2x+3y)^2}\mathrm{d}A$, $D=[0,1]\times[1,2]$;

(5) $\iint_D x\mathrm{e}^{xy}\mathrm{d}A$, $D=[-1,2]\times[0,1]$.

Solution (1) $\iint_D 6xy^2\mathrm{d}A$, $D=[2,4]\times[1,2]$.

It doesn't matter which variable we integrate with respect to first, we will get the same answer regardless of the order of integration. To prove that, let's work this one with each order to make sure that we do get the same answer.

Solution 1 In this case we will integrate with respect to y first. So, the iterated integral that we need to compute is

$$\iint_D 6xy^2 \mathrm{d}A = \int_2^4 \int_1^2 6xy^2 \mathrm{d}y \mathrm{d}x$$

Since we are integrating with respect to y first, we need to have y limits. Also we need to regard x as constants and keep the integrals with respect to x.

To compute this we will do the inner integral first and typically keep the outer integral around as follows,

$$\iint_D 6xy^2 \mathrm{d}A = \int_2^4 (2xy^3) \Big|_1^2 \mathrm{d}x = \int_2^4 (16x - 2x) \mathrm{d}x = \int_2^4 14x \mathrm{d}x$$

Remember that we treat the x as a constant when doing the first integral and don't do any integration with it yet. Now, we have a normal single integral, so let's finish the integral by computing this

$$\iint_D 6xy^2 \mathrm{d}A = 7x^2 \Big|_2^4 = 84$$

Solution 2 In this case we'll integrate with respect to x first and then to y. Sure enough the same answer as the first solution.

$$\iint_D 6xy^2 \mathrm{d}A = \int_1^2 (3x^2y^2) \Big|_2^4 \mathrm{d}y = \int_1^2 36y^2 \mathrm{d}x = 84$$

So, remember that we can do the integration in any order.

(2) $\iint_D (2x - 4y^3) \mathrm{d}A$, $D = [-5,4] \times [0,3]$.

For this integral we'll integrate with respect to y first.

$$\begin{aligned}\iint_D (2x - 4y^3) \mathrm{d}A &= \int_{-5}^4 \int_0^3 (2x - 4y^3) \mathrm{d}y \mathrm{d}x \\ &= \int_{-5}^4 (2xy - y^4) \Big|_0^3 \mathrm{d}x = \int_{-5}^4 (6x - 81) \mathrm{d}x \\ &= (3x^2 - 81x) \Big|_{-5}^4 = -756\end{aligned}$$

(3) $\iint_D [x^2y^2 + \cos(\pi x) + \sin(\pi y)] \mathrm{d}A$, $D = [-2, -1] \times [0,1]$.

In this case we'll integrate with respect to x first.

$$\begin{aligned}&\iint_D [x^2y^2 + \cos(\pi x) + \sin(\pi y)] \mathrm{d}A \\ &= \int_0^1 \int_{-2}^{-1} \left[\frac{1}{3}x^3y^2 + \frac{1}{\pi}\sin(\pi x) + x\sin(\pi y)\right] \mathrm{d}x \mathrm{d}y \\ &= \int_0^1 \left[\frac{7}{3}y^2 + \sin(\pi y)\right] \mathrm{d}y = \left[\frac{7}{9}y^3 - \frac{1}{\pi}\cos(\pi y)\right]\Big|_0^1 \\ &= \frac{7}{9} + \frac{2}{\pi}\end{aligned}$$

(4) $\iint\limits_D \frac{1}{(2x+3y)^2}\mathrm{d}A$, $D=[0,1]\times[1,2]$.

In this case because the limits for x are simple (i. e. they are zero and one which are often nice for evaluation), let's integrate with respect to x first.

$$\begin{aligned}\iint\limits_D \frac{1}{(2x+3y)^2}\mathrm{d}A &= \int_1^2\left[-\frac{1}{2}(2x+3y)^{-1}\right]\Big|_0^1\mathrm{d}y\\ &= -\frac{1}{2}\int_1^2\left(\frac{1}{2+3y}-\frac{1}{3y}\right)\mathrm{d}y\\ &= -\frac{1}{2}\left(\frac{1}{3}\ln|2+3y|-\frac{1}{3}\ln y\right)\Big|_1^2\\ &= -\frac{1}{6}(\ln 8-\ln 2-\ln 5)\end{aligned}$$

(5) $\iint\limits_D x\mathrm{e}^{xy}\mathrm{d}A$, $D=[-1,2]\times[0,1]$.

Now, while we can technically integrate with respect to either variable first, sometimes one way is significantly easier than the other one. In this case it will be much easier to integrate with respect to y first as we will see.

$$\begin{aligned}\iint\limits_D x\mathrm{e}^{xy}\mathrm{d}A &= \int_{-1}^2\left(x\frac{1}{x}\mathrm{e}^{xy}\right)\Big|_0^1\mathrm{d}x=\int_{-1}^2(\mathrm{e}^x-1)\,\mathrm{d}x\\ &= \mathrm{e}^2-2-(\mathrm{e}^{-1}+1)\\ &= \mathrm{e}^2-\frac{1}{\mathrm{e}}-3\end{aligned}$$

Now let's see what would happen if we had integrated with respect to x first.

$$\begin{aligned}\iint\limits_D x\mathrm{e}^{xy}\mathrm{d}A &= \int_0^1\int_{-1}^2 x\mathrm{e}^{xy}\,\mathrm{d}x\,\mathrm{d}y=\int_0^1\left(\frac{x}{y}\mathrm{e}^{xy}-\frac{1}{y^2}\mathrm{e}^{xy}\right)\Big|_{-1}^2\mathrm{d}y\\ &= \int_0^1\left[\left(\frac{2}{y}\mathrm{e}^{2y}-\frac{1}{y^2}\mathrm{e}^{2y}\right)-\left(-\frac{1}{y}\mathrm{e}^{-y}-\frac{1}{y^2}\mathrm{e}^{-y}\right)\right]\mathrm{d}y\end{aligned}$$

We see that the calculation gets more complicated, so choosing the right order of integration may make the problem simpler.

When the integrand is as the following, a double integral can be reduced directly to the product of two definite integrals.

Theorem 7.2 If $f(x,y)=g(x)h(y)$ are integrating over the rectangle $D=[a,b]\times[c,d]$, then

$$\iint\limits_D f(x,y)\mathrm{d}A=\iint\limits_D g(x)h(y)\mathrm{d}A=\left(\int_a^b g(x)\mathrm{d}x\right)\left(\int_c^d h(y)\mathrm{d}y\right)$$

Example 7.2 Evaluate $\iint\limits_D x\cos^2 y\,\mathrm{d}A$, $D=[-2,3]\times\left[0,\frac{\pi}{2}\right]$.

Solution Since the integrand is a function of x times a function of y, we can use the conclusion above. We have

$$\iint_D x\cos^2 y\,dA = \int_{-2}^{3} x\,dx\int_0^{\frac{\pi}{2}}\cos^2 y\,dy$$

$$=\left(\frac{1}{2}x^2\right)\Big|_{-2}^{3}\int_0^{\frac{\pi}{2}}\frac{1+\cos 2y}{2}dy$$

$$=\frac{5}{2}\times\frac{1}{2}\left(y+\frac{1}{2}\sin 2y\right)\Big|_0^{\frac{\pi}{2}}=\frac{5}{8}\pi$$

7.2 Double Integrals over General Regions

In the previous section we studied double integrals over rectangular regions. The problem with this is that most of the regions are not rectangular, so now we need to consider the following double integral,

$$\iint_D f(x,y)\,dA$$

where D can be any region.

7.2.1 Calculation of Double Integrals over Two Types of Domains

There are two types of regions that we need to look at. Here is a sketch of both of them, see figure 7 - 4.

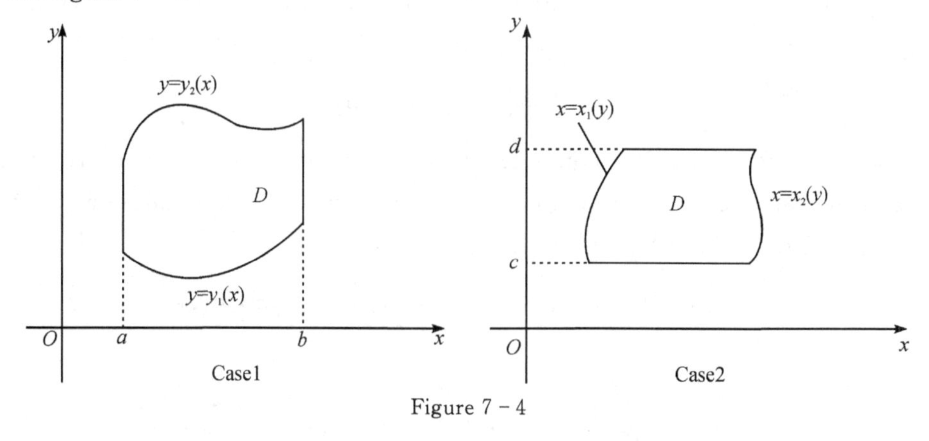

Figure 7 - 4

We will often use set notation to describe these regions, which is called **inequalities of the boundary about the region.** Here is the definition for the region in case 1, it is called x-type region.

$$D=\{(x,y)\,|\,a\leqslant x\leqslant b, y_1(x)\leqslant y\leqslant y_2(x)\}$$

and here is the definition for the region in case 2, it is called y-type region.

$$D=\{(x,y)\,|\,x_1(y)\leqslant x\leqslant x_2(y), c\leqslant y\leqslant d\}$$

Next, we will discuss the calculation over x-type region by the interpretation of double integral. This method is similar to that over y-type region. See figure 7 - 5, assume that $f(x,y)\geqslant 0$, $(x,y)\in D$, then the volume of the cylinder with a surface roof $z=f(x,y)$,

$(x, y) \in D$ is $V = \iint\limits_D f(x, y)\mathrm{d}x\mathrm{d}y$.

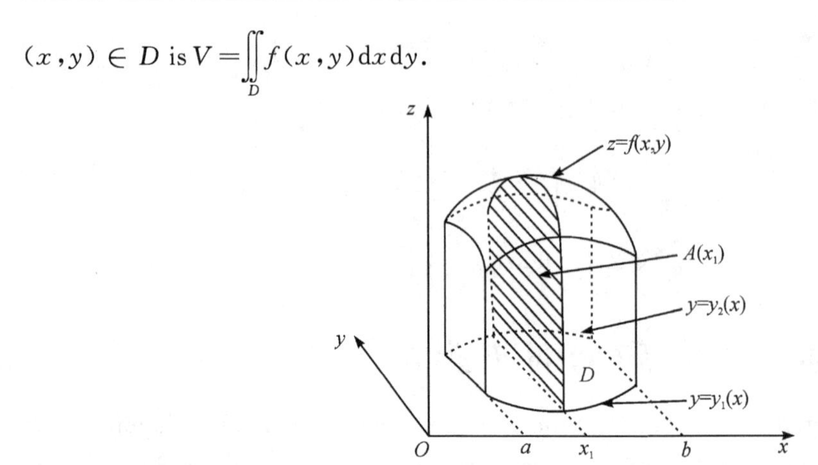

Figure 7 - 5

Choosing any point x_1 on the interval $[a, b]$, consider the area $A(x_1)$ of the section generated by the cylinder with a surface roof and the plane $x = x_1$. It is easy to see that the section is a trapezoid with a curved top $z = f(x_1, y)$ and based on the interval $[y_1(x_1), y_2(x_1)]$. As the shadow shown in figure 7 - 5. Its area is given by the definite integral

$$A(x_1) = \int_{y_1(x_1)}^{y_2(x_1)} f(x_1, y)\mathrm{d}y$$

Now rewrite x_1 as x so that the area $A(x)$ of cross section changes as varies on the interval $[a, b]$, and is thus a function over $[a, b]$

$$A(x) = \int_{y_1(x)}^{y_2(x)} f(x, y)\mathrm{d}y$$

It should be noted that when we compute the above integral, x should be regarded as a constant and the variable of integration is y. When the area of the section $A(x)$ is obtained, we can easily find the volume V of the cylinder

$$V = \iint\limits_D f(x, y)\mathrm{d}x\mathrm{d}y = \int_a^b A(x)\mathrm{d}x = \int_a^b \left[\int_{y_1(x)}^{y_2(x)} f(x, y)\mathrm{d}y\right]\mathrm{d}x$$

Therefore

$$\iint\limits_D f(x, y)\mathrm{d}x\mathrm{d}y = \int_a^b \left[\int_{y_1(x)}^{y_2(x)} f(x, y)\mathrm{d}y\right]\mathrm{d}x = \int_a^b \mathrm{d}x \int_{y_1(x)}^{y_2(x)} f(x, y)\mathrm{d}y \tag{7.1}$$

The right hand of (7.1) is called the **iterated integral** with respect to y and then to x. That means, firstly, set x as a constant regarding the function $f(x, y)$ as s function of one variable y. Next, find the definite integral with respect to y from $y_1(x)$ to $y_2(x)$ and then find the definite integral for the expression obtained previously from $x = a$ to $x = b$.

In the discussion above, we assume that $f(x, y) \geqslant 0$ when $(x, y) \in D$. Actually, it works for cases where $f(x, y)$ can be positive or negative.

According to the analysis above, the following theorem can be obtained.

Here we give the calculation method of double integrals over the region of two types of integrals without proof.

Theorem 7.3 Iterated integral over a general region.

(1) When the region is a x-type region, the double integration can be reduced to iterated integral with respect to y first and then to x.

$$\iint_D f(x,y)\mathrm{d}A = \int_a^b \left[\int_{y_1(x)}^{y_2(x)} f(x,y)\mathrm{d}y\right]\mathrm{d}x$$

(2) When the region is a y-type region, the double integration can be reduced to iterated integral with respect to x first and then to y.

$$\iint_D f(x,y)\mathrm{d}A = \int_c^d \left[\int_{x_1(y)}^{x_2(y)} f(x,y)\mathrm{d}x\right]\mathrm{d}y$$

According to the conclusion above, we can use a ray threading a x-region in order to determine the limits of y, that is, draw any ray in an interval which x lies in and thread the region from the bottom to the top, the first intersect is the lower limit, then the second intersect is the upper limit. Similarly, this way is suitable to determine the limits of x in y-type region.

Let's take a look at some examples of double integrals over general regions.

Example 7.3 Evaluate each of the following integrals over the given region D.

(1) $\iint_D e^{\frac{x}{y}}\mathrm{d}A$, $D=\{(x,y)\mid 1\leqslant y\leqslant 2, y\leqslant x\leqslant y^3\}$.

(2) $\iint_D (4xy-y^3)\mathrm{d}A$, D is the region bounded by $y=\sqrt{x}$ and $y=x^3$.

(3) $\iint_D (6x^2-40y)\mathrm{d}A$, D is the triangle with vertices $A(0,3)$, $B(1,1)$, and $C(5,3)$.

Solution

(1) Here is a sketch of the given integral domain as shown in figure 7 - 6 .

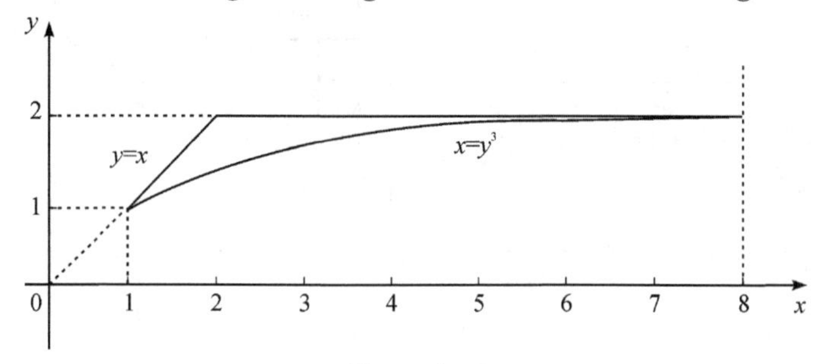

Figure 7 - 6

It is y-type region. We choose the order with respect to x and then to y.

$$\iint_D e^{\frac{x}{y}}\mathrm{d}A = \int_1^2\left(\int_y^{y^3} e^{\frac{x}{y}}\mathrm{d}x\right)\mathrm{d}y = \int_1^2 y e^{\frac{x}{y}}\Big|_y^{y^3}\mathrm{d}y = \int_1^2 y e^{y^2} - y e \mathrm{d}y$$

$$= \left(\frac{1}{2}e^{y^2} - \frac{1}{2}y^2 e\right)\Big|_1^2 = \frac{1}{2}e^4 - 2e$$

(2) $\iint\limits_D (4xy - y^3)\mathrm{d}A$, D is the region bounded by $y=\sqrt{x}$ and $y=x^3$.

The integral domain is shown in figure 7 - 7. Since the given region is both x-type domain and y-type domain, both types of integral orders are acceptable. Here, we consider the integral domain as the x-type domain.

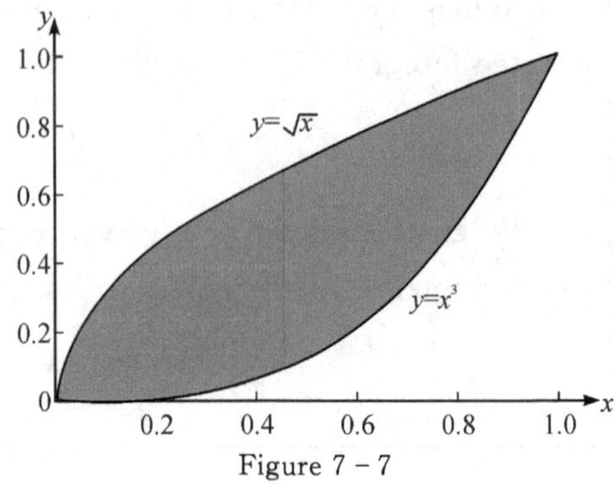

Figure 7 - 7

In this case we need to determine the two inequalities for x and y, the best way is to graph the two curves. Here is a sketch as shown in figure 7 - 7.

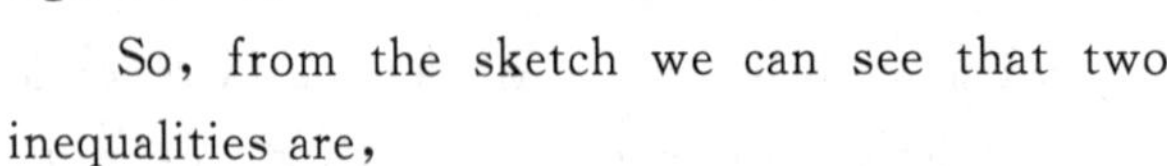
So, from the sketch we can see that two inequalities are,

$$0 \leqslant x \leqslant 1, \quad x^3 \leqslant y \leqslant \sqrt{x}$$

We can now do the integral,

$$\iint\limits_D (4xy - y^3)\mathrm{d}A = \int_0^1 \left\{\int_{x^3}^{\sqrt{x}} [(4xy - y^3)\mathrm{d}y]\right\} \mathrm{d}x = \int_0^1 \left(2xy^2 - \frac{1}{4}y^4\right)\Bigg|_{x^3}^{\sqrt{x}} \mathrm{d}x$$

$$= \int_0^1 \left(\frac{7}{4}x^2 - 2x^7 + \frac{1}{4}x^{12}\right)\mathrm{d}x$$

$$= \left(\frac{7}{12}x^3 - \frac{1}{4}x^8 + \frac{1}{52}x^{13}\right)\Bigg|_0^1 = \frac{55}{156}$$

(3) $\iint\limits_D (6x^2 - 40y)\mathrm{d}A$, D is the triangle with vertices $A(0,3)$, $B(1,1)$ and $C(5,3)$. We got even less information about the region this time. Let's start with sketching the triangle as shown in figure 7 - 8.

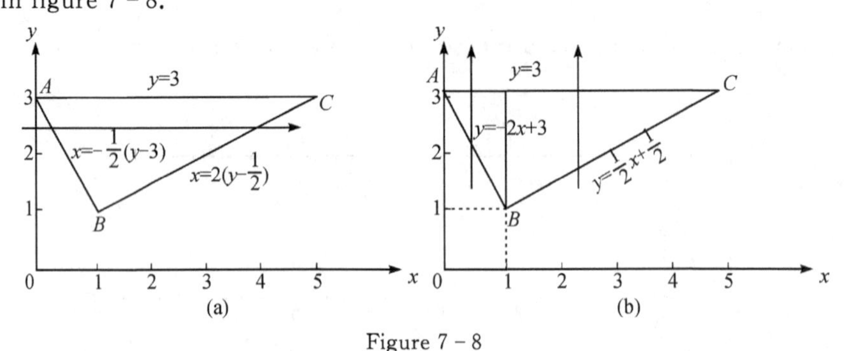

Figure 7 - 8

Solution 1　The integral domain can be regarded as the y-type domain. Since the equation of the line can be determined by two points, the equation of each side of the triangle is first solved, and expressed x as a function in terms of y.

The equation of line AB : $\quad x = -\dfrac{1}{2}(y-3)$

The equation of line BC : $x = 2(y - \frac{1}{2})$

Thus, the inequalities of the boundary about the region is:

$$D = \left\{ (x, y) \mid 1 \leqslant y \leqslant 3, -\frac{1}{2}(y-3) \leqslant x \leqslant 2(y - \frac{1}{2}) \right\}$$

Therefore, the double integration can be reduced to iterated integral with respect to x first and then to y.

$$\iint_D (6x^2 - 40y)\mathrm{d}A = \int_1^3 \left[\int_{-\frac{1}{2}y+\frac{3}{2}}^{2y-1} (6x^2 - 40y)\mathrm{d}x \right] \mathrm{d}y = \int_1^3 (2x^3 - 40xy) \Big|_{-\frac{1}{2}y+\frac{3}{2}}^{2y-1} \mathrm{d}y$$

$$= \int_1^3 100y - 100y^2 + 2(2y-1)^3 - 2\left(-\frac{1}{2}y + \frac{3}{2}\right)^3 \mathrm{d}y$$

$$= \left[50y^2 - \frac{100}{3}y^3 + \frac{1}{4}(2y-1)^4 + \left(-\frac{1}{2}y + \frac{3}{2}\right)^4 \right] \Big|_1^3$$

$$= -\frac{935}{3}$$

Solution 2 If we regard region as x-type, as shown in the figure 7 - 8(b), we will have to separate the region into two different pieces since the lower function is different depending upon the value of x. In this case the region would be given by $D = D_1 \cup D_2$ where,

$$D_1 = \{(x, y) \mid 0 \leqslant x \leqslant 1, -2x + 3 \leqslant y \leqslant 3\}$$

$$D_2 = \left\{ (x, y) \mid 1 \leqslant x \leqslant 5, \frac{1}{2}x + \frac{1}{2} \leqslant y \leqslant 3 \right\}$$

Note the $\cup$ is the "union" symbol and just means that D is the region combing the two regions. If we do so then we'll need to do two separate integrals, one for each of the regions, and expressed y as a function in terms of x.

$$x = -\frac{1}{2}y + \frac{3}{2} \quad \Rightarrow \quad y = -2x + 3$$

$$x = 2y - 1 \quad \Rightarrow \quad y = \frac{1}{2}x + \frac{1}{2}$$

Then the double integral is

$$\iint_D (6x^2 - 40y)\mathrm{d}A = \iint_{D_1} (6x^2 - 40y)\mathrm{d}A + \iint_{D_2} (6x^2 - 40y)\mathrm{d}A$$

$$= \int_0^1 \left[\int_{-2x+3}^3 (6x^2 - 40y)\mathrm{d}y \right] \mathrm{d}x + \int_1^5 \left[\int_{\frac{1}{2}x+\frac{1}{2}}^3 (6x^2 - 40y)\mathrm{d}y \right] \mathrm{d}x$$

$$= \int_0^1 (6x^2 y - 20y^2) \Big|_{-2x+3}^3 \mathrm{d}x + \int_1^5 (6x^2 y - 20y^2) \Big|_{\frac{1}{2}x+\frac{1}{2}}^3 \mathrm{d}x$$

$$= \int_0^1 \left[12x^3 - 180 + 20(3-2x)^2 \right] \mathrm{d}x + \int_1^5 \left[-3x^3 + 15x^2 - 180 + 20\left(\frac{1}{2}x + \frac{1}{2}\right)^2 \right] \mathrm{d}x$$

$$= \left[3x^4 - 180x - \frac{10}{3}(3-2x)^3 \right] \Big|_0^1 + \left[-\frac{3}{4}x^4 + 5x^3 - 180x + \frac{40}{3}\left(\frac{1}{2}x + \frac{1}{2}\right)^3 \right] \Big|_1^5$$

$$= -\frac{935}{3}$$

Obviously, the solution 1 is simpler than the solution 2. Therefore, when choosing the order of integration, seperating region must be avoided.

7.2.2 Exchange the Order of Double Integrals

In fact, sometimes you can't integrate in one order, and have to reverse the order of integration. Let's look at a couple of examples.

Example 7.4 Evaluate the following integrals by first reversing the order of integration.

(1) $\int_0^3\int_{x^2}^9 x^3 e^{y^3}\,dy\,dx$; (2) $\int_0^8\int_{\sqrt[3]{y}}^2 \sqrt{x^4+1}\,dx\,dy$.

Solution

(1) Since the antiderivative of the integrand cannot be expressed in terms of an elementary function in the given order of integration, we are hoping that if reverse the order of integration we will get an integral that can be integrated.

From the given condition, we can obtain the boundary inequality of the region.

$$D=\{(x,\ y)\mid 0\leqslant x\leqslant 3,\ x^2\leqslant y\leqslant 9\}$$

The sketch is shown in figure 7 - 9.

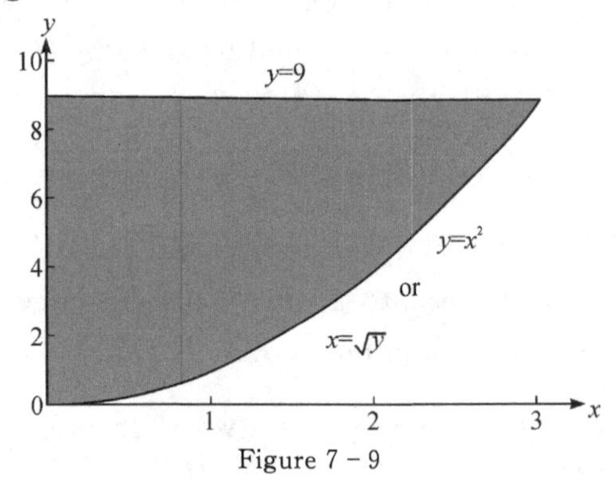

Figure 7 - 9

Accoding to the sketch, we rewite the region as y-region.

$$D=\{(x,\ y)\mid 0\leqslant y\leqslant 9,\ 0\leqslant x\leqslant \sqrt{y}\}$$

The integral, with the order reversed, is now,

$$\int_0^3\int_{x^2}^9 x^3 e^{y^3}\,dy\,dx=\int_0^9\int_0^{\sqrt{y}} x^3 e^{y^3}\,dx\,dy=\int_0^9 \frac{1}{4}x^4 e^{y^3}\Big|_0^{\sqrt{y}}\,dy$$

$$=\int_0^9 \frac{1}{4}y^2 e^{y^3}\,dy=\frac{1}{12}e^{y^3}\Big|_0^9=\frac{1}{12}(e^{729}-1)$$

Note: When the order of integration is need to be reversed, we need to determine the limits under the new order of integration according to the graph of the region. Just changing the letter is not right.

(2) Same as the first integral, in the original order of integration, we cannot find the antiderivative of the integrand, we need to reverse the order of integration.

Given the conditions, the boundary inequalities about the region is

$$D=\{(x,\ y)\,|\,0\leqslant y\leqslant 8,\ \sqrt[3]{y}\leqslant x\leqslant 2\}$$

Sketch the region according to the boundary inequalities, as shown in figure 7 - 10.

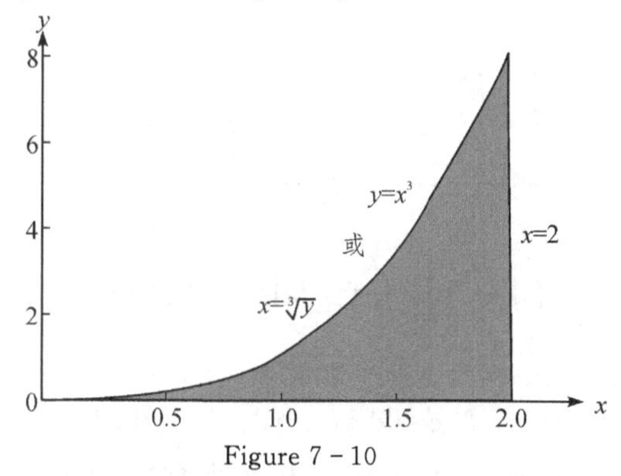

Figure 7 - 10

According to the sketch, the corresponding boundary inequalitis is written in the x-type region.

$$D=\{(x,\ y)\,|\,0\leqslant x\leqslant 2,\ 0\leqslant y\leqslant x^3\}$$

The integral is then,

$$\int_0^8\int_{\sqrt[3]{y}}^2\sqrt{x^4+1}\,\mathrm{d}x\,\mathrm{d}y=\int_0^2\int_0^{x^3}\sqrt{x^4+1}\,\mathrm{d}y\,\mathrm{d}x=\int_0^2 y\sqrt{x^4+1}\,\Big|_0^{x^3}\,\mathrm{d}x$$

$$=\int_0^2 x^3\sqrt{x^4+1}\,\mathrm{d}x=\frac{1}{6}\left(17^{\frac{3}{2}}-1\right)$$

7.3 Double Integrals in Polar Coordinates

7.3.1 Poalr Coordinate

In two dimensional space, the Cartesian coordinates (x, y) specify the location of a point P in the plane. Another two-dimensional coordinate system is polar coordinate system. Polar coordinates specify the location of a point P in the plane by the distance of OP and the angle θ between the line segment from the origin to P and the polar axis. The polar coordinates (r,θ) of a point P are illustrated in figure 7 - 11.

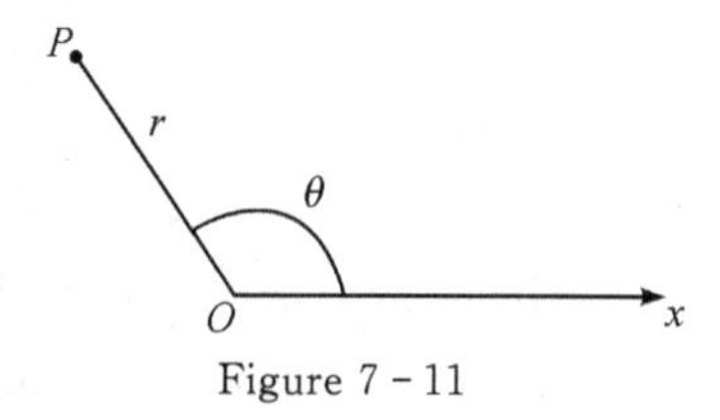

Figure 7 - 11

As r ranges from 0 to infinity and θ ranges from 0 to 2π, the polar coordinates (r,θ) covers every point in the plane, i. e. , every point (x,y) in the plane corresponds to (r,θ).

It is easy to get the conversion formulas between the Cartesian coordinates and the polar coordinates from figure 7 - 12.

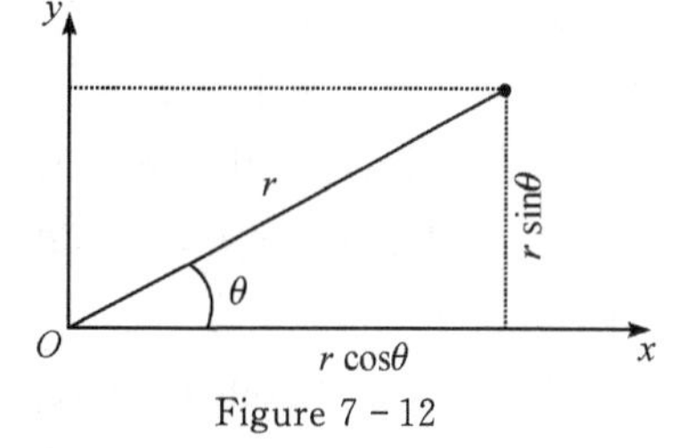

Figure 7 - 12

Conversion Formulas

$$\begin{cases} x = r\cos\theta \\ y = r\sin\theta \end{cases} \qquad \begin{cases} r = \sqrt{x^2 + y^2} \\ \theta = \arctan \dfrac{y}{x} \end{cases}$$

7.3.2 Calculation of Double Integrals in Polar Coordinates

To this point we've seen quite a few double integrals. However, in every case we've seen to this point that the region D could be easily described in terms of simple functions in Cartesian coordinates. In this section we want to look at some regions that are much easier to describe in terms of polar coordinates. For instance, we might have a region that seems like disk, ring, or a portion of a disk or ring. In these cases, using Cartesian coordinates could be somewhat cumbersome. For instance, let's suppose we wanted to do the following integral,

$$\iint_D f(x, y)\mathrm{d}A\text{, } D \text{ is a disk of radius 2}$$

To this we would have to determine a set of inequalities for x and y that describe this region. These would be,

$$-2 \leqslant x \leqslant 2, \; -\sqrt{4-x^2} \leqslant y \leqslant \sqrt{4-x^2}$$

With these limits the integral would become

$$\iint_D f(x, y)\mathrm{d}A = \int_{-2}^{2}\int_{-\sqrt{4-x^2}}^{\sqrt{4-x^2}} f(x, y)\mathrm{d}y\mathrm{d}x$$

Due to the limits on the inner integral, this is liable to be an unpleasant integral to compute.

However, a disk of radius 2 can be defined in polar coordinates by the following inequalities.

$$0 \leqslant \theta \leqslant 2\pi, \; 0 \leqslant r \leqslant 2$$

These are very simple limits and, in fact, are constant limits of integration which almost always makes integrals somewhat easier.

So, if we could convert the double integral formula into one involving polar coordinates, it would be in pretty good shape. The problem is that we can't just convert the $\mathrm{d}x$ and the $\mathrm{d}y$ into a $\mathrm{d}r$ and a $\mathrm{d}\theta$. In computing double integrals we have been using the fact that $\mathrm{d}A = \mathrm{d}x\mathrm{d}y$, and this really does require Cartesian coordinates to use. Once we've moved into polar coordinates, $\mathrm{d}A \neq \mathrm{d}r\mathrm{d}\theta$, and so we're going to determine just what $\mathrm{d}A$ is under polar coordinates.

Here is a sketch of a region in polar coordinates as shown in figure 7 - 13. So, the general region will be defined by inequalities,

$$\alpha \leqslant \theta \leqslant \beta, \; r_1(\theta) \leqslant r \leqslant r_2(\theta)$$

Now, to find dA in polar coodcinates, we'll separate the region into a mesh of radial lines and arcs, see figure 7 - 14. Now, if we pull one of the pieces of the mesh out as shown, we have something that is almost, but not quite a rectangle. The area of this piece is ΔA.

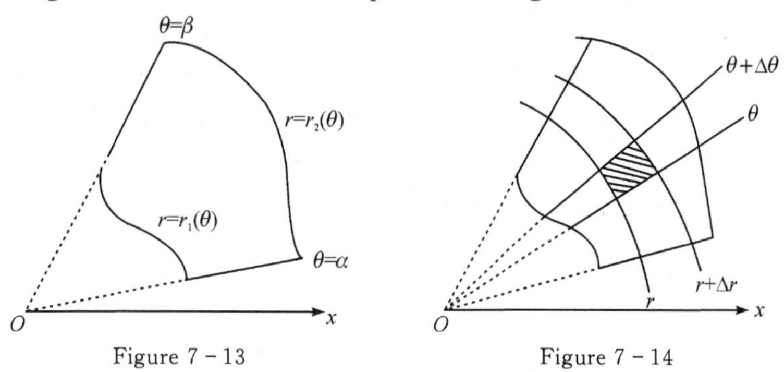

Figure 7 - 13 Figure 7 - 14

Because the seperated region is very small, we can take each sub-region approximately as a rectangle. So the area of the shaded part is approximately $\Delta A \approx r\Delta r\Delta\theta$. We must assume that each sub-region is small enough, this is not an unreasonable assumption, because the double integral definition is composed of two limits, and that illustrate the finer the partition, the bettter the approximation.

Now, if we were to use the transformation formula to convert a double integral in rectangular coordinates to a double integral in polar coordinates

$$\iint\limits_D f(x,y)\mathrm{d}A = \int_\alpha^\beta \int_{r_1(\theta)}^{r_2(\theta)} f(r\cos\theta, r\sin\theta)\, r\mathrm{d}r\mathrm{d}\theta$$

It is important to not forget the added r and convert the Cartesian coordinates in the function over to polar coordinates.

Let's look at a couple of examples of these kinds of integrals.

Example 7.5 Evaluate the following integrals by converting them into polar coordinates.

(1) $\iint\limits_D 2xy\mathrm{d}A$, D is the portion of the region between the circles of radius 2 and radius 5 centered at the origin that lies in the first quadrant.

(2) $\iint\limits_D \mathrm{e}^{x^2+y^2}\mathrm{d}A$, D is the unit disk centered at the origin.

Solution

(1) We want the region between the two circles, so we will have the following inequality for r.

$$2 \leqslant r \leqslant 5$$

Also, since we only want the portion that is in the first quadrant, we get the following range of θ.

$$0 \leqslant \theta \leqslant \frac{\pi}{2}$$

Now we can do the integral.

$$\iint_D 2xy\,\mathrm{d}A=\int_0^{\frac{\pi}{2}}\int_2^5 r^3\sin(2\theta)\,\mathrm{d}r\mathrm{d}\theta=\int_0^{\frac{\pi}{2}}\frac{1}{4}r^4\sin(2\theta)\Big|_2^5\mathrm{d}\theta$$

$$=\int_0^{\frac{\pi}{2}}\frac{609}{4}\sin(2\theta)\,\mathrm{d}\theta=-\frac{609}{8}\cos(2\theta)\Big|_0^{\frac{\pi}{2}}=\frac{609}{4}$$

(2) In this case we can't do this integral in terms of Cartesian coordinates. However, we will be able to do it in polar coordinates. First, the region D is defined by

$$0\leqslant\theta\leqslant 2\pi,\ 0\leqslant r\leqslant 1$$

In terms of polar coordinates the integral is

$$\iint_D \mathrm{e}^{x^2+y^2}\,\mathrm{d}A=\int_0^{2\pi}\int_0^1 r\mathrm{e}^{r^2}\,\mathrm{d}r\mathrm{d}\theta=\int_0^{2\pi}\frac{1}{2}\mathrm{e}^{r^2}\Big|_0^1\mathrm{d}\theta$$

$$=\int_0^{2\pi}\frac{1}{2}(\mathrm{e}-1)\,\mathrm{d}\theta=\pi(\mathrm{e}-1)$$

Example 7.6 Determine the volume of the region that lies under the sphere $x^2+y^2+z^2=9$, above the plane $z=0$ and inside the cylinder $x^2+y^2=5$.

Solution The solid is in figure 7 - 15.

So, the solid that we want the volume for is really a cylinder with a cap that comes from the sphere. And the cap is the surface described by $z=\sqrt{9-x^2-y^2}$, the base is the region described by $x^2+y^2\leqslant 5$.

We know that the formula for finding the volume of a region is

$$V=\iint_D f(x,y)\,\mathrm{d}A$$

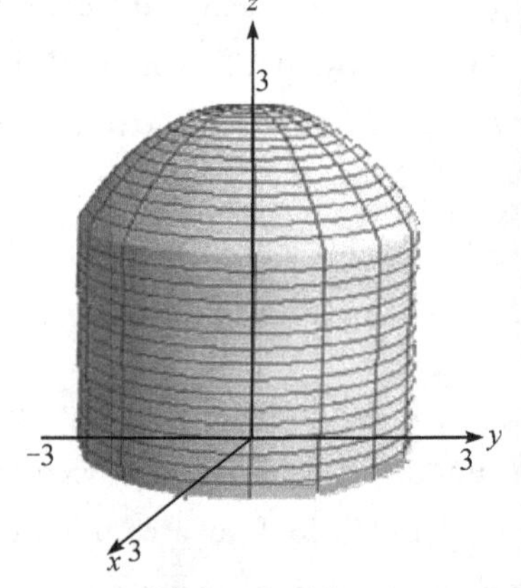

Figure 7 - 15

We are going to do this integral in terms of polar coordinates, so here are the limits (in polar coordinates) for the region,

$$0\leqslant\theta\leqslant 2\pi,\ 0\leqslant r\leqslant\sqrt{5}$$

and we'll need to convert the function to polar coordinates as well.

$$z=\sqrt{9-(x^2+y^2)}=\sqrt{9-r^2}$$

The volume is then,

$$V=\iint_D\sqrt{9-x^2-y^2}\,\mathrm{d}A=\int_0^{2\pi}\int_0^{\sqrt{5}} r\sqrt{9-r^2}\,\mathrm{d}r\mathrm{d}\theta$$

$$=\int_0^{2\pi}-\frac{1}{3}(9-r^2)^{\frac{3}{2}}\Big|_0^{\sqrt{5}}\mathrm{d}\theta=\int_0^{2\pi}\frac{19}{3}\mathrm{d}\theta=\frac{38\pi}{3}$$

Example 7.7 Find the volume of the region that lies inside $z=x^2+y^2$ and below the plane $z=16$.

Solution Let's start with a quick sketch of the region as shown in figure 7 - 16.

Now, in this case the standard formula is not going to work.

First, notice that

$$V_1 = \iint_D 16\mathrm{d}A$$

will be the volume under $z = 16$ (of course we'll need to determine D eventually) while

$$V_2 = \iint_D x^2 + y^2 \mathrm{d}A$$

is the volume under $z = x^2 + y^2$, using the same D.

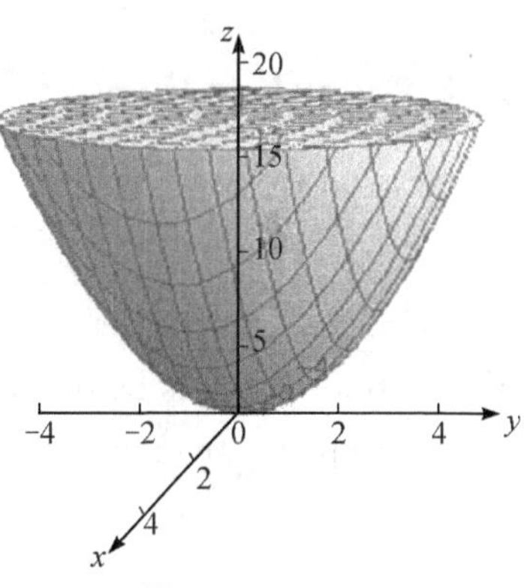

Figure 7 - 16

The volume that we're after is really the difference between these two or,

$$V = \iint_D 16\mathrm{d}A - \iint_D (x^2 + y^2)\mathrm{d}A = \iint_D [16 - (x^2 + y^2)]\mathrm{d}A$$

Now all that we need to do is to determine the region D and then convert everything over to polar coordinates.

Here are the inequalities for the region and the function we'll be integrating in terms of polar coordinates.

$$0 \leqslant \theta \leqslant 2\pi,\ 0 \leqslant r \leqslant 4,\ z = 16 - r^2$$

The volume is then,

$$V = \iint_D 16 - (x^2 + y^2)\,\mathrm{d}A = \int_0^{2\pi}\int_0^4 r(16 - r^2)\,\mathrm{d}r\mathrm{d}\theta$$

$$= \int_0^{2\pi} \left(8r^2 - \frac{1}{4}r^4\right)\Big|_0^4 \mathrm{d}\theta = \int_0^{2\pi} 64\mathrm{d}\theta = 128\pi$$

Example 7.8 Evaluate the following integral by first converting it to polar coordinates.

$$\int_{-1}^{1}\int_{-\sqrt{1-x^2}}^{0} \cos(x^2 + y^2)\,\mathrm{d}y\mathrm{d}x$$

Solution First, we sketch the region as shown in figure 7 - 17.

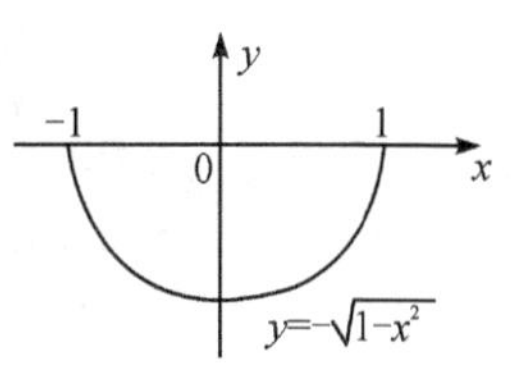

Figure 7 - 17

Notice that we cannot do this integral in Cartesian coordinates, and so converting to polar coordinates may be the only option we have for actually doing the integral. Notice that the function will be converted to polar coordinates nicely and the region is a semi-disk, so it is suitable to solve in polar coordinates.

Let's first determine the region in terms of polar coordinates as

$$\pi \leqslant \theta \leqslant 2\pi,\ 0 \leqslant r \leqslant 1$$

Finally, we just need to remember that,

$$\mathrm{d}x\mathrm{d}y = \mathrm{d}A = r\mathrm{d}r\mathrm{d}\theta$$

and so the integral becomes,

$$\int_{-1}^{1}\int_{-\sqrt{1-x^2}}^{0} \cos(x^2 + y^2)\,\mathrm{d}y\mathrm{d}x = \int_{\pi}^{2\pi} \frac{1}{2}\sin(r^2)\Big|_0^1 \mathrm{d}\theta$$

$$= \int_{\pi}^{2\pi} \frac{1}{2}\sin(1)\mathrm{d}\theta = \frac{\pi}{2}\sin(1)$$

7.4 Application of Double Integrals in Geometry

7.4.1 Volume of Solid

We have known that the geometrical interpretation of the double integral is the volume of the solid that was below the surface of the function $z = f(x, y) \geqslant 0$ and over the region D in the xy-plane given by

$$V = \iint_D f(x, y) \mathrm{d}A$$

Example 7.9 Find the volume of the solid that lies below the surface given by $z = 16xy + 200$ and above the region in the xy-plane bounded by $y = x^2$ and $y = 8 - x^2$.

Solution Here is the graph of the surface and we've tried to show the region in the xy-plane below the surface as shown in figure 7 - 18. Here is a sketch of the region in the xy-plane, see figure 7 - 19.

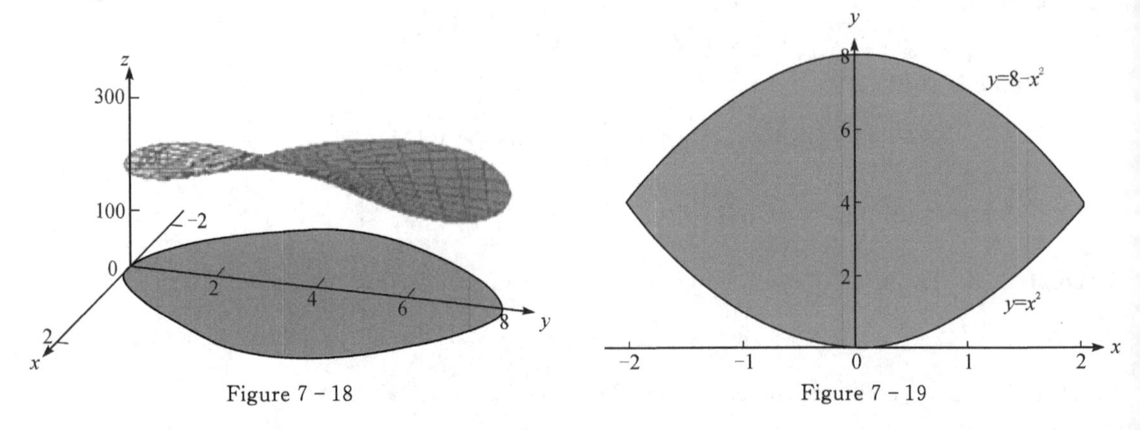

Figure 7 - 18　　Figure 7 - 19

By setting the two bounding equations equal we can see that they will intersect at $x = 2$ and $x = -2$. So, the inequalities that define the region D in the xy-plane are

$$-2 \leqslant x \leqslant 2,\ x^2 \leqslant y \leqslant 8 - x^2$$

The volume is then given by,

$$V = \iint_D 16xy + 200 \mathrm{d}A = \int_{-2}^{2} \int_{x^2}^{8-x^2} 16xy + 200 \mathrm{d}y \mathrm{d}x$$

$$= \int_{-2}^{2} (8xy^2 + 200y) \Big|_{x^2}^{8-x^2} \mathrm{d}x$$

$$= \int_{-2}^{2} (-128x^3 - 400x^2 + 512x + 1600) \mathrm{d}x$$

$$= \left(-32x^4 - \frac{400}{3}x^3 + 256x^2 + 1600x\right) \Big|_{-2}^{2}$$

$$= \frac{12800}{3}$$

Example 7.10 Find the volume of the solid enclosed by the planes $4x+2y+z=10$, $y=3x$, $z=0$, $x=0$.

Solution This example is a little different from the previous one. Here the region D is not explicitly given so we have to find it.

The first plane, $4x+2y+z=10$, is the top of the solid and so we are really looking for the volume under $z=10-4x-2y$, and above the region D in the xy-plane. The second plane, $y=3x$ (yes, that is a plane), gives one of the sides of the solid as shown in figure 7 - 20.

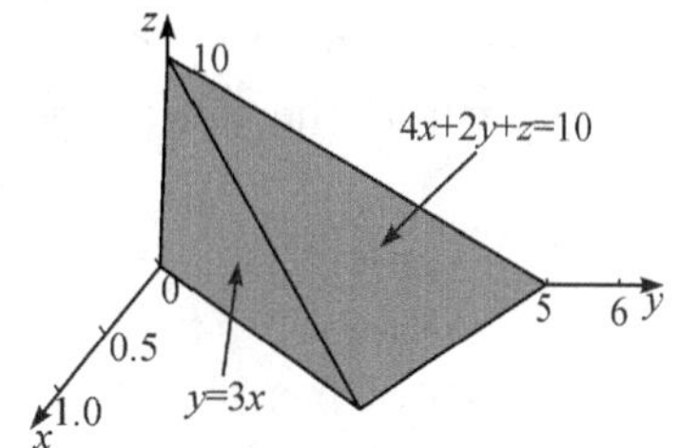

Figure 7 - 20

The region D will be the region in the xy-plane (i. e. $z=0$) that is bounded by $y=3x$, $x=0$, and the line where $z+4x+2y=10$ intersects the xy-plane. We can determine where $z+4x+2y=10$ intersects the xy-plane by plugging $z=0$ into it.

$$0+4x+2y=10 \quad \Rightarrow \quad 2x+y=5 \quad \Rightarrow \quad y=-2x+5$$

So, here is a sketch of the region D as shown in figure 7 - 21.

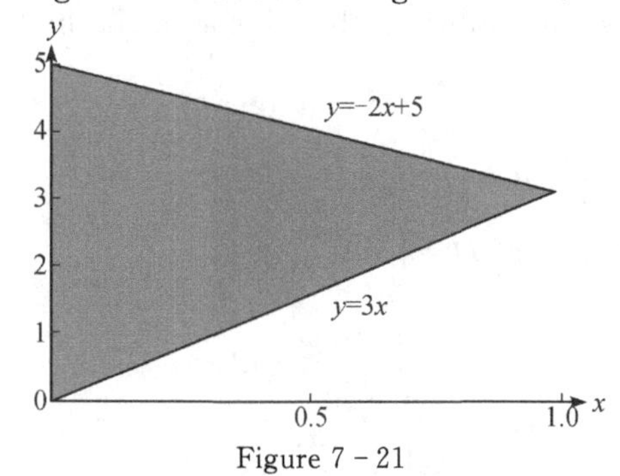

Figure 7 - 21

The region D is really where this solid will sit on the xy-plane and here are the inequalities that define the region

$$0 \leqslant x \leqslant 1, \; 3x \leqslant y \leqslant -2x+5$$

Here is the volume of this solid

$$V=\iint_D 10-4x-2y\mathrm{d}A=\int_0^1\int_{3x}^{-2x+5}(10-4x-2y)\mathrm{d}y\mathrm{d}x$$

$$=\int_0^1(10y-4xy-y^2)\Big|_{3x}^{-2x+5}\mathrm{d}x=\int_0^1 25x^2-50x+25\mathrm{d}x$$

$$=\left(\frac{25}{3}x^3-25x^2+25x\right)\Big|_0^1=\frac{25}{3}$$

7.4.2 Area of Region on Plane

We saw a similar idea in definite integral,

$$A=\int_a^b f(x)\mathrm{d}x$$

gives the net area between the curve given by $y=f(x)$ and the interval $[a,b]$ on the x-axis.

The second geometric interpretation of a double integral is as following,

$$\text{area of } D=\iint_D \mathrm{d}A$$

Suppose that we want to find the area of the region as shown in figure 7 - 22. The region is enclosed by the curve $y=y_1(x)$, $y=y_2(x)$ and the line $x=a$, $x=b$.

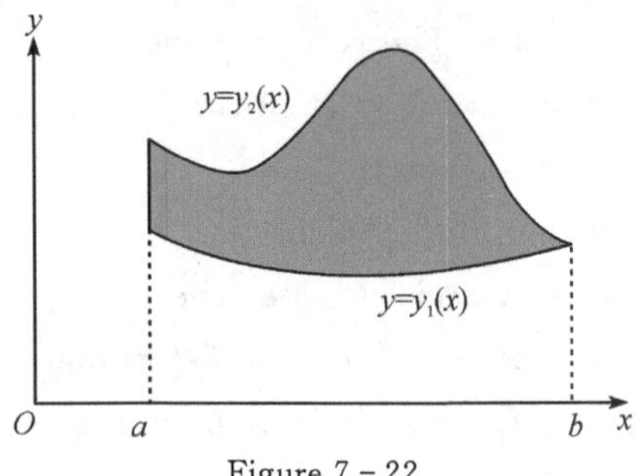

Figure 7 - 22

From definite integral we have known that this area can be found by the integral

$$A=\int_a^b [y_2(x)-y_1(x)]\,\mathrm{d}x$$

in terms of double integral we have,

$$\begin{aligned}\text{area of } D&=\iint_D \mathrm{d}A\\&=\int_a^b\int_{y_1(x)}^{y_2(x)}\mathrm{d}y\,\mathrm{d}x\\&=\int_a^b y\Big|_{y_1(x)}^{y_2(x)}\mathrm{d}x\\&=\int_a^b[y_2(x)-y_1(x)]\,\mathrm{d}x\end{aligned}$$

This is exactly the same formula we had in chapter 5.

7.4.3 The Mass and Centroid of Gravity of a Lamina

Suppose a lamina with a continuous density function $\mu(x,y)$ occupies the region D, divide the region into n sub-regions, then the mass of the sub-region ΔA_i is approximately $\mu(x_i,y_i)\Delta A_i$. If we add up all the approximations, we have

$$m\approx\sum_{i=1}^{n}\mu(x_i,y_i)\Delta A_i$$

The exact value of the mass can be obtained by taking the limit of the above equation, i. e

$$m=\iint_D \mu(x,y)\,\mathrm{d}A$$

Now, we focus on the concepts **center of gravity of a lamina.**

(1) Assume there are n mass points on the xy-plane with mass $m_1, m_2, \cdots, m_n$ and locate at the points $(x_1, y_1), (x_2, y_2), \cdots, (x_n, y_n)$. This is called a discrete particle system. Let its coordinate of certer of mass is$(\bar{x}, \bar{y})$, then the total mass of this system by physics is $M = \sum_{i=1}^{n} m_i$ and its center of gravity is given by

$$\bar{x} = \frac{M_y}{M} = \frac{\sum_{i=1}^{n} m_i x_i}{\sum_{i=1}^{n} m_i}, \quad \bar{y} = \frac{M_x}{M} = \frac{\sum_{i=1}^{n} m_i y_i}{\sum_{i=1}^{n} m_i}$$

where $M_y = \sum_{i=1}^{n} m_i x_i$, $M_x = \sum_{i=1}^{n} m_i y_i$ are called the **static moment** about y-axis and x-axis respectively.

(2) If a lamina with a continuous density function $\mu(x, y)$ occupies a region D in the xy-plane, then its center of gravity is given by

$$\bar{x} = \frac{M_y}{M} = \frac{\iint_D x\mu(x, y)\,\mathrm{d}A}{\iint_D \mu(x, y)\,\mathrm{d}A}, \quad \bar{y} = \frac{M_x}{M} = \frac{\iint_D y\mu(x, y)\,\mathrm{d}A}{\iint_D \mu(x, y)\,\mathrm{d}A}$$

where $M_y = \iint_D x\mu(x, y)\,\mathrm{d}A$, $M_x = \iint_D y\mu(x, y)\,\mathrm{d}A$ are the static moment about y-axis and x-axis respectively.

In the special of a homogeneous lamina, the center of gravity is called **the centroid of the lamina** or sometimes the **centroid of the region** D. Because the density function $\mu(x, y)$ is constant for a homogeneous lamina, the factor μ may be moved through the integral signs and canceled. Thus, the centroid is given by the following formulas which is called the **formulas of centroid.**

$$\bar{x} = \frac{1}{A}\iint_D x\,\mathrm{d}A, \quad \bar{y} = \frac{1}{A}\iint_D y\,\mathrm{d}A$$

where $A = \iint_D \mathrm{d}\sigma$ is area of the region D.

Example 7.11 Assume a lamina is enclosed by x-axis, $x=2$ and $y=x^2$, its density is $\mu = xy$, find its mass.

Solution　$m=\iint\limits_D xy\,\mathrm{d}x\,\mathrm{d}y=\int_0^2\int_0^{x^2}xy\,\mathrm{d}y\,\mathrm{d}x=\int_0^2\left(\frac{1}{2}xy^2\right)\Big|_0^{x^2}\mathrm{d}x=\frac{16}{3}$

Example 7.12　Find the center of gravity of the homogeneous lamina which lies between two circles $r=2\sin\theta$ and $r=4\sin\theta$ (Figure 7 - 23).

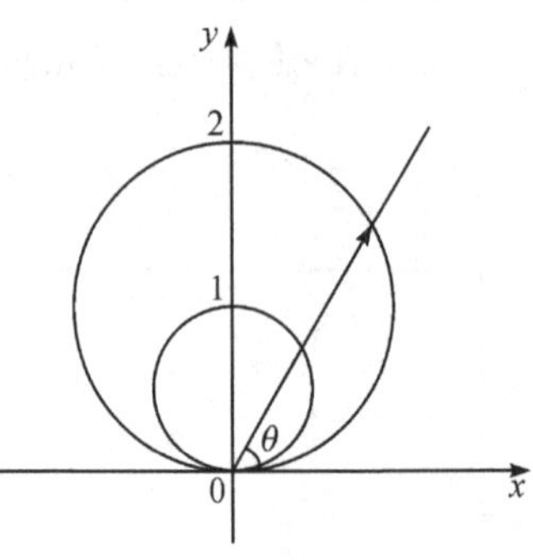

Figure 7 - 23

Solution　Because the region D is symmetric about y-axis, the center of gravity locate on y-axis, i. e. $\bar{x}=0$. The area of D is $A=3\pi$, and

$$\iint\limits_D y\,\mathrm{d}A=\iint\limits_D r\sin\theta\cdot r\,\mathrm{d}r\,\mathrm{d}\theta=\int_0^{\pi}\sin\theta\,\mathrm{d}\theta\int_{2\sin\theta}^{4\sin\theta}r^2\,\mathrm{d}r$$

$$=\frac{56}{3}\int_0^{\pi}\sin^4\theta\,\mathrm{d}\theta=7\pi$$

Therefore, using the formulas of centroid, yields

$$\bar{y}=\frac{7\pi}{3\pi}=\frac{7}{3}$$

So, the coordinate of the gravity is $(0,\frac{7}{3})$.

Example 7.13　Locate the centroid of a homogeneous semi-circle.

Solution　Place the semi-circle D above x-axis and its diameter on x-axis, then D may be expressed as

$$0\leqslant y\leqslant\sqrt{R^2-x^2},\ -R\leqslant x\leqslant R$$

D is symmetric about y-axis, so $\bar{x}=0$, such that we only need to evaluate $\bar{y}$. And the area of D is $A=\frac{1}{2}\pi R^2$,

$$\iint\limits_D y\,\mathrm{d}A=\int_{-R}^{R}\mathrm{d}x\int_0^{\sqrt{R^2-x^2}}y\,\mathrm{d}y=\int_{-R}^{R}\frac{R^2-x^2}{2}\mathrm{d}x=\frac{2}{3}R^3$$

Thus

$$\bar{y}=\frac{1}{A}\iint\limits_D y\,\mathrm{d}A=\frac{2R^3/3}{\pi R^2/2}=\frac{4R}{3\pi}$$

Therefore, the centroid coordinate of the region is $(0,4R/3\pi)$.

7.5　Triple Integrals

7.5.1　Integral by First Single and then Double (Strip Method)

Now let's consider a mass problem of space solid shown as figure 7 - 24(a). Assume that the solid occupies the space region

$$V=\{(x,y,z)\,|\,z_1(x,y)\leqslant z\leqslant z_2(x,y),(x,y)\in D\subset\mathbf{R}^2\}$$

It is also called the xy-type region and its mass is unevenly distributed and the volume density function is $\mu=\mu(x,y,z)$. We will evaluate the mass of this solid by the similar method used in calculating double integral.

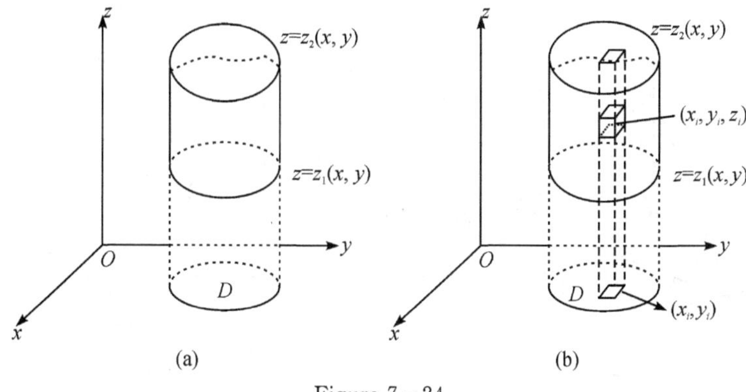

Figure 7 - 24

First, we partition the block into several cuboid small blocks, the mass of which is

$$\Delta m_k \approx \mu(x_i, y_i, z_i)\Delta V_i$$

where ΔV_i is the volume of the i-th small block, see figure 7 - 24(b), fix this small block firstly, integrating about z yields the mass of the strip,

$$\int_{z_1(x,y)}^{z_2(x,y)} \mu(x,y,z)\mathrm{d}z$$

And then add up all the mass of the strips, i. e. integrating on the projection region, yields the mass of the block

$$m = \iint_D \left[\int_{z_1(x,y)}^{z_2(x,y)} \mu(x,y,z)\mathrm{d}z\right]\mathrm{d}A$$

This method is called the **cutting strip method.**

Generally,

If $f(x,y,z)$ is continous on

$$V = \{(x,y,z) \mid z_1(x,y) \leqslant z \leqslant z_2(x,y), (x,y) \in D \subset \mathbf{R}^2\}$$

then $$\iiint_V f(x,y,z)\mathrm{d}V = \iint_D \left[\int_{z_1(x,y)}^{z_2(x,y)} f(x,y,z)\mathrm{d}z\right]\mathrm{d}A$$

The triple integral is called the iterated integral first single then double.

Specifically, the volume of the space region $= \iiint_V f(x,y,z)\mathrm{d}V$ when $f(x,y,z)=1$.

Example 7. 14 Evaluate the triple integral $\iiint_V x\,\mathrm{d}V$, where V is a tetrahedron with the vertices $(0,0,0)$, $(1,0,0)$, $(0,1,0)$ and $(0,0,1)$, as shown in figure 7 - 25(a).

Solution According to the shape of the solid, it is obviously a xy-type region. The projection on xy-plane is a triangle, see figure 7 - 25(b), consequently, the triple integral may be converted into the iterated integral first single and then double.

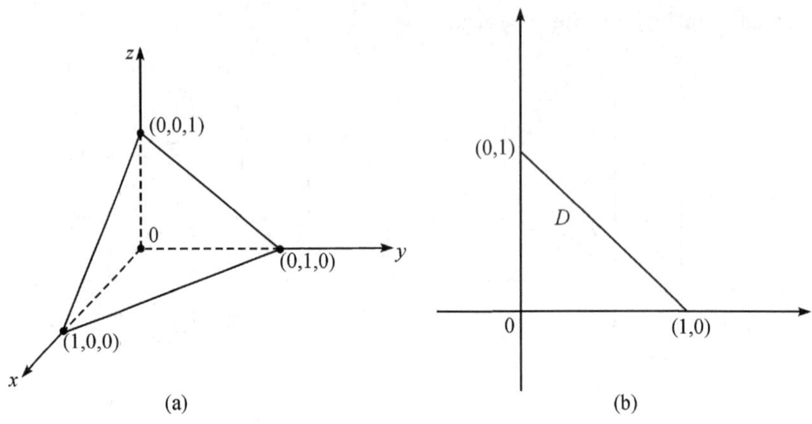

Figure 7 - 25

$$\iiint_V x\,\mathrm{d}V = \iint_D x\left(\int_0^{1-x-y}\mathrm{d}z\right)\mathrm{d}A = \iint_D xz\Big|_0^{1-x-y}\mathrm{d}x\,\mathrm{d}y = \iint_D x(1-x-y)\,\mathrm{d}x\,\mathrm{d}y$$

$$= \int_0^1 \mathrm{d}x\int_0^{1-x} x(1-x-y)\,\mathrm{d}y = \frac{1}{24}$$

Example 7.15 Evaluate $\iiint_V z\,\mathrm{d}V$, where $V = \left\{(x,y,z)\,\middle|\, 0 \leqslant z \leqslant \sqrt{1-x^2-y^2}\right\}$.

Solution Let's project the upper semi-sphere on xy-plane according to the shape of the region in space, the projection region on xy-plane is

$$D = \{(x,y)\,|\,x^2+y^2 \leqslant 1\}$$

So, the triple integral may be converted into the integral by first single and then double.

$$\iiint_V z\,\mathrm{d}V = \iint_D \left(\int_0^{\sqrt{1-x^2-y^2}} z\,\mathrm{d}z\right)\mathrm{d}A = \frac{1}{2}\iint_D (1-x^2-y^2)\,\mathrm{d}x\,\mathrm{d}y$$

$$= \frac{1}{2}\int_0^{2\pi}\mathrm{d}\theta\int_0^1 (1-r^2)\,r\,\mathrm{d}r = \frac{\pi}{4}$$

7.5.2 Integral by First Double and then Single (Slice Method)

If we regard the material block as being stacked on top of each other as shown in figure 7 - 26, taking any value $z\,(c \leqslant z \leqslant d)$, fixing it, so the mass of this thin slice is $\iint_{D_z} \mu(x,y,z)\,\mathrm{d}A$, and then add up along vertical direction to obtain the mass of the whole material block

$$m = \int_c^d \left[\iint_{D_z} \mu(x,y,z)\,\mathrm{d}x\,\mathrm{d}y\right]\mathrm{d}z$$

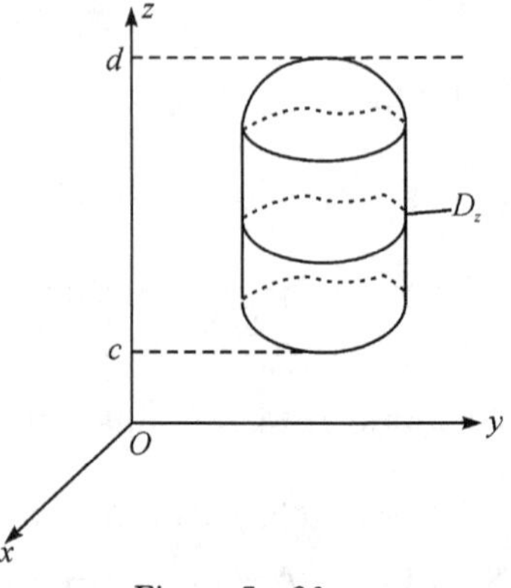

Figure 7 - 26

Generally,

If $f(x,y,z)$ is continuous on V , then

$$\iiint_V f(x,y,z)\mathrm{d}V = \int_c^d \left[\iint_{D_z} f(x,y,z)\mathrm{d}A\right]\mathrm{d}z$$

This type of integral is called integral first double and then single or slice method.

Specially, if $f(x,y,z)=f(z)$, then

$$\iiint_V f(x,y,z)\mathrm{d}V = \int_c^d f(z)A_z\mathrm{d}z$$

where A_z is area of D_z.

Example 7.16 Using slice method to evaluate example 7.15.

Solution

$$\iiint_V z\mathrm{d}V = \int_0^1 zA_z\mathrm{d}z = \int_0^1 z\pi(1-z^2)\,\mathrm{d}z = \frac{\pi}{4}$$

7.5.3 Wedge-Shape Method

If you think of the material blocks as adding up one wedge after another (for example, when you buy a watermelon, the seller may cut a wedge for the customer to taste), the mass of one of the wedges can be calculated first. To do this, you need to know the spherical coordinate system. As shown in figure 7 - 27, a point in space can be represented by the intersection of three surfaces: A sphere with radius r, a conical surface with vertex angle φ, a half plane passing through x-axis and rotation angle θ to xy-plane. (r,φ,θ) is called the spherical coordinates of the point $P(x,y,z)$.

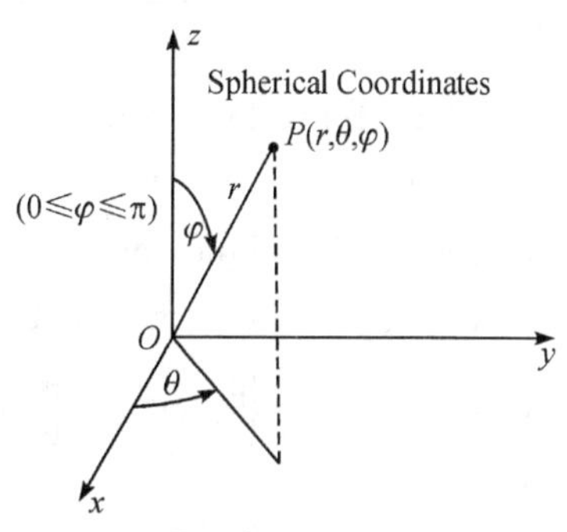

Figure 7 - 27

It is easy to see that, the transformation formula between Cartesian coordinates and spherical coordinates is

$$\begin{cases} x = r\sin\varphi\cos\theta \\ y = r\sin\varphi\sin\theta \\ z = r\cos\varphi \end{cases}$$

We first partition the material block into many small wedge-shaped blocks with the three sets of coordinate planes in spherical coordinates and one of the small blocks is shown in figure 7 - 28(a). The volume of the small block is $r^2\sin\varphi\mathrm{d}r\,\mathrm{d}\varphi\mathrm{d}\theta$ with the aid of figure 7 - 28(b).

Thus the mass of the small wedge-shaped block can be expressed as

$$f(r\sin\varphi\cos\theta, r\sin\varphi\sin\theta, r\cos\varphi)\,r^2\sin\varphi\mathrm{d}r\,\mathrm{d}\varphi\mathrm{d}\theta$$

Using triple integral yields the formula of the mass of the whole block. Generally,

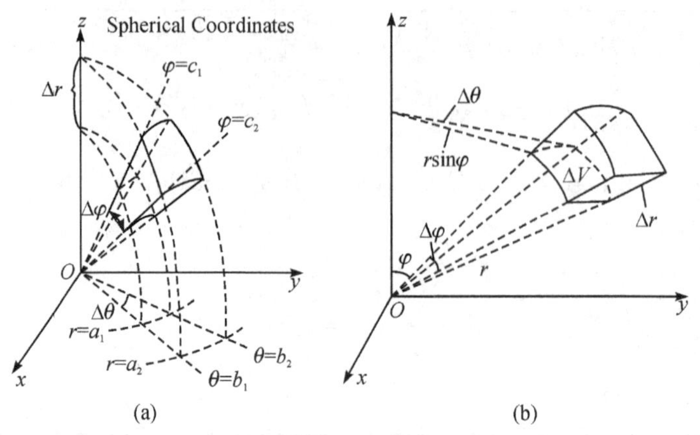

Figure 7 - 28

> If $f(x,y,z)$ is continuous on V, then
> $$\iiint_V f(x,y,z)\mathrm{d}V=\iiint_V f(r\sin\varphi\cos\theta, r\sin\varphi\sin\theta, r\cos\varphi)r^2\sin\varphi\mathrm{d}r\mathrm{d}\varphi\mathrm{d}\theta$$

Note: The centroid of a mass in space is similar to the centroid of a lamina, which will not be repeated here.

Example 7.17 Find the mass of a sphere with radius a and its density is proportional to the distance from the point on the sphere to the origin.

Solution $$m=\iiint_V \mu\mathrm{d}V=k\iiint_V \sqrt{x^2+y^2+z^2}\mathrm{d}V=k\int_0^{2\pi}\mathrm{d}\theta\int_0^{\pi}\mathrm{d}\varphi\int_0^a r\cdot r^2\sin\varphi\mathrm{d}r$$
$$=2\pi\frac{ka^4}{4}\int_0^{\pi}\sin\varphi\mathrm{d}\varphi=k\pi a^4$$

Example 7.18 Find the volume and centroid of a solid bounded by a cone and a sphere of radius 1 centered at the origin.

Solution $$V=\iiint_V \mathrm{d}V=\int_0^{2\pi}\mathrm{d}\theta\int_0^{\frac{\pi}{4}}\mathrm{d}\varphi\int_0^1 r^2\sin\varphi\mathrm{d}r=\frac{2}{3}\pi\int_0^{\frac{\pi}{4}}\sin\varphi\mathrm{d}\varphi=\frac{2}{3}\pi\left(1-\frac{\sqrt{2}}{2}\right)$$

The sphere is symmetric about z-axis, so $\bar{x}=\bar{y}=0$. Denote the moment of the sphere about xy-plane by M_{xy}, then

$$\bar{z}=\frac{M_{xy}}{V}=\frac{\iiint_V z\mathrm{d}V}{V}=\frac{\int_0^{2\pi}\mathrm{d}\theta\int_0^{\frac{\pi}{4}}\mathrm{d}\varphi\int_0^1 r\cos\varphi\cdot r^2\sin\varphi\mathrm{d}r}{\frac{2}{3}\pi\left(1-\frac{\sqrt{2}}{2}\right)}$$

$$=\frac{\frac{\pi}{8}}{\frac{2}{3}\pi\left(1-\frac{\sqrt{2}}{2}\right)}=\frac{2}{8}\left(1+\frac{\sqrt{2}}{2}\right)$$

Thus, the centroid of the sphere is $\left(0,0,\frac{2}{8}\left(1+\frac{\sqrt{2}}{2}\right)\right)$.

Exercise 7

1. Compute the following double integral over the indicated rectangle: (a) by integration with respect to x first; (b) by integrating with respect to y first.

 (1) $\iint\limits_D (16xy - 9x^2 + 1)\,dA$, $D = [2,3]\times[-1,1]$;

 (2) $\iint\limits_D \cos x \sin y\,dA$, $D = \left[\frac{\pi}{6},\frac{\pi}{4}\right]\times\left[\frac{\pi}{4},\frac{\pi}{3}\right]$.

2. Compute the given double integral over the indicated rectangle.

 (1) $\iint\limits_D (6y\sqrt{x} - 2y^3)\,dA$, $D = [1,4]\times[0,3]$;

 (2) $\iint\limits_D \left(\sin 2x - \frac{1}{1+6y}\right)dA$, $D = \left[\frac{\pi}{4},\frac{\pi}{2}\right]\times[0,1]$;

 (3) $\iint\limits_D (ye^{y^2-4x})\,dA$, $D = [0,2]\times[0,2\sqrt{2}]$;

 (4) $\iint\limits_D xy\cos(x^2y)\,dA$, $D = [-2,3]\times[-1,1]$.

3. Determine the volume that lies under $f(x,y) = 9x^2 + 4xy + 4$ and above the rectangle given by $[-1,1]\times[0,2]$ in the xy-plane.

4. Explain each of the following equalities by means of the geometric meaning of double integral:

 (1) $\iint\limits_D k\,dA = kA$, where k is a constant, A is the area of the region D;

 (2) $\iint\limits_D \sqrt{R^2 - x^2 - y^2}\,dA = \frac{2}{3}\pi R^3$, where D is a circle with radius R and center at the origin;

 (3) If the domain of integration is symmetric about y-axis, then

 ① $\iint\limits_D f(x,y)dA = 0$, if $f(x,y)$ is an odd function with respect to x, i. e. $f(x,y) = -f(-x,y)$;

 ② $\iint\limits_D f(x,y)dA = 2\iint\limits_{D_1} f(x,y)dA$, if $f(x,y)$ is an even function with respect to y, i. e. $f(x,y) = f(-x,y)$, where D_1 is the part of the region D located in the right-half plane $x \geqslant 0$;

 (4) If the domain of integration is symmetric with x-axis, what conditions can guarantee respectively the following equalities?

$$\iint\limits_D f(x,y)dA = 0,\quad \iint\limits_D f(x,y)dA = 2\iint\limits_{D_1} f(x,y)dA$$

where D_1 is the part of region D located in the upper-half plane $y \geqslant 0$.

5. Calculation the following double integrals:

(1) $\iint\limits_D (7x^2 + 14y)\, dA$, where D is bounded by $x = 2y^2$ and $x = 8$;

(2) $\iint\limits_D x(y-1)\, dA$, where D is bounded by $y = 1 - x^2$ and $y = x^2 - 3$;

(3) $\iint\limits_D 5x^3\cos(y^3)\, dA$, where D is bounded by $y = 2, y = \frac{1}{4}x^2$ and y;

(4) $\iint\limits_D \frac{1}{\sqrt[3]{y}(x^3+1)} dA$, where D is bounded by $x = -\sqrt[3]{y}, x = 3$ and x-axis.

6. Exchange the order of the following integrals:

(1) $\int_0^2 dx \int_0^{\sqrt{x}} f(x,y) dy$; (2) $\int_0^4 dy \int_{2y}^8 f(x,y) dx$;

(3) $\int_0^2 dy \int_1^{e^y} f(x,y) dx$; (4) $\int_1^e dx \int_0^{\ln x} f(x,y) dy$;

(5) $\int_0^1 dy \int_{\arcsin y}^{\frac{\pi}{2}} f(x,y) dx$; (6) $\int_0^1 dy \int_{y^2}^{\sqrt{y}} f(x,y) dx$.

7. Evaluate the following integrals by reversing the order of the integrations.

(1) $\int_0^1 dx \int_{4x}^4 e^{-y^2} dy$; (2) $\int_0^2 dy \int_{\frac{y}{2}}^1 \cos x^2 dx$;

(3) $\int_0^4 dy \int_{\sqrt{y}}^2 e^{x^3} dx$; (4) $\int_1^3 dx \int_0^{\ln x} x dy$.

8. Evalute the following iterated integrals:

(1) $\int_0^{\frac{\pi}{2}} d\theta \int_0^{\sin\theta} r\cos\theta dr$; (2) $\int_0^{\pi} \left[\int_0^{1+\cos\theta} r dr\right] d\theta$;

(3) $\int_0^{\frac{\pi}{2}} d\theta \int_0^{a\sin\theta} r dr$; (4) $\int_0^{\frac{\pi}{6}} \left[\int_0^{\cos 3\theta} r dr\right] d\theta$.

9. Calculate the following double integrations in polar coordinates.

(1) $\iint\limits_D (y^2 + 3x)\, dA$, where D is the portion of the region between the circles $x^2 + y^2 = 1$ and $x^2 + y^2 = 9$ that lies in the first quadrant;

(2) $\iint\limits_D \sqrt{1 + 4x^2 + 4y^2}\, dA$, where D is the lower half of the circle $x^2 + y^2 = 16$;

(3) $\iint\limits_D (4xy - 7)\, dA$, where D is portion of the circle $x^2 + y^2 = 2$ that lies in the first quadrant;

(4) $\int_0^3 \int_{-\sqrt{9-x^2}}^0 e^{x^2+y^2} dy dx$.

10. Solving the following probelms by double integrals.

(1) Find the area enclosed by the curve $y = 1 - x^2$ and $y = x^2 - 3$;

(2) Find the area of a region enclosed by the cardioid $r = 1 - \cos\theta$;

(3) Find the area of a region in the first quadrant bounded by $r=1$ and $r=\sin 2\theta$;

(4) Find the volume of the solid enclosed by the surface $z=x^2+y^2$ and the planes $y=2x$, $y=2$, $x=0$ as well as xy-plane;

(5) Find the volume of the solid enclosed by the surface $z=2x^2+2y^2$, the cylinder $x^2+y^2=16$ and xy-plane;

(6) Find the volume of the solid enclosed by two surfaces $z=8-x^2-y^2$ and $z=3x^2+3y^2-4$.

11. Solve the following problems by triple integral.

(1) Find the volume of the solid enclosed by the paraboloid $z=x^2+y^2$ and the plane $z=4$;

(2) Find the volume of the solid enclosed by the surface $z=4-x^2-y^2$, the cylinder $x^2+y^2=2x$ and xy-plane;

(3) Find the volume of the solid enclosed by the cone $z=\sqrt{x^2+y^2}$ and the shpere $x^2+y^2+z^2=16$;

(4) Find the mass of the solid with density $\mu=3-z$ that is bounded by the cone $z=\sqrt{x^2+y^2}$ and the plane $z=0$;

(5) Find the centroid of the solid enclosed by two paraboloids $z=12-2x^2-2y^2$ and $z=x^2+y^2$;

(6) Find the center of gravity of the shpere of radius a and its density is proportional to the distance from the center to a point on the sphere;

(7) Find the center of gravity of the solid bounded by the paraboloid $z=1-x^2-y^2$ and the xy-plane, assuming the density to be $\mu=x^2+y^2+z^2$.

Chapter 8 Curve Integral and Surface Integral

8.1 Curve Integral

One kind of generalization of the definite integral $\int_a^b f(x)\mathrm{d}x$ is obtained by replacing the set $[a,b]$ over which we integrate by two-and three-dimensional sets. This led us to the double and triple integrals of Chapter 7. A very different generalization is obtained by replacing $[a,b]$ with a curve C in the xy-plane. The resulting integral $\int_C f(x,y)\mathrm{d}s$ is called a **curve integral** .

8.1.1 The Definition of Curve Integral

Let C be a smooth plane curve, it is given parametrically by

$$x=x(t),\ y=y(t),\ a\leqslant t\leqslant b$$

where $x(t)$ and $y(t)$ are continuous and not simultaneously zero on (a,b). Consider the partion of the parameter interval $[a,b]$ obtained by inserting the points

$$a=t_0<t_1<t_2<\cdots<t_n=b$$

This partition of $[a,b]$ results in a division of the curve C into n subarcs $\widehat{P_{i-1}P_i}$ in which the point P_i corresponds to t_i. Let Δs_i denote the length of the arc and let $d=\max\limits_{a\leqslant t\leqslant b}\{\Delta s_i\}$ be the largest norm of the partition. Finally, choose a sample point $Q_i(\bar{x}_i,\bar{y}_i)$ on the subarc $\widehat{P_{i-1}P_i}$, see figure 8 - 1.

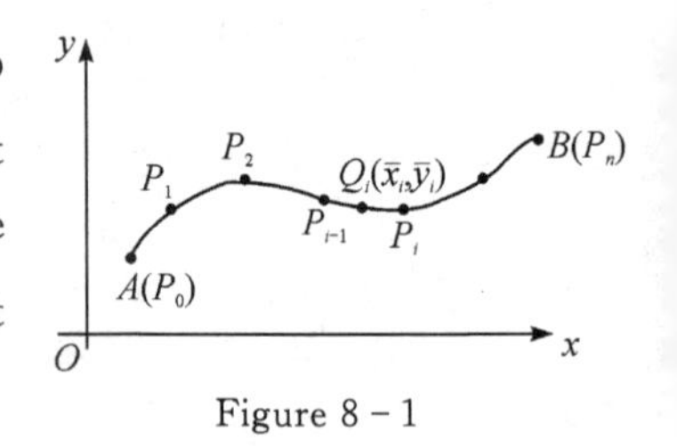

Figure 8 - 1

Now consicer the Riemann sum

$$\sum_{i=1}^{n} f(\bar{x}_i,\bar{y}_i)\Delta s_i$$

If $f(x,y)$ is nonnegative, this sum approximates the area of the curved vertical curtain shown in figure 8 - 2.

If $f(x,y)$ is continuous on a region D containing the curve C, then this Riemann sum has a limit as $d\to 0$, this limit is called the **curve integral of** $f(x,y)$ **along** C **from** A to B **with respect to arc length**, that is

$$\int_C f(x,y)\mathrm{d}s=\lim_{d\to 0}\sum_{i=1}^{n} f(\bar{x}_i,\bar{y}_i)\Delta s_i$$

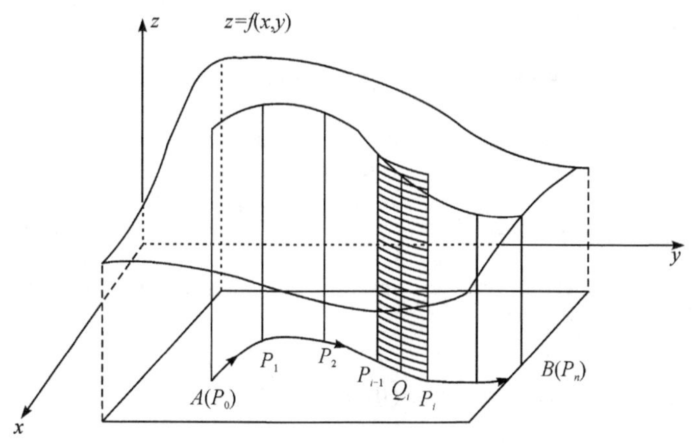

Figure 8 - 2

8.1.2 Calculation of Curve Integral

The definition of curve integral does not provide a very good way of evaluating. That is best accomplished by expressing everything in terms of the parameter and leads to an odinary definite integral. Using $\widehat{P_{i-1}P_i}$ (see section 2. 3) we have

$$\int_C f(x,y)\mathrm{d}s=\int_a^b f[x(t),y(t)]\sqrt{[x'(t)]^2+[y'(t)]^2}\,\mathrm{d}t$$

Remark:

(1) If $f(x,y)=1$, then $\int\limits_C \mathrm{d}s=\int_a^b\sqrt{[x'(t)]^2+[y'(t)]^2}\,\mathrm{d}t$ gives the length of the curve C.

(2) The lower limit must be smaller than the upper limit for the nonnegative of the length of any arc.

All that above extends easily to a smooth curve C in three-dimensional space. In particular, if C is given parametrically by

$$x=x(t),\ y=y(t),\ z=z(t),\ a\leqslant t\leqslant b$$

then

$$\int_C f(x,y,z)\mathrm{d}s=\int_a^b f[x(t),y(t),z(t)]\sqrt{[x'(t)]^2+[y'(t)]^2+[z'(t)]^2}\,\mathrm{d}t$$

Example 8. 1 Evaluate $\int\limits_C x^2y\,\mathrm{d}s$, where C is determined by the parametric equation $x=\cos t$, $y=\sin t, 0\leqslant t\leqslant\frac{\pi}{2}$. Also show that the parameterization $x=\sqrt{1-y^2}$, $y=y, 0\leqslant y\leqslant 1$, gives the same value.

Solution　Using the first parametrization, we obtain

$$\int_C x^2 y \mathrm{d}s = \int_0^{\frac{\pi}{2}} \cos^2 t \sin t \sqrt{(-\sin t)^2 + \cos^2 t} \, \mathrm{d}t$$

$$= \int_0^{\frac{\pi}{2}} \cos^2 t \sin t \, \mathrm{d}t = \left(-\frac{1}{3}\cos^3 t\right)\Big|_0^{\pi/2} = \frac{1}{3}$$

For the second parametrization, we use another formula as given in section 2.3. This gives

$$\mathrm{d}s = \sqrt{(\mathrm{d}x)^2 + (\mathrm{d}y)^2} = \sqrt{1 + \left(\frac{\mathrm{d}x}{\mathrm{d}y}\right)^2} \, \mathrm{d}y$$

Thus

$$\int_C x^2 y \mathrm{d}s = \int_0^1 (1 - y^2) y \frac{1}{\sqrt{1 - y^2}} \mathrm{d}y = \int_0^1 \sqrt{1 - y^2} \, y \mathrm{d}y = \frac{1}{3}$$

Example 8.2　Find the lateral area of the part of the elliptic cylinder $\frac{x^2}{5} + \frac{y^2}{9} = 1$ cut by the planes $z = y$ and $z = 0$ located in the first and second octantsm, see figure 8 - 3.

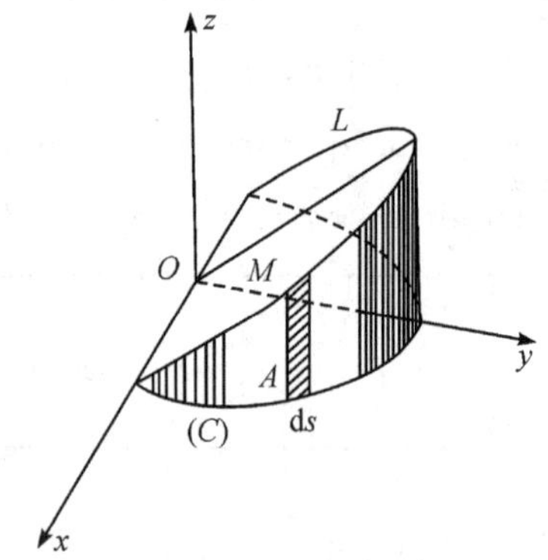

Figure 8 - 3

Solution　It is easy to see that the generator of the elliptic cylinder is the semi-ellipse in the xy-plane:

So that the lateral area we seek is

$$A = \int_C y \mathrm{d}s$$

Rewrite the equation of the curve C to the parametric equations

$$x = \sqrt{5}\cos t, \quad y = 3\sin t, \quad 0 \leqslant t \leqslant \pi$$

We have

$$A = \int_C y \mathrm{d}s = \int_0^{\pi} 3\sin t \sqrt{5\sin^2 t + 9\cos^2 t} \, \mathrm{d}t = \int_0^{\pi} 3\sin t \sqrt{5 + 4\cos^2 t} \, \mathrm{d}t$$

$$= -3\int_0^{\pi} \sqrt{4 + 5\cos^2 t} \, \mathrm{d}\cos t = 9 + \frac{15}{4}\ln 5$$

Example 8.3　Suppose $y = \frac{2}{3}x^{\frac{3}{2}}$, find the length of the curve from point $(0,0)$ to point $\left(4, \frac{16}{3}\right)$.

Solution　Rewrite the equation of the function as a paramatric form

$$x = x, \quad y = \frac{2}{3}x^{\frac{3}{2}}, \quad 0 \leqslant x \leqslant 4$$

then

$$L = \int_L \mathrm{d}s = \int_0^4 \sqrt{1 + \left[\left(\frac{2}{3}x^{\frac{3}{2}}\right)'\right]^2} \, \mathrm{d}x = \int_0^4 \sqrt{1 + x} \, \mathrm{d}x = \frac{2}{3}(5\sqrt{5} - 1)$$

Example 8.4 A thin wire is bent in the shape of the semicircle

$$x = a\cos t, \quad y = a\sin t, \quad 0 \leqslant t \leqslant \pi$$

If the density of the wire at a point is proportional to its distance from the x-axis, find the mass and center of mass of the wire.

Solution It is similar to the method of finding the mass of a lamina or a solid.

Let $\mu(x, y) = ky$ be the density at (x, y) (k is a constant), then the mass of the whole wire is

$$m = \int_C ky\,ds = \int_0^{\pi} ka\sin t\sqrt{a^2\sin^2 t + a^2\cos^2 t}\,dt = ka^2\int_0^{\pi}\sin t\,dt = 2ka^2$$

The static moment of the wire respect to x-axis is given by

$$M_x = \int_C y \cdot ky\,ds = \int_0^{\pi} ka^3\sin^2 t\,dt = \frac{ka^3}{2}\int_0^{\pi}(1 - \cos 2t)\,dt = \frac{ka^3}{2}\left(t - \frac{1}{2}\sin 2t\right)\Big|_0^{\pi} = \frac{ka^3\pi}{2}$$

Thus

$$\bar{y} = \frac{M_x}{m} = \frac{\frac{1}{2}ka^3\pi}{2ka^2} = \frac{1}{4}\pi a$$

Symmetrically, $\bar{x} = 0$, so the center of mass is at $\left(0, \frac{\pi a}{4}\right)$.

Example 8.5 Find the mass of a wire C of density $\mu = kz$, if it has the shape of the helix with parametrization

$$x = 3\cos t,\ y = 3\sin t,\ z = 4t,\ 0 \leqslant t \leqslant \pi$$

Solution $$m = \int_C \mu\,ds = \int_0^{\pi} k \cdot 4t\sqrt{9\sin^2 t + 9\cos^2 t + 16}\,dt = 20k\int_0^{\pi} t\,dt = 10k\pi^2$$

8.2 Surface Integral

8.2.1 Surface Area

We have known some special cases of surface area. For example, the surface area of a sphere is $4\pi r^2$. In this section, we develop a formula for the area of a surface defined by $z = f(x, y)$ over a region.

Suppose that Σ is such a surface over the closed and bounded region D in the xy-plane. Assume that $f_x(x, y)$ and $f_y(x, y)$ are continuous in D. We begin by creating a partition of D ,then take a representative region on D and make a cylinder upwards. Correspondingly, a small surface is produced by the intersection of the thin cylinder and the surface. The tangent plane is made at the point $P_k(x_k, y_k, z_k)$ and its area may approximate the area of the minor surface. In order to make a clear illustration, we magnify the local graphic as shown in figure 8 - 4.

We next find the area of the parallelogram whose sides are $|\boldsymbol{T}_x|$ and $|\boldsymbol{T}_y|$, then

$$\boldsymbol{T}_x = \Delta x_k\boldsymbol{i} + f_x(x_k, y_k)\Delta x_k\boldsymbol{k}$$

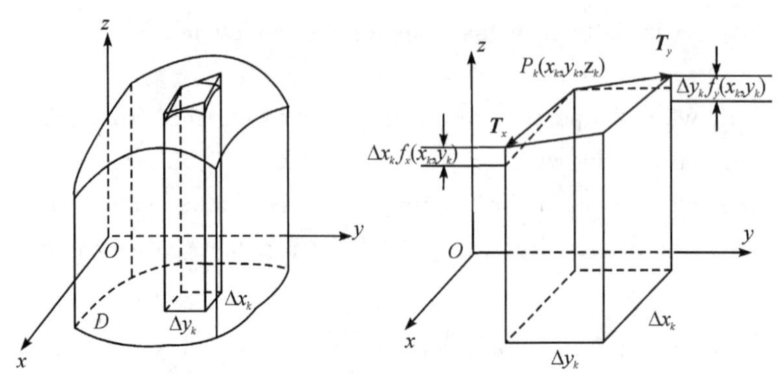

Figure 8 - 4

$$T_y = \Delta y_k \boldsymbol{j} + f_y(x_k, y_k)\Delta y_k \boldsymbol{k}$$

From Section 5. 2 of *Basic Linear Algbra*, we know that area of the tangent parallelogram is $|\boldsymbol{T}_x \times \boldsymbol{T}_y|$ where

$$\boldsymbol{T}_x \times \boldsymbol{T}_y = \begin{vmatrix} \boldsymbol{i} & \boldsymbol{j} & \boldsymbol{k} \\ \Delta x_k & 0 & f_x(x_k, y_k)\Delta x_k \\ 0 & \Delta y_k & f_y(x_k, y_k)\Delta y_k \end{vmatrix} = \Delta x_k \Delta y_k [-f_x(x_k, y_k)\boldsymbol{i} - f_y(x_k, y_k)\boldsymbol{j} + \boldsymbol{k}]$$

The area of the parallelogram is therefore

$$|\boldsymbol{T}_x \times \boldsymbol{T}_y| = \Delta x_k \Delta y_k \sqrt{1 + [f_x(x_k, y_k)]^2 + [f_y(x_k, y_k)]^2}$$

We then add the areas of these tangent paralleograms and take the limit to get the surface area of Σ.

$$\text{Area of } \Sigma = \lim_{n\to\infty} \sum_{k=1}^{n} \sqrt{1 + [f_x(x_k, y_k)]^2 + [f_y(x_k, y_k)]^2}\, \Delta x_k \Delta y_k$$

Therefore, the formula of a surface area is given by

$$\text{area of } \Sigma = \iint_D \sqrt{1 + [f_x(x, y)]^2 + [f_y(x, y)]^2}\, \mathrm{d}x\,\mathrm{d}y$$

Example 8.6 If D is the rectangular region in the xy-plane that is bounded by the lines $x = 0$, $x = 1$, $y = 0$ and $y = 2$, find the area of the part of the cylindrical surface $\Sigma : z = \sqrt{4 - x^2}$ that projects onto D.

Solution Let $f(x, y) = \sqrt{4 - x^2}$, then $f_x = \dfrac{-x}{\sqrt{4 - x^2}}$, $f_y = 0$, and

$$\text{area of } \Sigma = \iint_D \sqrt{1 + f_x^2 + f_y^2}\, \mathrm{d}x\,\mathrm{d}y = \iint_D \sqrt{\frac{x^2}{4 - x^2} + 1}\, \mathrm{d}x\,\mathrm{d}y = 2\iint_D \frac{1}{\sqrt{4 - x^2}} \mathrm{d}x\,\mathrm{d}y$$

$$= \int_0^1 \frac{1}{\sqrt{4 - x^2}} \mathrm{d}x \int_0^2 2\mathrm{d}y = 4\int_0^1 \frac{1}{\sqrt{4 - x^2}} \mathrm{d}x = 4\arcsin \frac{x}{2} \Big|_0^1 = \frac{2}{3}\pi$$

Example 8.7 Find the area of the surface $z = x^2 + y^2$ below the plane $z = \sqrt{20}$.

Solution Let $f(x,y)=x^2+y^2$, then $f_x=2x, f_y=2y$, and

$$\text{area of } \Sigma=\iint_D \sqrt{1+f_x^2+f_y^2}\,dx\,dy=\iint_D \sqrt{1+4x^2+4y^2}\,dx\,dy$$

$$=\int_0^{2\pi} d\theta \int_0^{\sqrt{20}} \sqrt{1+4r^2}\,r\,dr=2\pi\times\frac{1}{8}\times\frac{2}{3}(1+4r^2)^{\frac{3}{2}}\Big|_0^{\sqrt{20}}=\frac{243}{2}\pi$$

8.2.2 Surface Integral

Similar to the formula of the calculation of the line integral, we have the following formula.

> Let Σ be a surface given by $z=z(x,y)$, its projection in xy-plane is D. If $f(x,y,z)$, $\frac{\partial z}{\partial x}$ and $\frac{\partial z}{\partial y}$ are continuous, then
>
> $$\iint_\Sigma f(x,y,z)\,dS=\iint_D f[x,y,z(x,y)]\sqrt{1+z_x^2+z_y^2}\,dx\,dy$$

Example 8.8 Evaluate $\iint_\Sigma xz\,dS$, where Σ is lower part of cone $z=\sqrt{x^2+y^2}$ cut by $x^2+y^2=2ax\ (a>0)$.

Solution The projection of Σ in xy-plane is $D: x^2+y^2\leqslant 2ax$, then

$$\iint_\Sigma xz\,dS=\iint_D x\sqrt{x^2+y^2}\sqrt{1+z_x^2+z_y^2}\,dx\,dy$$

$$=\sqrt{2}\iint_D x\sqrt{x^2+y^2}\,dx\,dy=\sqrt{2}\int_{-\frac{\pi}{2}}^{\frac{\pi}{2}} d\theta\int_0^{2a\cos\theta} r^3\cos\theta\,dr$$

$$=4\sqrt{2}a^4\int_{-\frac{\pi}{2}}^{\frac{\pi}{2}}\cos^5\theta\,d\theta=\frac{64\sqrt{2}}{15}a^4$$

Example 8.9 The portion of the spherical surface Σ with equation

$$z=\sqrt{9-x^2-y^2}$$

where x and y satisfy $x^2+y^2\leqslant 4$ has a thin metal covering whose density at (x,y,z) is $\mu(x,y,z)=z$. Find the mass of this covering.

Solution The projection of coving metal in xy-plane is $D: x^2+y^2\leqslant 4$, then

$$m=\iint_\Sigma \mu(x,y,z)\,dS=\iint_D z\sqrt{1+z_x^2+z_y^2}\,dx\,dy$$

$$=\iint_D z\sqrt{1+\frac{x^2}{9-x^2-y^2}+\frac{y^2}{9-x^2-y^2}}\,dx\,dy$$

$$=3\iint_D dx\,dy=12\pi$$

Example 8.10 Calculate the surface area of a sphere of radius a.

Solution Suppose that the equation of the shpere is $x^2+y^2+z^2=a^2$, we only need to calculate the surface area of the upper semi-sphere because of the symmetry. Let $z = \sqrt{a^2 - x^2 - y^2}$,the projection of Σ in xy-plane is $D: x^2 + y^2 \leqslant a^2$, then

$$z_x = \frac{-x}{\sqrt{a^2 - x^2 - y^2}},\ z_y = \frac{-y}{\sqrt{a^2 - x^2 - y^2}}$$

Hence
$$\mathrm{d}S = \sqrt{1 + z_x^2 + z_y^2}\,\mathrm{d}x\,\mathrm{d}y = \frac{a}{\sqrt{a^2 - x^2 - y^2}}\mathrm{d}x\,\mathrm{d}y$$

Using the formula of a surface area yields the surface area of the sphere

$$\begin{aligned} S &= 2\iint_D \frac{a}{\sqrt{a^2 - x^2 - y^2}}\mathrm{d}x\,\mathrm{d}y = 2\iint_D \frac{a}{\sqrt{a^2 - r^2}} r\,\mathrm{d}r\,\mathrm{d}\theta \\ &= 2a\int_0^{2\pi}\mathrm{d}\theta\int_0^a \frac{r\,\mathrm{d}r}{\sqrt{a^2 - r^2}} \\ &= 4\pi a(-\sqrt{a^2 - r^2})\Big|_0^a \\ &= 4\pi a^2 \end{aligned}$$

Example 8.11 Find the area of the surface with equation $x^2 + y^2 + z^2 = 4a^2$ in the first octant, which is cut by the cylinder $x^2 + y^2 = 2ax\,(a > 0)$, see figure 8 - 5.

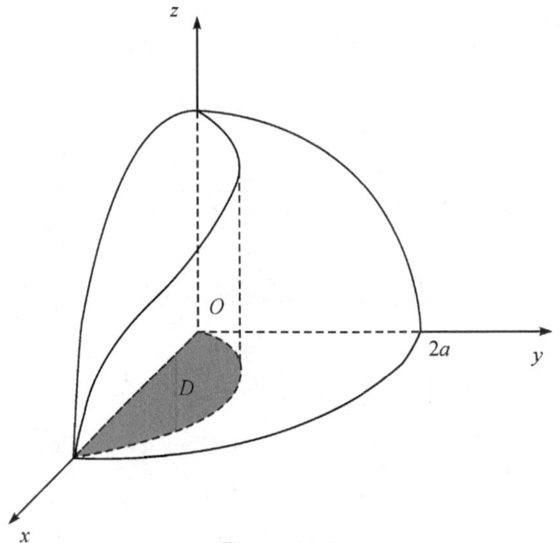

Figure 8 - 5

Solutioin By symmetry, the desired area is equal to 4 times the area of the first octant. Let $z = \sqrt{4a^2 - x^2 - y^2}$,its projection in xy-plane is $D: x^2 + y^2 \leqslant 2ax\,(y \geqslant 0)$, hence

$$\begin{aligned} S &= 4\iint_D \sqrt{1 + z_x^2 + z_y^2}\,\mathrm{d}\sigma = 4\iint_D \frac{2a}{\sqrt{4a^2 - x^2 - y^2}}\mathrm{d}x\,\mathrm{d}y \\ &= 8a\int_0^{\frac{\pi}{2}}\mathrm{d}\theta\int_0^{2a\cos\theta}\frac{r}{\sqrt{4a^2 - r^2}}\mathrm{d}r = -8a\int_0^{\frac{\pi}{2}}\sqrt{4a^2 - r^2}\Big|_0^{2a\cos\theta}\mathrm{d}\theta \\ &= -16a^2\int_0^{\frac{\pi}{2}}(\sin\theta - 1)\mathrm{d}\theta = 4(\pi - 4)a^2 \end{aligned}$$

Example 8.12 Find the centre coordinates of an equilateral triangle with vertices $(a, 0, 0)$,

$(0,a,0)$ and $(0,0,a)$.

Solutioin Let the centre coordinates be $(\bar{x},\bar{y},\bar{z})$. The euqation of the plane passing through the center is $x+y+z=a$.

Method 1
$$\bar{x}=\frac{\iint_{\Sigma} x\,\mathrm{d}S}{\iint_{\Sigma}\mathrm{d}S}=\frac{\iint_{D} x\sqrt{1+z_x^2+z_y^2}\,\mathrm{d}x\,\mathrm{d}y}{\frac{1}{2}\sqrt{2}a\cdot\sqrt{2}a\cdot\frac{\sqrt{3}}{2}}$$
$$=\frac{\sqrt{3}\int_0^a \mathrm{d}x\int_0^{a-x} x\,\mathrm{d}y}{\frac{\sqrt{3}}{2}a^2}=\frac{1}{3}a$$

Method 2 Because $\iint_{\Sigma} x\,\mathrm{d}S=\iint_{\Sigma} y\,\mathrm{d}S=\iint_{\Sigma} z\,\mathrm{d}S$, so $\iint_{\Sigma} x\,\mathrm{d}S=\frac{1}{3}\iint_{\Sigma}(x+y+z)\,\mathrm{d}S=\frac{a}{3}\iint_{\Sigma}\mathrm{d}S$, therefore
$$\bar{x}=\frac{\iint_{\Sigma} x\,\mathrm{d}S}{\iint_{\Sigma}\mathrm{d}S}=\frac{\frac{1}{3}a\iint_{\Sigma}\mathrm{d}S}{\iint_{\Sigma}\mathrm{d}S}=\frac{a}{3}$$

Because of the symmetry of the figure, we have
$$\bar{x}=\bar{y}=\bar{z}=\frac{a}{3}$$

Exercise 8

1. Evaluate the following line integral:

(1) $\int_C (x+y)\sqrt{x^2+y^2}\,\mathrm{d}s$, where $C: x^2+y^2=a^2(y\geqslant 0)$;

(2) $\int_C (x^3+y)\,\mathrm{d}s$, where $C: x=3t, y=t^3, 0\leqslant t\leqslant 1$;

(3) $\int_C (\sin x+\cos y)\,\mathrm{d}s$, where C is the line segment from $(0,0)$ to $(\pi,2\pi)$;

(4) $\int_C x\mathrm{e}^y\,\mathrm{d}s$, where C is the line segment from $(-1,2)$ to $(1,1)$;

(5) $\int_C (2x+9z)\,\mathrm{d}s$, where $C: x=t, y=t^2, z=t^3, 0\leqslant t\leqslant 1$;

(6) $\int_C (x^2+y^2+z^2)\,\mathrm{d}s$, where $C: x=4\cos t, y=4\sin t, z=3t, 0\leqslant t\leqslant 2\pi$.

2. Evaluate $I=\oint_L (x+y)\,\mathrm{d}s$, where L is the boundry curve of the region enclosed by $y=2x$, $y=2$ and $x=0$.

3. Calculate the mass and the coordinates of the uniform circular arc with radius R and center angle α (with the line density as μ).

4. A wire of constant density has the shape of the helix

$$x = a\cos t, y = a\sin t, z = bt, 0 \leqslant t \leqslant 3\pi$$

Find its mass and center of mass.

5. Evaluate $\iint\limits_{\Sigma} g(x,y,z)\mathrm{d}S$ when

(1) $g(x,y,z) = x^2 + y^2 + z$, $\Sigma: z = x + y + 1, 0 \leqslant x \leqslant 1, 0 \leqslant y \leqslant 1$;

(2) $g(x,y,z) = x$, $\Sigma: x + y + 2z = 4, 0 \leqslant x \leqslant 1, 0 \leqslant y \leqslant 1$;

(3) $g(x,y,z) = x + y$, $\Sigma: z = \sqrt{4 - x^2}, 0 \leqslant x \leqslant \sqrt{3}, 0 \leqslant y \leqslant 1$;

(4) $g(x,y,z) = 2y^2 + z$, $\Sigma: z = x^2 - y^2, 0 \leqslant x^2 + y^2 \leqslant 1$;

(5) $g(x,y,z) = y$, $\Sigma: z = 4 - y^2, 0 \leqslant x \leqslant 3, 0 \leqslant y \leqslant 2$.

(6) $g(x,y,z) = x + y$, Σ is the surface of the cuboid: $0 \leqslant x \leqslant 1, 0 \leqslant y \leqslant 1, 0 \leqslant z \leqslant 1$.

6. Evaluate $\iint\limits_{G} (xy + z)\mathrm{d}S$, where G is the part of the plane $2x - y + z = 3$ above the triangle R sketched in figure 8 - 6.

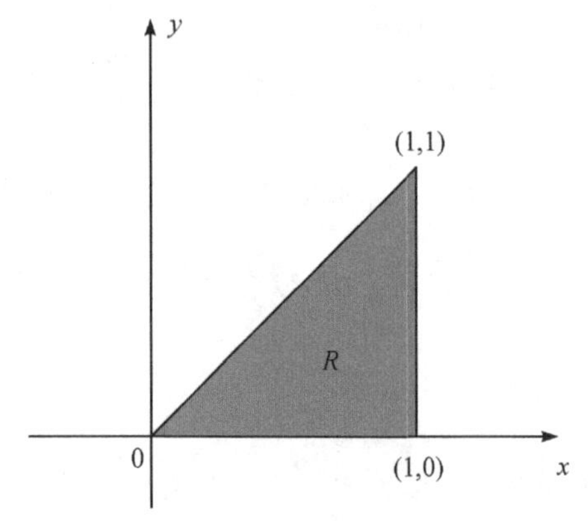

Figure 8 - 6

7. Evaluate $\iint\limits_{G} xyz\mathrm{d}S$, where G is the portion of the cone $z^2 = x^2 + y^2$ between the planes $z = 1$ and $z = 4$, see figure 8 - 7.

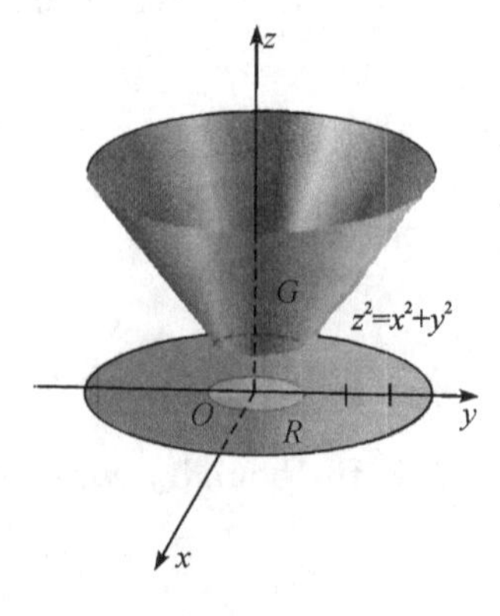

Figure 8 - 7

8. Evaluate the following surface area:

(1) the part of the surface $z=\sqrt{4-y^2}$ in the first octant that above the circle $x^2+y^2=4$ in the xy-plane.

(2) the part of the paraboloid $z=x^2+y^2$ that is cut off by the plane $z=4$.

9. Find the center of mass of the homogeneous triangle with vertices $(a,0,0)$, $(0,a,0)$, $(0,0,a)$ and density of $\mu(x,y,z)=kx^2$.

10. Find the mass of the surface $z=1-(x^2+y^2)/2$ over $0\leqslant x\leqslant 1, 0\leqslant y\leqslant 1$, if $\mu(x,y,z)=kxy$.

Chapter 9 Ordinary Differential Equation

In scientific research and practical problems, it is often necessary to find a function to describe those problems . However, the function of concern is not easily established directly. Usually we can only establish a relationship between the unknown function and its derivatives or differentials. We call such relationship differential equation.

9.1 Some Fundamental Concepts

9.1.1 Cite Example

To introduce some terminologies of differential equations we consider the following example firstly.

Example 9.1 Suppose that a plane curve pass through the point (1,2) in xy-plane. The slope of the tangent at any point (x, y) of the curve is $2x$. Find the equation of the curve.

Solution From the geometric interpretation of derivatives, the required curve should satisfy

$$\frac{dy}{dx}=2x \quad \text{or} \quad dy=2x\,dx \tag{9.1}$$

(9.1) is an equation involving derivatives of an unknown function which is called differential equation. In order to find the function, integrating on both sides of equation (9.1) with respect to x, we obtain

$$y=\int 2x\,dx=x^2+C \tag{9.2}$$

where C is an arbitrary constant.

Because the curve pass through the point (1,2) , that is, $y=y(x)$ satisfy

$$y\Big|_{x=1}=2 \tag{9.3}$$

Substituting(9.3) into (9.2), then $C=1$. Therefore, the equation of the required curve is

$$y=x^2+1 \tag{9.4}$$

9.1.2 Terminology

Definition 9.1 Differential Equation

An equation is called a **differential equation** if it contains the derivative or differential of an unknown function.

Definition 9.2 Order

The **order** of a differential equation is the order of the highest derivative that it contains. The order of the differential equation in example 9.1 is one.

Definition 9.3 Solution

A function $y = y(x)$ is a **solution** of a differential equation if the equation is satisfied identically when $y = y(x)$ and its derivatives are substituted into the equation.

Definition 9.4 General Solution

A solution of a differential equation from which all solutions can be derived by substituting values for arbitrary constants is called a **general solution** of the equation. For example, (9.2) is a general solution of the differential equation in example 9.1.

Definition 9.5 Particular Solution

A solution of a differential equation is called the **particular solution** if all the arbitrary constants in a solution have been determined. For example, (9.4) is a particular solution of the differential equation in example 9.1.

Definition 9.6 Initial-Value Problems

When an applied problem leads to a differential equation, there are usually conditions in the problem that determine specific values of the arbitrary constants. As a rule of thumb, it requires n conditions to determine values for all n arbitrary constants in the general solution of n-th-order differential equation (one condition for each constant). For a first-order equation, the single arbitrary constant can be determined by specifying the value of the unknown function $y(x)$ at an arbitrary x-value x_0, say $y(x_0) = y_0$. This is called an **initial condition**, and the problem of solving a first-order equation subject to an initial condition is called a **first-order initial-value problem.** Geometrically, the initial condition has the effect of isolating the integral curve that passes through the point from the complete family of integral curves.

In example 9.1, $\frac{\mathrm{d}y}{\mathrm{d}x} = 2x$, $y\big|_{x=1} = 2$ is an initial-value problem and $y\big|_{x=1} = 2$ is an initial condition. The general solution $y = x^2 + C$ represents complete family of integral curves and the particular solution $y = x^2 + 1$ is realized as the integral that passes through the point (1,2), see figure 9 - 1.

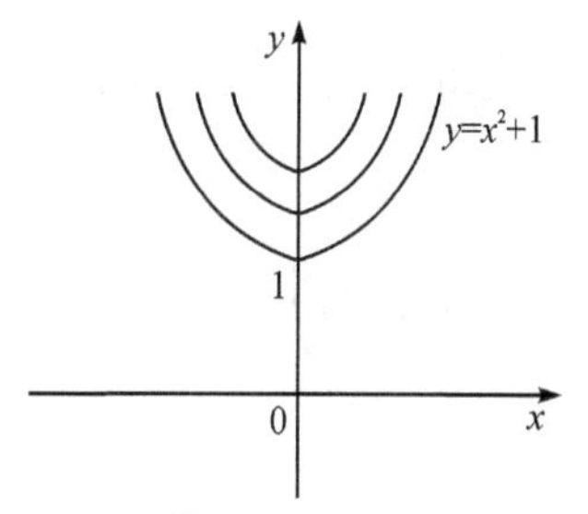

Figure 9 - 1

9.2 First-Order Equations

9.2.1 First-Order Separable Equations

If first-order equation can be written in the form of

$$\frac{\mathrm{d}y}{\mathrm{d}x} = h(x)g(y) \tag{9.5}$$

then such first-order equations are said to be separable. This name arises from the fact that this equation can be rewritten in the differential form when $g(y) \neq 0$,

$$\frac{1}{g(y)}\mathrm{d}y = h(x)\mathrm{d}x \tag{9.6}$$

in which the expressions involving x appear on one side and those involving y appear on the other. This process is called **separating variables.** When $g(y)=0$, (9.5) has one solution $y=C$ (C is a constant).

The method of solving the separable equations is to integrate both side of (9.6). We will illustrate the method by some examples.

Example 9.2 Find the general solution of the differential equation $\frac{dy}{dx}=2xy$.

Solution When $y\neq 0$, separating variables

$$\frac{dy}{y}=2x\,dx \tag{9.7}$$

Integrate both side of (9.7)

$$\int\frac{dy}{y}=\int 2x\,dx$$

yields

$$\ln|y|=x^2+C_1$$

Then

$$y=\pm e^{x^2+C_1}=\pm e^{C_1}e^{x^2}=Ce^{x^2}\quad (C=\pm e^{C_1}\neq 0)$$

Obviously, $y=0$ is the solution of the given differential equation.

Therefore, the general solution of the equation is

$$y=Ce^{x^2}\ (\text{where } C \text{ is an arbitrary constant})$$

Example 9.3 Solve the initial-value problem

$$\frac{dy}{dx}=-\frac{xy^2}{1+x^2},\ y\Big|_{x=0}=1$$

Solution For $y\neq 0$, we can write this equation in form of

$$\frac{1}{y^2}\frac{dy}{dx}=-\frac{x}{1+x^2} \tag{9.8}$$

Integrate both side of (9.8)

$$-\int\frac{1}{y^2}dy=\int\frac{x}{1+x^2}dx$$

yields

$$\frac{1}{y}=\frac{1}{2}\ln(1+x^2)+C$$

The initial condition requires that $y=1$ when $x=0$. Substituting these values into the solution yields $C=1$. Thus, a solution to the initial-value problem is

$$\frac{1}{y}=\frac{1}{2}\ln(1+x^2)+1$$

Example 9.4 Find a curve in the xy-plane that passes through (0,3) and with the slope $2x/y^2$ of the tangent at point (x,y).

Solution Since the slope of the tangent line is $\frac{dy}{dx}$, we have

$$\frac{dy}{dx}=\frac{2x}{y^2} \tag{9.9}$$

and, since the curve passes through (0,3) , we have the initial condition

$$y(0)=3 \tag{9.10}$$

Equation (9.9) is separable and can be written as

$$y^2 dy = 2x dx$$

so

$$\int y^2 dy = \int 2x dx \quad \text{or} \quad \frac{1}{3}y^3 = x^2 + C$$

It follows from the initial condition (9.10) that $y=3$ if $x=0$. Substituting these values into the last equation yields $C=9$, so the equation of the desired curve is

$$\frac{1}{3}y^3 = x^2 + 9 \quad \text{or} \quad y=(3x^2+27)^{\frac{1}{3}}$$

9.2.2 First-Order Linear Equations

If a first-order equation can be written in the form of

$$\frac{dy}{dx}+P(x)y=Q(x) \tag{9.11}$$

and if $Q(x)\equiv 0$, then equation (9.11) becomes

$$\frac{dy}{dx}+P(x)y=0 \tag{9.12}$$

which is called a **homogeneous linear differential equation of first order**, and if $Q(x)\neq 0$, (9.11) is called a **nonhomogeneous linear differential equation of first order.**

For instance, $\frac{dy}{dx}+x^2y=e^x$, $\frac{dy}{dx}+5y=2$ and $\frac{dy}{dx}+(\sin x)y+x^3=0$ are all nonhomogeneous linear differential equation of first order.

9.2.3 Method of Variation of Constants for the First-Order Linear Equations

It is easy to see that (9.12) may be solved by separating variables

$$\frac{dy}{y}=-P(x)dx \tag{9.13}$$

Integrate both side of (9.13) yields

$$\ln y = -\int P(x)dx + \ln C$$

then

$$y=Ce^{-\int P(x)dx} \quad (C \text{ is an arbitray constant})$$

It is a general solution of (9.13) and it is reasonable to deduce that (9.11) has a solution in the form of

$$y=C(x)e^{-\int P(x)dx} \tag{9.14}$$

Hence, to find the solution, we only need to determine $C(x)$.

Substituting (9.14) into (9.11), we have

$$C'(x)e^{-\int P(x)dx} - C(x)P(x)e^{-\int P(x)dx} + P(x)C(x)e^{-\int P(x)dx} = Q(x)$$

or

$$C'(x) = Q(x)e^{\int P(x)dx}$$

So that

$$C(x) = \int Q(x)e^{\int P(x)dx}dx + C$$

Substitution this expression of $C(x)$ into (9.14), we obtain

$$y = e^{-\int P(x)dx}\left(\int Q(x)e^{\int P(x)dx}dx + C\right) \qquad (9.15)$$

or

$$y = Ce^{-\int P(x)dx} + e^{-\int P(x)dx}\int Q(x)e^{\int P(x)dx}dx \qquad (9.16)$$

This is the general solution of (9.11) because it is a solution which contains an arbitrary constant. From (9.16) we can see that **the general solution of the nonhomogeneous linear equation equals to the sum of a particular solution of it and the general solution of the corresponding homogeneous linear equation.**

Example 9.5 Solve the differential equation

$$\frac{dy}{dx} + y\cos x = e^{-\sin x}$$

Solution Substituting $P(x) = \cos x$, $Q(x) = e^{-\sin x}$ into the formula (9.15) yields the general solution of the desired equation

$$\begin{aligned} y &= e^{-\int \cos x dx}\left(\int e^{-\sin x}e^{\int \cos x dx}dx + C\right) \\ &= e^{-\sin x}\left(\int e^{-\sin x}e^{\sin x}dx + C\right) \\ &= e^{-\sin x}(x + C) = xe^{-\sin x} + Ce^{-\sin x} \end{aligned}$$

Example 9.6 Solve the initial-value problem

$$xdy + ydx = xe^x, \qquad y\Big|_{x=1} = 1$$

Solution Change the given equation into a proper form

$$y' + \frac{1}{x}y = e^x$$

Using the formula (9.15) yields the general solution of the differential equation

$$\begin{aligned} y &= e^{-\int \frac{1}{x}dx}\left(\int e^x e^{\int \frac{1}{x}dx}dx + C\right) \\ &= \frac{1}{x}\left(\int xe^x dx + C\right) \end{aligned}$$

$$=\frac{1}{x}(x\mathrm{e}^{x}-\mathrm{e}^{x}+C)$$

It follows from the initial condition that $y=1$ if $x=1$. Substituting these values into the last equation yields $C=1$, so the solution of the desired equation is

$$y=\frac{1}{x}(x\mathrm{e}^{x}-\mathrm{e}^{x}+1)$$

Example 9.7 Find the general solution of the differential equation

$$\frac{\mathrm{d}y}{\mathrm{d}x}=\frac{1}{x+y} \tag{9.17}$$

Solution The given equation is not a linear differential equation with respect to x.

If we change (9.17) into a proper form

$$\frac{\mathrm{d}x}{\mathrm{d}y}-x=y \tag{9.18}$$

then (9.18) is a linear differential equation with respect to y.

$$x=\mathrm{e}^{\int 1\mathrm{d}y}\left(\int y\mathrm{e}^{-\int 1\mathrm{d}y}\mathrm{d}y+C\right)$$

$$=\mathrm{e}^{y}\left(\int y\mathrm{e}^{-y}\mathrm{d}y+C\right)$$

$$=\mathrm{e}^{y}(-y\mathrm{e}^{-y}-\mathrm{e}^{-y}+C)=C\mathrm{e}^{y}-y-1$$

9.3 Differential Equations of Second-Order Solvable

There are three types of second-order differential equations will be introduced in this section.

9.3.1 $y''=f(x)$

This type of equations is very simple and we only need to integrate $f(x)$ successively twice.

Example 9.8 Find the general solution of the equation

$$y'''=\mathrm{e}^{-x}-\sin x$$

Solution Integrating the given equation successively three times

$$y''=-\mathrm{e}^{-x}+\cos x+C_1$$

$$y'=\mathrm{e}^{-x}+\sin x+C_1x+C_2$$

Then, the general solution of the equation is

$$y=-\mathrm{e}^{-x}-\cos x+\frac{C_1}{2}x^2+C_2x+C_3$$

9.3.2 $y''=f(x,y')$

The character of the equation is that the function f does not contain the unknown function y.

Solution　Let $y' = p(x) = p$, then $y'' = p' = \frac{dp}{dx}$. Substituting it into $y'' = f(x, y')$ yields

$$p' = f(x, p) \tag{9.19}$$

If we find the general solution of (9.19) which is denoted by

$$p = \frac{dy}{dx} = \varphi(x, C_1) \tag{9.20}$$

Integrating (9.20) again, we obtain the desired solution

$$y = \int \varphi(x, C_1) dx + C_2$$

Example 9.9　Find the general solution of the equation

$$y'' = y' + x$$

Solution　Let $y' = p$, then $y'' = \frac{dp}{dx}$, substituting it into the given function, yields

$$p' - p = x$$

From the formula (9.15)

$$p = e^{\int dx}\left[\int x e^{-\int dx} dx + C_1\right] = e^x\left(\int x e^{-x} dx + C_1\right)$$

$$= e^x(-xe^{-x} - e^{-x} + C_1) = C_1 e^x - x - 1$$

i. e.

$$y' = C_1 e^x - x - 1$$

Integrating again, we obtain the general solution of the desired equation

$$y = C_1 e^x - \frac{1}{2}x^2 - x + C_2$$

Example 9.10　Solve the initial-value problem

$$y'' = \frac{2x}{1+x^2}y', \quad y\big|_{x=0} = 1, \ y'\big|_{x=0} = 3$$

Solution　Let $y' = p$, then $y'' = \frac{dp}{dx}$. Substituting it into the given equation, yields

$$\frac{dp}{dx} = \frac{2x}{1+x^2}p$$

This is separable equation. Separating variables and integrating, yields

$$\int \frac{dp}{p} = \int \frac{2x}{1+x^2} dx$$

$$\ln p = \ln(1+x^2) + \ln C_1$$

i. e.

$$y' = p = C_1(1+x^2)$$

It follows from the initial condition that $y' = 3$ if $x = 0$. Substituting these values into the last equation yields $C_1 = 3$, then

$$\frac{dy}{dx} = 3(1+x^2)$$

Integrating again,

$$y = x^3 + 3x + C_2$$

It follows from another initial condition that $y = 1$ if $x = 0$. Substituting these values into the last equation, yields $C_2 = 1$, then the desired particular solution is

$$y = x^3 + 3x + 1$$

9.3.3 $y'' = f(y, y')$

The character of the equation is that the function f does not contain the independent variable x.

Solution Let $y' = p(x) = p$, then using the chain rule, yields

$$y'' = \frac{dy'}{dx} = \frac{dp}{dx} = \frac{dp}{dy} \cdot \frac{dy}{dx} = p \frac{dp}{dy} \tag{9.21}$$

Substituting (9.21) into $y'' = f(y, y')$, we have a differential equation of first order.

$$p \frac{dp}{dy} = f(y, p) \tag{9.22}$$

If the general solution of (9.22) is

$$p = \frac{dy}{dx} = \varphi(y, C_1)$$

Separating the variables and integrating, yields the general solution of $y'' = f(y, y')$

$$\int \frac{dy}{\varphi(y, C_1)} = x + C_2$$

Example 9.11 Find the general solution of the equation

$$yy'' - y'^2 = 0$$

Solution Let $y' = p$, then $y'' = p \frac{dp}{dy}$. Substituting it into the given equation, yields

$$yp \frac{dp}{dy} - p^2 = 0$$

If $y \neq 0$ and $p \neq 0$, we have

$$\frac{dp}{p} = \frac{dy}{y} \tag{9.23}$$

Integrating both side of (9.23), yields

$$\ln p = \ln y + \ln C_1$$

i. e.

$$\frac{dy}{dx} = p = C_1 y$$

Separating the variables and integrating, yields the general solution of the given equation

$$\ln |y| = C_1 x + C$$

or

$$y = C_2 e^{C_1 x} \quad (C_2 = \pm e^C)$$

9.4 The Application of Differential Equations of First-Order

We have seen that the application of differential equations in geometry in example 9.4.

The following examples show some applications in other problems.

9.4.1 Mixing Problems

Example 9.11 At time $t=0$ a tank contains 10 L salt solution with 1 kg salt. Suppose that water is allowed to enter the tank at a rate of 3 L/min and that the mixed solution is drained from the tank at the same rate. Find the amount of salt in the tank after one hour.

Solution Let $y(t)$ be the amount of salt after t minutes. We are given that $y(0)=1$, and we want to find $y(60)$. We will begin by finding a differential equation that is satisfied by $y(t)$. To do so, observe that $\frac{dy}{dt}$, which is the rate at which the amount of salt in the tank changes with time, can be expressed as

$$\frac{dy}{dt}=\text{rate in}-\text{rate out} \tag{9.24}$$

where rate in means the rate at which salt enters the tank, and rate out means the rate at which salt leaves the tank. Because the water is allowed enter the tank, the rate in is 0.

Since liquid enters and drains from the tank at the same rate, the volume of brine in the tank stays constant at 10 L. Thus, after t minutes, the tank contains $y(t)$ kg of salt, and the rate at which salt leaves the tank at that instant is

$$\text{rate out}=\frac{y(t)}{10}\cdot 3 \quad \text{L/min}$$

Therefore, (9.24) can be written as

$$\frac{dy}{dt}=-\frac{y(t)}{10}\cdot 3$$

Which is the first-order separable differential equation. Since we are given that $y(0)=1$, the function can be obtained by solving the initial-value problem

$$\frac{dy}{dt}=-\frac{y(t)}{10}\cdot 3, \quad y(0)=1 \text{ kg}$$

Seperating and integrating, yields

$$y(t)=e^{-0.3t}$$

At time $t=60$ the amount of salt in the tank is

$$y(60)=e^{-18} \text{ kg}$$

9.4.2 A Model of Free-Fall Motion Retarded by Air Resistance

Example 9.12 Assume that a parachute of mass m falls from the tower and the air resistance is directly proportional to the landing speed. If the speed is zero when the parachute leaves the tower, find the falling velocity function with respect to time t.

Solution In the case of free-fall notion retarded by air resistance, the net force acting on the parachute is

$$F=F_G-F_R$$

where F_G is the gravity and F_R is the resistance.

From **Newton's second law**, we have

$$ma = F_G - F_R \quad \text{or} \quad m\frac{dv}{dt} = mg - kv \tag{9.25}$$

and since the speed is zero when the parachute leaves the tower, we have the initial condition $v(0)=0$.

(9.25) is a first order linear equation (also separable) and can be rewritten as

$$\frac{dv}{dt} + \frac{k}{m}v = g$$

Using the general solution formula (9.15), yields the general solution

$$v(t) = e^{-\int \frac{k}{m}dt}\left(\int g e^{\int \frac{k}{m}dt}dt + C\right) = e^{-\frac{k}{m}t}\left(\int g e^{\frac{k}{m}t}dt + C\right)$$

$$= e^{-\frac{k}{m}t}\left(\frac{mg}{k}e^{\frac{k}{m}t} + C\right) = \frac{mg}{k} + Ce^{-\frac{k}{m}t}$$

It follows from the initial condition that $v=0$ if $t=0$. Substituting the value into the last expression yields $C = -\frac{mg}{k}$, then the desired solution is

$$v(t) = \frac{mg}{k}(1 - e^{-\frac{k}{m}t}) \tag{9.26}$$

(9.26) indicates that the velocity tends to the constant $\frac{mg}{k}$ as time goes on.

In other words, the jump starts with an accelerated motion, but gradually approaches an uniform motion. So it's theoretically safe for a skydiver to jump from high altitude to the ground.

9.5 Second-Order Linear Homogeneous Differential Equations with Constant Coefficients

9.5.1 A General Form of Second-Order Linear Differential Equation

A **second-order linear differential equation** is one of the form

$$\frac{d^2y}{dx^2} + p(x)\frac{dy}{dx} + q(x)y = Q(x)$$

or, in alternative notation,

$$y'' + p(x)y' + q(x)y = Q(x) \tag{9.27}$$

If $Q(x)$ is identically 0, then (9.27) can be reduces to

$$y'' + p(x)y' + q(x)y = 0 \tag{9.28}$$

(9.28) is called the **second-order linear homogeneous differential equation.**

9.5.2 Linearly Dependent and Independent of Functions

In order to discuss the solutions of (9.27), it will be useful to introduce some terminology. Two functions f and g which are defined on the interval I are said to be **linearly**

dependent if one is a constant multiple of the other. If neither is constant multiple of the other, then they are called **linearly independent**. Thus,

$$f(x)=\mathrm{e}^x \quad \text{and} \quad g(x)=2\mathrm{e}^x$$

are linearly dependent, but

$$f(x)=x \quad \text{and} \quad g(x)=x^2$$

are linearly independent. The following theorem is central to the study of second-order linearly homogeneous differential equations.

Theorem 9.1 If $y_1(x)$ and $y_2(x)$ are two linearly independent solutions to (9.28) on an interval I, then a general solution on I is given by

$$y(x)=C_1y_1(x)+C_2y_2(x) \tag{9.29}$$

That is, every solution of (9.28) on I can be obtained by choosing appropriate values of the constants C_1 and C_2; conversely, (9.29) produces all the solutions of (9.28) for all choices of C_1 and C_2.

9.5.3 Method of Solving Second-Order Linear Homogeneous Differential Equations with Constant Coefficients

We will restrict our attention to second-order linear homogeneous equations of the form

$$y''+py'+qy=0 \tag{9.30}$$

where p and q are constants. It follows from theorem 9.1 that to determine a general solution to (9.30) we only need to find two linearly independent solutions, $y_1(x)$ and $y_2(x)$. The general solution will then be given by $y(x)=C_1y_1(x)+C_2y_2(x)$, where C_1 and C_2 are arbitrary constants.

We will start by looking for solutions to (9.30) of the form $y=\mathrm{e}^{rx}$. This is motivated by the fact that the first and second derivatives of this function are still exponential function, suggesting that a solution of (9.30) might result by choosing r appropriately. To find such a r, we substitute

$$y=\mathrm{e}^{rx}, \quad y'=r\mathrm{e}^{rx}, \quad y''=r^2\mathrm{e}^{rx}$$

into (9.30) to obtain

$$(r^2+pr+q)\mathrm{e}^{rx}=0$$

which is satisfied if and only if

$$r^2+pr+q=0 \tag{9.31}$$

since $\mathrm{e}^{rx}\neq 0$ for every x.

(9.31) is called the **characteristic equation** to the differential equation (9.30). The solutions, r_1 and r_2 of (9.31) can be obtained by factoring or by the quadratic formula. These solutions are

$$r_{1,2}=\frac{-p\pm\sqrt{p^2-4q}}{2} \tag{9.32}$$

Depending on whether p^2-4q is positive, zero, or negative, these roots will be distinct and real, equal and real, or complex conjugates. We will consider each of these cases separately.

1. Distinct Real Roots

If r_1 and r_2 are distinct real roots, then (9.30) has two solutions

$$y_1=e^{r_1x} \quad \text{and} \quad y_2=e^{r_2x}$$

Neither of the solutions e^{r_1x} and e^{r_2x} is a constant multiple of the other, so the general solution of (9.30) in this case is

$$y(x)=C_1e^{r_1x}+C_2e^{r_2x}$$

2. Equal Real Roots

If r_1 and r_2 are equal real roots, say $r_1=r_2(=r)$, then the characteristic equation yields only one solution of (9.30)

$$y_1(x)=e^{rx}$$

We may verify that

$$y_2(x)=xe^{rx}$$

is another solution of (9.30) and they are linearly independent solution. Thus the general solution of (9.30) in this case is

$$y=C_1e^{rx}+C_2xe^{rx}=(C_1+C_2x)e^{rx}$$

3. Complex Roots

If the characteristic equation has complex roots $r_1=\alpha+i\beta$ and $r_2=\alpha-i\beta$ (α, β are real numbers, and $\beta\neq 0$), then we may verify $y_1=e^{\alpha x}\cos\beta x$ and $y_2=e^{\alpha x}\sin\beta x$ are linearly independent solutions of (9.30). Thus the general solution of (9.30) is

$$y=e^{\alpha x}(C_1\cos\beta x+C_2\sin\beta x)$$

To sum up, the steps to solve the general solution of (9.30) are as follows:

Step 1 Write out the characteristic equation of (9.30).

$$r^2+pr+q=0$$

Step 2 Find the roots of the characteristic equation, r_1, r_2.

Step 3 Determine the general solution of (9.30) by the table 9 - 1.

Table 9 - 1

The Roots of $r^2+pr+q=0$	The General Solution of $y''+py'+qy=0$
Two distinct real roots $r_1\neq r_2$	$y(x)=C_1e^{r_1x}+C_2e^{r_2x}$
Two eaqual real roots $r_1=r_2=r$	$y=(C_1+C_2x)e^{rx}$
A pair of conjugate complex roots $r_{1,2}=\alpha\pm i\beta$	$y=e^{\alpha x}(C_1\cos\beta x+C_2\sin\beta x)$

Example 9.13 Find the general solution of $y''-2y'-3y=0$.

Solution The characteristic equation is

$$r^2-2r-3=0$$

So its roots are $r_1=-1, r_2=3$, which are two distinct real roots. So from table 9-1, the general solution of the differential equation is

$$y=C_1e^{-x}+C_2e^{3x}$$

Example 9.14 Find the general solution of $y''-8y'+16y=0$.

Solution The characteristic equation is

$$r^2-8r+16=0$$

So its roots are $r_1=r_2=4$, which are two equal real roots. So from table 9-1, the general solution of the differential equation is

$$y=C_1e^{4x}+C_2xe^{4x}$$

Example 9.15 Solve the initial-value problem

$$\frac{d^2s}{dt^2}+2\frac{ds}{dt}+s=0, \quad s|_{t=0}=4, \ s'|_{t=0}=-2$$

Solution The characteristic equation is

$$r^2+2r+1=0$$

So the roots are $r_1=r_2=-1$, which are two distinct real roots. So from table 9-1, the general solution of the differential equation is

$$s=(C_1+C_2t)e^{-t} \tag{9.33}$$

and the derivative of this solution is

$$s'=(C_2-C_1-C_2t)e^{-t} \tag{9.34}$$

Substituting $t=0$ in (9.33) and (9.34), and using the initial conditions $s|_{t=0}=4$ and $s'|_{t=0}=-2$ yields

$$C_1=4 \text{ and } C_2=2$$

So the particular solution of the differential equation is

$$s=(4+2t)e^{-t}$$

Example 9.16 Find the general solution of $y''+6y'+13y=0$.

Solution The characteristic equation is

$$r^2+6r+13=0$$

So its roots are $r_{1,2}=-3\pm 2i$, which is a pair of conjugate complex roots. From table 9-1, the general solution of the differential equation is

$$y=e^{-3x}(C_1\cos 2x+C_2\sin 2x)$$

Exercise 9

1. Find the general solutions of the following differential equations.

(1) $xy'-y\ln y=0$; (2) $3x^2+5x-5y'=0$;

(3) $\sqrt{1-x^2}\,y'=\sqrt{1-y^2}$; (4) $y\,dx+(x^2-4x)dy=0$.

2. Solve the following initial-value problems.

(1) $y' = e^{2x-y}, y|_{x=0} = 0$;

(2) $xy' = y\ln y, y|_{x=\frac{\pi}{2}} = e$;

(3) $\cos y dx + (1 + e^{-x})\sin y dy = 0$, $y|_{x=0} = \frac{\pi}{4}$.

3. Find the general solutions of the following linear differential equations of first order.

(1) $\frac{dy}{dx} + y = e^{-x}$; (2) $xy' + y = x^2 + 3x + 2$;

(3) $y' + y\tan x = \sin 2x$; (4) $(y^2 - 6x)\frac{dy}{dx} + 2y = 0$.

4. A curve passes through $(0,0)$ and the slope of the tangent line at point (x,y) is $2x + y$, find the equation of this curve.

5. A particle with a mass of 1 g moves in a straight line under an external force, which is proportional to time and inversely proportional to the velocity of the particle. At time $t = 10$ s, the velocity is 50 cm/s, the external force is 4 g · cm/s², what is the velocity after one minute from the beginning of the motion?

6. An upward convex arc $\overset{\frown}{OA}$ connecting two points $O(0,0)$ and $A(1,1)$ is provided. For any point $P(x,y)$ on $\overset{\frown}{OA}$, the area of the graph enclosed by the curve arc $\overset{\frown}{OP}$ and the straight line segment $\overline{OP}$ is x^2, find the equation of the curve arc $\overset{\frown}{OA}$.

7. Find the general solutions of the following differential equations.

(1) $\frac{d^2y}{dx^2} - \frac{9}{4}x = 0$; (2) $y''' = xe^x$;

(3) $(1 + x^2)y'' = 2xy'$; (4) $y'' - \frac{2}{1-y}y'^2 = 0$

8. Solve the following initial-value problems?

(1) $y''' = e^x, y|_{x=1} = y'|_{x=1} = y''|_{x=1} = 0$;

(2) $y'' = 3\sqrt{y}, y|_{x=0} = 1, y'|_{x=0} = 2$;

(3) $y'' - e^{2y} = 0, y|_{x=0} = y'|_{x=0} = 0$;

(4) $y^3 y'' + 1 = 0, y|_{x=1} = 1, y'|_{x=1} = 0$.

9. Which set of functions is linearly independent on its domain?

(1) $\cos x, x^2$; (2) $x^2, 5x^2$;

(3) $2x, x^3$; (4) $e^{2x}, 3e^{2x}$.

10. Verify $y_1 = e^{-2x}$ and $y_2 = e^{-6x}$ are both the solutions of the equation $y'' + 8y' + 12y = 0$ and write out its general solution.

11. Verify $y_1 = \sin x$ and $y_2 = \cos x$ are both the solutions of the equation $y'' + y = 0$, and write out its general solution.

12. Find the general solution of the following differential equations.

(1) $y'' - 3y' - 10y = 0$; (2) $y'' - 4y' = 0$;

(3) $y'' + 2y = 0$; (4) $y'' + 8y' + 16y = 0$;

(5) $\frac{d^2x}{dt^2}-6\frac{dx}{dt}+9x=0$;　　(6) $y''+2y'+2y=0$.

13. Solve the following initial-value problems.

(1) $y''-6y'+8y=0, y|_{x=0}=1, y'|_{x=0}=6$;

(2) $4y''+4y'+y=0, y|_{x=0}=2, y'|_{x=0}=0$.

14. If the volume of a workshop is 10800 m^3, and at the time $t=0$, the bulk concentration of CO_2 in the air is equal to 0.12%. In order to reduce the content of CO_2, the fresh air containing the bulk concentration of CO_2 of 0.04% is blowed by a blower with air volume of 1500 m^3/min. Assume that the incoming fresh air and the original air in the workshop can be quickly mixed evenly and discharged at the same speed, what is the concentration of CO_2 in the workshop after 10 min?

Answers

Chapter 1

1. (1) 0; (2) $\frac{1}{2}$; (3) $\frac{1}{2}$; (4) $\frac{2}{3}a^{-\frac{1}{3}}$; (5) $-\frac{1}{3}$; (6) $\frac{1}{4\sqrt{3a}}$; (7) $\sqrt{2}$; (8) $\frac{1}{2}$;

   (9) 2.

2. (1) $\frac{3}{4}$; (2) $\frac{5}{3}$; (3) 0; (4) $\frac{a}{2}$.

3. (1) $\frac{m}{n}$; (2) $\frac{a}{b}$; (3) $\frac{a}{c}$; (4) 2; (5) $a-b$; (6) $\frac{1}{2}$; (7) 2; (8) $\cos y$; (9) $-\sqrt{2}$;

   (10) $\frac{\sec c}{2\sqrt{c}}$.

4. (1) 6; (2) 1; (3) $a-b$; (4) $\ln ab$; (5) 1; (6) -1.

5. (1) continuous; (2) discontinuous; (3) discontinuous; (4) discontinuous.

6. (1) continuous; (2) discontinuous.

7. (2) No, $f(x)=\begin{cases} 2x-3, & x<2 \\ 1, & x=2. \\ 3x-5, & x>2 \end{cases}$

8. (1) 1; (2) 12.

Chapter 2

1. (1) $2x$; (2) $-\frac{1}{x^2}$; (3) $\frac{1}{2\sqrt{x}}$; (4) $2x^{\frac{3}{4}}+3x^{\frac{1}{2}}+x^{\frac{1}{4}}+x^{-\frac{1}{4}}$; (5) $4x^3(27x^5+25x-9)$;

   (6) $\frac{x^2+2x-2}{(x+1)^2}$.

2. (1) $4x(8x^2-3)$; (2) $\frac{4}{3}(4x+5)^{-\frac{2}{3}}$; (3) $8(3x^2+2x-1)^3(3x+1)$;

   (4) $\frac{-5}{2\sqrt{8-5x}}$; (5) $\frac{2ax+b}{2(ax^2+bx+c)}$; (6) $\sqrt{\frac{x^2+a^2}{x^2-a^2}}$.

3. (1) 14; (2) $36x^2-2$; (3) $\frac{18}{x^4}$; (4) $\frac{8}{(2x+1)^3}$.

4. (1) $-\frac{x}{y}$; (2) $\frac{a^2}{b^2}\frac{x}{y}$; (3) $\frac{2x(1+2y)}{3y^2-2x^2}$; (4) $\frac{y}{x}$.

5. (1) $-2\sin 4x$; (2) $\dfrac{\cos 2x}{\sqrt{\sin 2x}}$; (3) $10x\sec^2(5x^2+6)$; (4) $-\dfrac{2}{x^2}\sec^2\dfrac{1}{x}\tan\dfrac{1}{x}$;

(5) $-5\sec^2(\cos 5x)\sin 5x$; (6) $12\csc^3(\cot 4x)\cot(\cot 4x)\csc^2 4x$;

(7) $\dfrac{1}{\sqrt{x}}\sec^2(\tan\sqrt{x})\tan(\tan\sqrt{x})\sec^2\sqrt{x}$; (8) $-6\sin(2\cos 6x)\sin 6x$.

6. (1) $(2x+3)\sin 5x+5(x^2+3x)\cos 5x$; (2) $(1+2\cos 2x)\sec 3x^2+6x(x+\sin 2x)$;

(3) $\dfrac{1}{2x}\left(\cos\sqrt{x}-\dfrac{1}{\sqrt{x}}\sin\sqrt{x}\right)$; (4) $\dfrac{n(ax-b)\sec nx\tan x-a\sec nx}{ax-b}$;

(5) $(m+n)\sin 2(m+n)x-(m-n)\sin 2(m-n)x$; (6) $4\cos 8x-\cos 2x$.

7. (1) $\sec x\tan x$; (2) $2n\sec^2 nx$; (3) $\tan\dfrac{1}{2}x\sec^2\dfrac{1}{2}x$; (4) $-\dfrac{1}{2}\sec^2\left(\dfrac{\pi}{4}-\dfrac{x}{2}\right)$;

(5) $\sec^2\left(\dfrac{\pi}{4}+x\right)$; (6) $\cot\dfrac{1}{2}x\csc^2\dfrac{1}{2}x$; (7) $2\sec x(\sec x+\tan x)^2$; (8) $\sec^2\left(\dfrac{\pi}{4}+x\right)$.

8. (1) $\dfrac{3}{\sqrt{1-(3x-4)^2}}$; (2) $\dfrac{-2\sqrt{3}}{\sqrt{4-3x^2}}$; (3) $\dfrac{2}{1+x^2}$; (4) $\dfrac{2x}{1+(1-x^2)^2}$; (5) 1;

(6) $\dfrac{1}{\sqrt{1-x^2}}$.

9. (1) $\dfrac{1+\sin(x-y)}{\sin(x-y)-1}$; (2) $\dfrac{2x-y\cos xy}{x\cos xy-2y}$; (3) $\dfrac{2xy^2-a\sec^2(ax+by)}{b\sec^2(ax+by)-2x^2y}$;

(4) $\dfrac{y-2x\sec^2(x^2+y^2)}{2y\sec^2(x^2+y^2)-x}$.

10. (1) $-\dfrac{b}{a}$; (2) $-\dfrac{b}{a}\tan 2t$; (3) $\tan\theta$; (4) $\tan t$; (5) $-\dfrac{1}{t}$.

11. (1) $\cot x$; (2) $\dfrac{1+\sec^2 x}{x+\tan x}$; (3) $\dfrac{5e^{5x}}{1+e^{5x}}$; (4) $\dfrac{1}{x\ln x}$; (5) $\tan x$; (6) $\dfrac{\sin 2x}{1+\sin^2 x}$.

12. (1) $-\dfrac{e^{\sqrt{\cos x}}\sin x}{2\sqrt{\cos x}}$; (2) $\dfrac{e^{1+\ln x}}{x}$; (3) $\dfrac{1}{x}e^{\sin(\ln x)}\cos(\ln x)$; (4) $\dfrac{1}{x}\sec^2(\ln x)$;

(5) $-\tan x\cdot\sin(\ln\sec x)$; (6) $\dfrac{\sec(\ln\tan x)\tan(\ln\tan x)\sec^2 x}{\tan x}$.

13. (1) $\dfrac{y^2}{x(1-y\ln x)}$; (2) $-\dfrac{e^{\sin x}\cos x}{e^{\sin y}\cos y}$; (3) $-\dfrac{y(y+x\ln y)}{x(y\ln x+x)}$; (4) $\dfrac{y(x\cos x\ln y-\sin y)}{x(y\cos y\ln x-\sin x)}$.

Chapter 3

1. (1) Increasing on $(1,+\infty)$ and decreasing on $(-\infty,1)$;

(2) Increasing on $\left(\dfrac{1}{4},+\infty\right)$ and decreasing on $\left(-\infty,\dfrac{1}{4}\right)$;

(3) Decreasing on $(-\infty,-1)\cup(2,\infty)$ and increasing on $(-1,2)$;

(4) Increasing on $[-3,-2)\cup(2,5]$ and decreasing on $(-2,2)$.

2. (1) Global max. $=20$, Global min. $=0$;

(2) Global max. $=42$, Global min. $=33$.

3. (1) Local min. value $=0$ at $x=1$, No point of inflection;

   (2) Local max. value $=16$ at $x=-1$, Local min. value $=-109$ at $x=4$;

   (3) Local max. value $=\frac{7}{2}$ at $x=-\frac{1}{2}$, Local min. value $=-\frac{25}{2}$ at $=\frac{3}{2}$, the point of inflection is at $x=\frac{3}{2}$;

   (4) Local max. value $=15$ at $x=10$, Local min. value $=-25$ at $=-10$, no point of inflection.

5. (1) Concave for $x>1$, concave for $x<0$ and convex for $0<x<1$.

   (2) Concave for $x>0$; convex for $x<0$.

6. 1296 m^2.

8. The area of the surface has the minimum value when the radium is $\left(\frac{26}{\pi}\right)^{1/3}$ and the height is $2\left(\frac{26}{\pi}\right)^{1/3}$.

9. 5, 5.

10. 29 m/s, 4 m/s^2.

11. (1) $S'(t)=\frac{9000t}{(t^2+50)^2}$;

    (2) $S(10)=60, S'(10)=4$;

    (3) About 64000 CD.

12. (1) $C'(t)=\frac{0.14-0.14t^2}{(t^2+1)^2}$;

    (2) $C'(0.5)=0.0672$, $C'(3)=-0.0112$.

13. $C'(x)=0.02x+10$, $L'(x)=-0.02x+20$, 1000 units.

14. $R'(x)=\frac{1}{5}x+100$, When the sales volume is less than 500, additional sales can increase the total revenue, but when the sales volume is more than 500, the revenue will decrease.

15. (1) 693 cm^3/min; (2) $\frac{9}{32\pi}$ cm/min; (3) $\frac{8}{27\pi}$ cm/min.

16. (1) $\frac{9}{128\pi}$ cm/s.   17. 10 cm/s.

18. (1) 10.08 km/h; (2) 9 cm/s.

Chapter 4

1. (1) $2x^3+\frac{7}{2}x^2+2x+C$;   (2) $\frac{1}{3}x^3+\frac{1}{x}+C$;

   (3) $\frac{2}{3}x^{3/2}-2x^{1/2}+C$;   (4) $\frac{3}{2}x^2-5x+2\ln|x|+C$;

(5) $\frac{3}{8}x^{8/3}+\frac{9}{5}x^{5/3}+\frac{15}{2}x^{2/3}+C$;
(6) $-\frac{(a-bx)^6}{6b}+C$;

(7) $\sqrt{2x+7}+C$;
(8) $3x+5\ln|x-2|+C$;

(9) $\frac{1}{3a}[(x+a)^{3/2}+(x-a)^{3/2}]+C$;
(10) $\frac{2}{25}[(5x+3)^{3/2}+(5x+3)^{1/2}]+C$;

(11) $\frac{1}{2}x^2-\frac{1}{x+3}+C$;
(12) $\frac{1}{2}x^2+2x+\ln|x+1|+C$;

(13) $\frac{e^{px}}{p}-\frac{e^{-qx}}{q}+C$;
(14) $\frac{1}{3}e^{3x}+e^x+C$.

2. (1) $\frac{\sin(a^2x+b)}{a^2}+C$;
(2) $\frac{1}{2}\tan(2x+3)+C$;

(3) $\frac{1}{2}\left(x-\frac{\sin 2ax}{2a}\right)+C$;
(4) $\frac{1}{a}\tan ax-x+C$;

(5) $\frac{1}{32}(12x-8\sin 2x+\sin 4x)+C$;
(6) $\tan x-\cot x+C$;

(7) $-\cot x-\frac{1}{2}x-\frac{1}{4}\sin 2x+C$;
(8) $\frac{1}{a}(\sin ax-\cos ax)+C$;

(9) $\frac{1}{a}(\tan ax+\sec ax)+C$;
(10) $\frac{\sin 2x}{4}-\frac{\sin 12x}{24}+C$.

3. (1) $\frac{1}{4}(x^3+1)^4+C$;
(2) $-\frac{1}{4(3x^2+4x+1)^2}+C$;

(3) $\frac{2}{3}\sqrt{x^3+3x+4}+C$;
(4) $\frac{1}{2}(\ln x)^2+C$;

(5) $\frac{1}{8}\cos^8 x-\frac{1}{6}\cos^6 x+C$;
(6) $\frac{1}{4a}(a\sin x-b)^4+C$;

(7) $\frac{1}{4}[\ln(\sin x)]^4+C$;
(8) $\frac{1}{3}\tan^3\theta+\frac{1}{5}\tan^5\theta+C$;

(9) $\frac{1}{4}\tan^4 x+\frac{1}{6}\tan^6 x+C$;
(10) $\frac{1}{2}\tan^2 x+\ln|\cos x|+C$;

(11) $e^{\sin x\cos x}+C$;
(12) $e^{x+1/x}+C$;

(13) $-2\cos\sqrt{x}+C$;
(14) $e^x-\ln(1+e^x)+C$;

(15) $2\ln(e^{x/2}+e^{-x/2})+C$.

4. (1) $\frac{2}{5}(x+1)^{\frac{5}{2}}-\frac{2}{3}(x+1)^{\frac{3}{2}}+C$;
(2) $\frac{1}{10}(2x+3)^{\frac{5}{2}}+\frac{1}{6}(2x+3)^{\frac{3}{2}}+C$;

(3) $\frac{x}{a^2\sqrt{a^2-x^2}}+C$;
(4) $\frac{a^2}{2}\arcsin\frac{x}{a}-\frac{1}{2}x\sqrt{a^2-x^2}+C$;

(5) $\ln\left|x+\sqrt{x^2-4}\right|+C$;
(6) $-\frac{\sqrt{x^2+1}}{x}+C$;

(7) $a\arcsin\frac{x}{a}-\sqrt{a^2-x^2}+C$;
(8) $a\arcsin\sqrt{\frac{x}{a}}-\sqrt{x(a-x)}+C$.

5. (1) $\frac{1}{4}x^2(2\ln x-1)+C$; (2) $\frac{1}{25}(5x-1)e^{5x}+C$;
(3) $x\tan x+\ln\cos x+C$; (4) $\sin x-x\cos x+C$;
(5) $\frac{1}{2}\sec x\tan x+\frac{1}{2}\ln|\sec x+\tan x|+C$; (6) $x\arcsin x+\sqrt{1-x^2}+C$;
(7) $x\sec x-\ln|\sec x+\tan x|+C$; (8) $\frac{1}{4}x^2-\frac{1}{4}x\sin 2x-\frac{1}{8}\cos 2x+C$.

Chapter 5

1. (1) $b-\frac{b^2}{2}$; (2) $\frac{4c^3}{3}$; (3) e^a-1.

2. (1) $\frac{b^3-a^3}{12b}$; (2) 240; (3) 57; (4) $2(9-2\sqrt{6})$; (5) $\frac{1}{a}(e^{ca}-e^{ba})$; (6) $2\ln 2-1$.

3. (1) $\frac{32\sqrt{2}}{3}a^2$; (2) $\frac{4}{3}\sqrt{a}(h-a)^{3/2}$; (3) $\frac{b^2}{12}$.

4. (1) $5\frac{1}{3}$; (2) $\frac{1}{4}$; (3) $25\frac{3}{5}$; (4) $\frac{16}{3}a^2$; (5) $\frac{4}{3}$.

5. (1) $13\frac{1}{6}$; (2) $-\ln 2$; (3) $\frac{1}{12}(3\sqrt{3}-1)$; (4) $\frac{1}{4}e(e^6-1)$; (5) $\frac{\sqrt{2}-1}{\sqrt{2}a}$; (6) $\ln\frac{4}{3}$;
(7) $1-\frac{\pi}{4}$; (8) $\frac{1}{2}$; (9) $\frac{2}{3}$; (10) $\sqrt{2}$; (11) 2; (12) $\frac{3}{5}$; (13) $\frac{7}{120}\sqrt{2}$;
(14) $\frac{1}{2}(1-\ln 2)$; (15) $\frac{8}{15}$; (16) $-\frac{\pi}{6}$; (17) $\frac{\pi}{2}-1$; (18) $\frac{1}{4}(e^2+1)$.

6. $\frac{\pi}{4}\ln 5$. 7. $\frac{8}{5}\pi$. 10. (1) $\frac{1}{4}\pi$; (2) $\frac{8}{35}\pi$. 11. 2π.

Chapter 6

1. (1) $2x+4y\geqslant 1$; (2) $x>y$ (3) $-|y|\leqslant x\leqslant |y|, x\neq 0, y\neq 0$;
(4) $x+y\geqslant 0, x\geqslant 3$.

3. (1) 0; (2) $-\frac{4}{93}$; (3) Does not exist; (4) Does not exist.

4. (1) $\frac{\partial f}{\partial x}=y+\frac{1}{y}, \frac{\partial f}{\partial y}=x-\frac{x}{y^2}$; (2) $\frac{\partial f}{\partial x}=\frac{y}{1+x^2y^2}, \frac{\partial f}{\partial y}=\frac{x}{1+x^2y^2}$;
(3) $\frac{\partial f}{\partial x}=(1+y^2)^x\ln(1+y^2)$, $\frac{\partial f}{\partial y}=2xy(1+y^2)^{x-1}$;
(4) $\frac{\partial f}{\partial x}=\frac{x}{x^2+y^2+z^2}, \frac{\partial f}{\partial y}=\frac{y}{x^2+y^2+z^2}, \frac{\partial f}{\partial z}=\frac{z}{x^2+y^2+z^2}$.

5. (1) $f_x(1,1)=2, f_x(-1,-1)=-2, f_y(1,2)=\frac{3}{4}, f_y(2,1)=0$;
(2) $f_x(1,1)=\frac{1}{2}, f_x(-1,-1)=-\frac{1}{2}, f_y(1,2)=\frac{1}{5}, f_y(2,1)=\frac{2}{5}$;

(3) $f_x(1,1)=4\ln 2, f_x(-1,-1)=-\ln 2, f_y(1,2)=1, f_y(2,1)=4$.

6. 1.

7. 3.

8. (1) $\frac{\partial^2 f}{\partial x^2}=0, \frac{\partial^2 f}{\partial y^2}=0, \frac{\partial^2 f}{\partial x\partial y}=0, \frac{\partial^2 f}{\partial y\partial x}=0$;

(2) $\frac{\partial^2 f}{\partial x^2}=2, \frac{\partial^2 f}{\partial y^2}=2, \frac{\partial^2 f}{\partial x\partial y}=0, \frac{\partial^2 f}{\partial y\partial x}=0$;

(3) $\frac{\partial^2 f}{\partial x^2}=6x, \frac{\partial^2 f}{\partial y^2}=6y, \frac{\partial^2 f}{\partial x\partial y}=1, \frac{\partial^2 f}{\partial y\partial x}=1$;

(4) $\frac{\partial^2 f}{\partial x^2}=8+12y, \frac{\partial^2 f}{\partial y^2}=6y+12x^2, \frac{\partial^2 f}{\partial x\partial y}=24xy, \frac{\partial^2 f}{\partial y\partial x}=24xy$;

(5) $\frac{\partial^2 f}{\partial x^2}=0, \frac{\partial^2 f}{\partial y^2}=0, \frac{\partial^2 f}{\partial x\partial y}=1, \frac{\partial^2 f}{\partial y\partial x}=1$.

9. (1) 0,0,0; (2) 2,2,0; (3) 2, −12,1; (4) −28,36, −24.

11. (1) $\frac{\partial f}{\partial x}=\sin(xt)+2xt+xt\cos(xt), \frac{\partial^2 f}{\partial x\partial t}=2x\cos(xt)-x^2 t\sin(xt)+2x$;

(2) $\frac{\partial f}{\partial x}=zt-t\mathrm{e}^{xt}, \frac{\partial^2 f}{\partial x\partial t}=z-\mathrm{e}^{xt}-xt\mathrm{e}^{xt}$;

(3) $\frac{\partial f}{\partial x}=-6x\sin(t+x^2), \frac{\partial^2 f}{\partial x\partial t}=-6x\cos(t+x^2)$.

12. (1) (0,0) local minimum point, (0,2) saddle point;

(2) (0,0) saddle point, (−10,0) local maximum point;

(3) (0,0) local maximum point;

(4) (−1,3) saddle point;

(5) (0,0) saddle point, (1,0) local minimum point, (−1,0) local minimum point.

(6) $f(x,y)=(x^2+y^2+1)^2$, local minimum point of (0,0).

13. 2 millions of dollars on labors, 3 millions of dollars on equipments, minimize cost is $C(2,3)=15$ millions of dollars.

14. 6×4×2.

Chapter 7

1. (1) −135; (2) $\frac{3-2\sqrt{2}}{4}$.

2. (1) $\frac{9}{2}$; (2) $\frac{1}{2}-\frac{\pi}{24}\ln 7$; (3) $\frac{1}{8}(\mathrm{e}^8+\mathrm{e}^{-8}-2)$; (4) 0.

3. 28.

5. (1) 4096; (2) 0; (3) $\frac{20}{3}\sin 8$; (4) $-\frac{1}{2}\ln 28$.

6. (1) $\int_0^{\sqrt{2}}\mathrm{d}y\int_{y^2}^{x} f(x,y)\mathrm{d}x$; (2) $\int_0^{8}\mathrm{d}x\int_0^{\frac{x}{2}} f(x,y)\mathrm{d}y$;

(3) $\int_1^{e^2} dx \int_{\ln x}^{e^2} f(x,y)dy$;   (4) $\int_0^1 dy \int_{e^y}^{e} f(x,y)dx$;

(5) $\int_0^{\frac{\pi}{2}} dx \int_0^{\sin x} f(x,y)dy$;   (6) $\int_0^1 dx \int_{\sqrt{x}}^{x^2} f(x,y)dy$.

7. (1) $\frac{1}{8}(1-e^{-16})$;   (2) $\sin 1$;

(3) $\frac{1}{3}(e^8-1)$;   (4) $\frac{1}{2}\left(9\ln 3-\frac{9}{2}\right)$.

8. (1) 1/6; (2) $\frac{3\pi}{4}$; (3) $\frac{1}{8}\pi a^2$; (4) $\frac{\pi}{24}$.

9. (1) $5\pi-26$; (2) $\frac{\pi}{12}(65^{\frac{3}{2}}-1)$; (3) 256π; (4) $\frac{\pi}{4}(e^9-1)$.

10. (1) $\frac{16}{3}\sqrt{2}$; (2) $\frac{3\pi}{2}$; (3) $\frac{\pi}{16}$; (4) $\frac{31}{6}$; (5) 256π; (6) 18π.

11. (1) 8π; (2) $\frac{64\pi(2-\sqrt{2})}{3}$; (3) $\frac{5\pi}{2}$; (4) $\frac{27\pi}{4}$; (5) $(0,0,\frac{44}{45})$; (6) $(0,0,\frac{3a}{8})$;

(7) $(0,0,\frac{11}{30})$

Chapter 8

1. (1) $2a^3$; (2) $14(2\sqrt{2}-1)$; (3) $2\sqrt{5}$; (4) $4(e-e^2)$; (5) $\frac{1}{6}(14\sqrt{14}-1)$;

(6) $160\pi+120\pi^2$.

2. $I=I_1+I_2+I_3=\frac{3}{2}\sqrt{5}+\frac{5}{2}+2=\frac{3}{2}(3+\sqrt{5})$.

3. Mass: $2\mu\alpha R$; coordinates of the center of mass: $\left(0,\frac{R\sin\alpha}{\alpha}\right)$.

4. Mass: $3\pi k\sqrt{a^2+b^2}$; coordinates of the center of mass: $(0,\frac{2a}{3\pi},\frac{3\pi b}{2})$.

5. (1) $\frac{8\sqrt{3}}{3}$; (2) $\frac{\sqrt{6}}{4}$; (3) $\frac{\pi}{3}+2$; (4) $\frac{\pi(25\sqrt{5}+1)}{60}$; (5) $\frac{1}{4}(17\sqrt{17}-1)$; (6) 6.

6. $\frac{9\sqrt{6}}{8}$.

7. 0.

8. (1) 4; (2) $\frac{\pi}{6}(17\sqrt{17}-1)$.

9. $\frac{\sqrt{3}ka^4}{12}$.

10. $\frac{k(9\sqrt{3}-8\sqrt{2}+1)}{15}$.

Chapter 9

1. (1) $y = e^{Cx}$; (2) $y = \frac{1}{5}x^3 + \frac{1}{2}x^2 + C$;

   (3) $\arcsin y = \arcsin x + C$; (4) $(x-4)y^4 = Cx$.

2. (1) $e^y = \frac{1}{2}(e^{2x} + 1)$; (2) $y = e^{\frac{2}{\pi}x}$;

   (3) $(1 + e^x)\sec y = 2\sqrt{2}$.

3. (1) $y = e^{-x}(x + C)$; (2) $y = \frac{1}{3}x^2 + \frac{3}{2}x + 2 + \frac{C}{x}$;

   (3) $y = C\cos x - 2\cos^2 x$; (4) $x = Cy^3 + \frac{1}{2}y^2$.

4. $y = 2(e^x - x - 1)$.

5. $v = \sqrt{72500} \approx 269.3$ cm/s.

6. $y = x(1 - 4\ln x)$.

7. (1) $y = \frac{3}{8}x^3 + C_1 x + C_2$; (2) $y = xe^x - 3e^x + C_1 x^2 + C_2 x + C_3$;

   (3) $y = C_1(x + \frac{1}{3}x^3) + C_2$; (4) $(y-1)^3 = C_1 x + C_2$.

8. (1) $y = e^x - \frac{e}{2}x^2 - \frac{e}{2}$; (2) $y = \left(\frac{1}{2}x + 1\right)^4$;

   (3) $y = \ln\sec x$; (4) $y = \sqrt{2x - x^2}$.

9. (1) Linearly independent; (2) Linearly dependent;

   (3) Linearly independent; (4) Linearly dependent.

10. $y = C_1 e^{-2x} + C_2 e^{-6x}$.

11. $y = C_1 \sin x + C_2 \cos x$.

12. (1) $y = C_1 e^{-2x} + C_2 e^{5x}$; (2) $y = C_1 + C_2 e^{4x}$;

    (3) $y = C_1 \cos\sqrt{2}x + C_2 \sin\sqrt{2}x$; (4) $y = (C_1 + C_2 x)e^{-4x}$;

    (5) $x = (C_1 + C_2 t)e^{3t}$; (6) $y = e^{-x}(C_1 \cos x + C_2 \sin x)$.

13. (1) $y = -e^{2x} + 2e^{4x}$; (2) $y = (2 + x)e^{-\frac{x}{2}}$.

14. $\begin{cases} \frac{dx}{dt} = 1500 \times 0.04\% - 1500 \times \frac{x}{10800}, \\ x(0) = 10800 \times 0.12\% = 12.96 \end{cases}$ $\frac{x(10)}{10800} = 0.06\%$.

Reference

[1] STEWART J. Calculus Early Transcendentals[M]. Sixth edition. CA:Brooks Cole, 2003.

[2] 沃伯格,柏塞尔,里格登. 微积分[M]. 9 版. 北京:机械工业出版社,2009.

[3] 马知恩,王绵森,布劳尔. Fundamentals of Advanced Mathematics I [M]. 北京:高等教育出版社,2005.

[4] 王立冬,周文书. 微积分[M]. 北京:中国电力出版社,2010.

Reference

[illegible]